Analytic Aspects of Quantum Fields

Analytic Aspects of Quantum Fields

A. A. Bytsenko
State University of Londrina, Brazil
Sankt-Petersburg State Polytechnical University, Russia

G. Cognola
University of Trento, Italy

E. Elizalde
IEEC/CSIC and University of Barcelona, Spain

V. Moretti
University of Trento, Italy

S. Zerbini
University of Trento, Italy

World Scientific

NEW JERSEY • LONDON • SINGAPORE • SHANGHAI • HONG KONG • TAIPEI • BANGALORE

Published by

World Scientific Publishing Co. Pte. Ltd.

5 Toh Tuck Link, Singapore 596224

USA office: Suite 202, 1060 Main Street, River Edge, NJ 07661

UK office: 57 Shelton Street, Covent Garden, London WC2H 9HE

Library of Congress Cataloging-in-Publication Data
Analytic aspects of quantum fields / A.A. Bytsenko ... [et al.].
p. cm.
Includes bibliographical references and index.
ISBN 981-238-364-6 (alk. paper)
1. Quantum field theory. I. Title: Quantum fields. II. Bytsenko, Andrei A.

QC174.46.A53 2003
530.14'3--dc22 2003060130

British Library Cataloguing-in-Publication Data
A catalogue record for this book is available from the British Library.

This book is printed on acid-free paper.

Printed in Singapore by Mainland Press

to Elena and Anastasia
to Federica
to Maria Carme, Sergi and Aleix
to Francesca and Bianca
to Lucia, Livio and Giulia

Preface

As the title suggests, this is a book on some basic mathematical aspects of Quantum Field Theory, with special emphasis on new developments, many of which have only recently appeared as research articles in the specialized literature. In a way, this book is a follow up and an actualization of previous ones that have appeared not long ago on this and related subjects. The rapid development of this research field and the number of new results issued recently provide enough justification to write a new book that tries to be, on the one hand, self-contained on the subjects touched upon and, on the other, a documented guide to those flourishing new developments that cannot (all of them) be included in the ensuing few hundreds of pages.

We shall here deal mainly with basic analytic and geometric aspects of the mathematical formulation of Quantum Field Theory. We describe in some detail both old and new mathematical techniques that are commonplace in the language in which modern quantum physics is written. This language is, of course, mathematics, in one or several of its 'dialects' but, as it usually happens, the formulation of the concepts in the mathematical language is much too cold and abstract, what often prevents many members of the physical community from accessing the power and richness of a spectrum of quite beautiful and very useful techniques. That is why, one of the main purposes that have guided us on writing this book has been to try to explain in a basic manner the seemingly difficult issues of these mathematical structures, and we think that the best explanation comes from some specific and detailed examples, in each of the cases considered, corresponding to some comprehensible physical problems —to which the corresponding mathematical scheme is applied and its possibilities of further use duly demonstrated. We would be happy if the reader could loose in this way her/his probable initial prevention against the mathematical procedure itself, and at the same time obtain an adequate knowledge that will eventually lead to a mastering of the subject, so that he/she may be able to use, without further prevention, the techniques presented here in future work. In a few words, we try to close (or at least substantially narrow) the gap that always exists between the physics of Quantum Field Theories and the mathematical methods which are best suited for its formulation, which are more and more demanding on the mathematical abilities of the trained physicist.

All this said, this book comprises nine chapters, devoted to what we consider to be some basic subjects in quantum physics that make important use of analytic and geometric structures of mathematics, in obtaining clear cut results or predictions, in some fundamental quantum physical settings. The problems considered necessarily involve higher mathematics for their precise formulation. Again, the pedagogical purpose has been a major guideline along all these chapters. The first one has an introductory purpose, and contains a summary of path integral and regularization techniques for functional determinants, touching upon such basic issues as Schwinger's regularization, and generalizations thereof, the heat-kernel expansion in the more general setting that includes logarithmic terms, and the one-loop renormalization group equations. The corresponding physical applications considered (following the scheme already explained before) are several discussions of finite-temperature effects, in static and ultrastatic spacetimes, considering thereby the free energy and the thermodynamic potential, and the regularization of the vacuum energy, culminating in a final generalized vacuum-energy formula.

Chapters 2 and 3 are of a more heavy mathematical nature, albeit keeping always a view to important applications of the methods considered therein. Chapter 2 provides a fresh summary and original update of the zeta-function regularization method, including the description of a number of new issues that have never appeared in previous books or monographies on this subject. In particular, after an introduction to the procedure, local and global aspects of the zeta function on general compact manifolds are discussed. The effective action and effective Lagrangian are related to the zeta function and a comparison with the point-splitting regularization procedure is carried out. Green's functions are obtained via the zeta function and some issues related to the smoothness of the heat kernel and local zeta functions are considered too. The more difficult case of non-compact manifolds and of manifolds with a boundary is briefly discussed. The last part of the chapter is devoted to new developments of the local zeta-function technique, as applied to the stress-energy tensor and to field fluctuation regularization. In Chapter 3, very recent developments on the extension of the zeta-function formulation for quadratic forms to the most general case in which the spectral values are arbitrary forms (in the spectral vector, the number of dimensions being arbitrary), inhomogeneous, in general, and of degree two and less, is described in detail. This includes some of the most important and difficult cases in any compendium of explicit zeta-function formulae, together with some celebrated, non-trivial generalizations, as an extension of the Chowla-Selberg formula for the homogeneous, two-dimensional Epstein zeta function, to the most general quadratic case already described, and also an analytically continued formula extending the Barnes case to situations of physical relevance, involving collections of harmonic oscillator-like operators. Use is done of these formulae, in the second half of the chapter, for the study of the structure of the effective potential corresponding to a generic Quantum Field Theory (QFT) on a non-commutative space. It turns out that —as was recently discovered in original work by the authors— an unexpected new kind of zeta function turns out to be of primordial relevance in this situation. This zeta function exhibits higher order poles where the ordinary zeta functions for pseudodifferential operators admit only a simple pole (marking the abscissa of convergence).

Chapter 4 deals with locally symmetric spaces of rank one, with the corresponding spectral zeta function and the asymptotics of the heat kernel, and applications to the product of Einstein manifolds and real hyperbolic manifolds. Here, the Selberg zeta function appears, together with its factorization formula and meromorphic continuation. The chapter ends with the consideration of the Laplacian on forms and the important example of a scalar field in an ultrastatic spacetime whose spatial section is $\Gamma\backslash\mathbb{H}^3$.

Chapter 5 marks a clear evolution of the book towards more physical situations. It is devoted to the treatment of spinor fields, and thus contains the rather involved case of the Dirac operator, which needs the consideration of both zeta and eta functions. The very non-trivial case of the massive Dirac operator is also studied, to conclude with a specific one-dimensional example, where all this powerful machinery is put at work. It also deals with spinor fields on manifolds with a boundary and a most fundamental issue, in contemporary quantum physics, that is the index theorem. To this end, concepts such as the Ray-Singer norm, the twisted operators, the holonomy formula, the concept of the Dirac index of hyperbolic manifolds, and those of Chern-Simons invariants and vanishing cohomologies are introduced and carefully explained to the reader.

In Chapter 6 we consider the stress energy tensor and field fluctuations, and its different sections are devoted to the first and second variations of the effective action, the deformation of elliptic operators, the relative variance, and some explicit physical examples which are solved in full detail, as the case of the Casimir slab and the D-dimensional torus.

In Chapter 7 we introduce the concept of the multiplicative anomaly, in the calculation of determinants defined by means of the zeta function. This is a subject which originated from work due mainly to Wodzicki who, some years ago, found an easy-to-use integral expression which provides the unique (up to an overall constant) extension of the Dixmier trace to the important class of pseudodifferential operators. This development is summarized in a section of the Chapter. Wodzicki also derived a remarkable formula for the multiplicative anomaly (or defect) in terms of the trace and the orders of the operators. This anomaly arises from the, by now well known fact, that the (zeta-regularized) determinant of a product of operators generically differs from the product of the determinants (even in the case when the two operators commute!). The difference of the determinant logs is defined to be the multiplicative anomaly. We start the Chapter with Ray and Singer's definition of the zeta determinant and provide a few, very basic examples of its calculation in practice. We already see in these cases how the anomaly appears. We provide a simple and original derivation in perturbation theory of the multiplicative anomaly formula and a first application of the same to the case of low dimensional systems. We also discuss the relation of the multiplicative anomaly with the heat-kernel coefficients and obtain an anomaly formula for locally symmetric spaces.

Chapter 8 is concerned with providing some physical applications of the multiplicative anomaly issue introduced in Chapter 7. It is devoted to the discussion of such anomaly in physically relevant situations involving Dirac-like operators, the interacting case in the one-loop approximation — in particular for $O(2)$ models— and the free charged bosonic model at finite temperature.

Finally, Chapter 9 is devoted to the Casimir effect. We provide there a detailed and up to date summary of the very important developments that have taken place in the last five years concerning this subject. It includes the present day experimental situation (a very hot subject), some new fundamental issues relating the Casimir effect with the cosmological constant, the first practical applications of the measurable fluctuations of the vacuum energy in nanotechnology (MEMS) and several further subjects, as the so-called lateral Casimir force, roughness, temperature effects, etc. The summary includes also new problems with the mathematical computation of the Casimir energy density (QFT infinities), which have given birth to a renovated and very strong controversy within the community of Casimir specialists. This involves, again, finer mathematics and provides another beautiful example of the interplay between mathematics and physics, which is one of the *leitmotivs* of this book. The chapter ends with two specific calculations, discussed in all details, of the Casimir energy density. The first is the case of the ball, most beautifully solved by using the zeta-function approach, which exemplifies a situation where the spectrum of the operator is not known explicitly. The second one is a braneworld computation, to illustrate the issue of the possible connection of the vacuum energy density (or pressure) with the cosmological constant (thus providing a plausible source for the observed acceleration of the universe). Another use one could also make of it —at least in principle— is in order to stabilize a standard class of braneworld models, by most naturally employing their inherent Casimir force only.

For the reader convenience two appendices have also been added. The first one, Appendix A, contains a collection of definitions and properties of special functions and the second, Appendix B, is entirely dedicated to pseudo-differential operators (ΨDOs) and Wodzicki Residue.

Barcelona, Londrina and Trento June 2003

Andrei A. Bytsenko
Guido Cognola
Emilio Elizalde
Valter Moretti
Sergio Zerbini

Contents

Chapter 1 Survey of Path Integral Quantization and Regularization Techniques **1**

1.1 Path Integral and Regularization Techniques for Functional Determinants . . . 1
1.2 Schwinger-Like Regularizations and Heat-Kernel Expansion . . 4
1.3 Logarithmic Terms in the Heat-Kernel Expansion 7
1.4 One-Loop Renormalization Group Equations 10
1.5 Static Spacetimes: Thermodynamic Effects 13
1.5.1 Static and ultrastatic spacetimes 13
1.5.2 Finite-temperature effects 16
1.5.3 The free energy 17
1.5.4 The thermodynamic potential 18
1.5.5 Regularization of the vacuum energy 20
1.5.6 A generalized vacuum energy formula 21

Chapter 2 The Zeta-Function Regularization Method **25**

2.1 Survey of the Chapter, Notation and Conventions 25
2.1.1 Feasibility of physical interpretation via Wick rotation . 26
2.1.2 Notation and general hypotheses 30
2.2 Heat-Kernel Expansion and Coefficients 32
2.2.1 The heat-kernel expansion on compact manifolds 32
2.2.2 The self-adjoint extension 37
2.2.3 Existence of the (differentiated) heat-kernel expansion . 38

2.2.4 The heat-kernel coefficients 46
2.3 Local and Global Spectral Zeta Functions on Compact Manifolds 48
2.3.1 Weyl's asymptotic formulae 49
2.3.2 Spectral zeta functions 50
2.4 Effective Action, Effective Lagrangian and Green Functions . . 59
2.4.1 Comparison with the point-splitting regularization procedure 62
2.4.2 Green functions and zeta functions 67
2.4.3 Differential calculus of the heat kernel and local zeta functions 73
2.5 Noncompact Manifolds and Manifolds with a Boundary 81
2.6 The Stress-Energy Tensor and Field-Fluctuation Regularization 85
2.6.1 The stress-energy tensor 85
2.6.2 Zeta-function regularization of the stress-energy tensor and the field fluctuation 88
2.6.3 The regularized stress tensor and its properties 91
2.6.4 On the physical interpretation 103

Chapter 3 Generalized Spectra and Spectral Functions on Non-commutative Spaces 109
3.1 Extended Chowla-Selberg Formulae and Arbitrary Spectral Forms 111
3.2 Barnes and Related Zeta Functions 119
3.2.1 The two-dimensional case 120
3.2.2 The D-dimensional case 122
3.3 Spectral Zeta Functions for Scalar and Vector Fields on a Space-time with a Non-commutative Toroidal Part 124
3.3.1 Poles of the zeta function 125
3.3.2 Explicit analytic continuation of $\zeta_\alpha(s)$ 127
3.4 Applications to Quantum Field Theory in Non-commutative Space 129
3.4.1 Finite-temperature partition function 130
3.4.2 The spectral zeta function and the regularized vacuum energy 132
3.4.3 The regularized vacuum energy 134
3.4.4 High-temperature expansion 136

Chapter 4 Spectral Functions of Laplace Operator on Locally Symmetric Spaces 139
4.1 Locally Symmetric Spaces of Rank One 139

4.2 The Spectral Zeta Function 142
4.3 Asymptotics of the Heat Kernel 144
4.4 Product of Einstein Manifolds 147
4.4.1 The Krönecker sum of Laplace operators 148
4.4.2 The Selberg zeta function. Factorization formula 151
4.4.3 Meromorphic continuation 153
4.5 Real Hyperbolic Manifolds 157
4.5.1 Laplacian on forms 158
4.5.2 Simple complex Lie group 161
4.5.3 An example of functional determinant evaluation 166
4.5.4 Scalar fields in spacetime with spatial section of the form $\Gamma\backslash\mathbb{H}^3$ 171

Chapter 5 Spinor Fields **175**
5.1 The Dirac Operator and Spectral Invariants 177
5.1.1 The eta invariant 180
5.1.2 Induced Chern-Simons terms by quantum effects 183
5.1.3 Another form for the eta invariant variation 185
5.2 The Massive Dirac Operator 187
5.3 One-Dimensional Example 190
5.4 The One-Loop Effective Action 193
5.5 Dirac Bundle and the Ray–Singer Norm 198
5.6 The Determinant Line Bundles 202
5.7 The Dirac Index of Hyperbolic Manifolds 204

Chapter 6 Field Fluctuations and Related Variances **209**
6.1 The First Variation of the Effective Action 211
6.1.1 Deformation of elliptic operators: the first variation . . 212
6.1.2 The vacuum expectation values 213
6.2 The Second Variation of the Effective Action 215
6.2.1 Deformation of elliptic operators: the second variation 216
6.2.2 The relative variance 216
6.3 Some Examples 218
6.3.1 The Casimir slab 218
6.3.2 The D-dimensional torus 220
6.4 Remarks 222

Chapter 7 The Multiplicative Anomaly 225
7.1 Introduction 225
7.2 Zeta Trace, Determinant and the Multiplicative Anomaly . . . 231
7.2.1 The zeta determinant 232
7.2.2 Simple examples of infinite determinants 233
7.3 Perturbative Derivation of the Multiplicative Anomaly 239
7.3.1 Explicit expression for the multiplicative anomaly from perturbation theory 242
7.4 The Multiplicative Anomaly Formula 244
7.4.1 The multiplicative anomaly formula in lower dimensions 246
7.4.2 Heat-kernel coefficients from the multiplicative anomaly 248
7.5 Multiplicative Anomaly for Locally Symmetric Spaces 248

Chapter 8 Applications of the Multiplicative Anomaly 253
8.1 Anomalies for Dirac-like Operators 253
8.1.1 The one-dimensional case 253
8.1.2 Generalization to arbitrary dimensions 255
8.1.3 Harmonic oscillator in D-dimensions 257
8.2 The Massive Dirac Operator 258
8.3 Consistent, Covariant and Multiplicative Anomalies 259
8.4 Interacting Charged Scalar Model 265
8.4.1 Chemical potential in the non-interacting case revised . 268
8.4.2 The interacting case in the one-loop approximation . . . 273
8.4.3 The interacting $O(2)$ model 275
8.4.4 The free charged bosonic model at finite temperature . 278
8.5 Concluding Remarks 285

Chapter 9 The Casimir Effect 287
9.1 Introduction 287
9.2 The Casimir Energy 294
9.3 The Casimir Energy in the Ball 295
9.3.1 The method 296
9.3.2 An explicit example: the scalar field with Dirichlet boundary conditions 300
9.4 A Braneworld Computation 302
9.4.1 Casimir energy density for a dS brane in a 5D AdS background 304
9.4.2 The one-brane case 306
9.4.3 The two-brane case 307

9.4.4 The massive case 310

Appendix A Useful Mathematical Relations **313**

A.1 The Poisson Summation Formula 313
A.2 The Mellin Transform 314
A.3 The McDonald Functions 314
A.4 The Riemann-Hurwitz Functions 315
A.5 The Epstein Z-Function 315

Appendix B The Wodzicki Residue **317**

Definitions and Conventions 323

Bibliography 327

Index 341

Chapter 1

Survey of Path Integral Quantization and Regularization Techniques

The path integral method [R.P. Feynman and A.R. Hibbs (1965)] is a very powerful approach to the quantization of physical systems. By means of it, it is possible to investigate, in an elegant and economic way, several issues including for example, quantization of gauge theory [L.D. Faddeev and V.N. Popov (1967)], finite-temperature effects [C. Bernard (1974)], quantum anomalies [K. Fujikawa (1980)], effective action [E.S. Abers and B.W. Lee (1973)] and quantum gravity [S.W. Hawking (1978); S.W. Hawking (1979)].

In the present chapter we make a brief introduction to the quantization, via path integral, of matter systems living on flat or curved background spaces. With regard to this issue, the evaluation of the propagator and heat kernel in curved spacetime by a path integral was already presented in [J.D. Bekenstein and L. Parker (1981)], where the Einstein universe has constant curvature spatial section. Most part of the issues discussed here are well known and they will be reconsidered in Chapter 2, with much more mathematical details.

1.1 Path Integral and Regularization Techniques for Functional Determinants

To start with, let ϕ be a quantum field living on a given curved background, that is a D-dimensional manifold with metric $g_{\mu\nu}$. As already mentioned, its physical properties are conveniently described by means of the path-integral functional

$$Z[g] = \int d[\phi]\, e^{iS[\phi,g]}, \tag{1.1}$$

where $S[\phi, g]$ is the classical action and the functional integral is taken over all matter fields satisfying suitable boundary or periodicity conditions. The quantity g is the metric tensor to be considered as a classical background field.

Recall that when the spacetime is asymptotically flat and the integral is over the fields infinitesimally closed to the classical vacuum at infinity, $Z[g]$ is supposed to give the vacuum-to-vacuum amplitude [B.S. DeWitt (1979)]. Feynman boundary conditions are implicitly assumed in the action functional. The related functional $W[g] = -i \log Z[g]$ plays also an important role, since it generates the effective field equations, namely the classical ones plus quantum corrections. For a non-asymptotically flat spacetime, the physical meaning of the functional $Z[g]$ as vacuum persistence amplitude is less clear, but the functional $W[g]$ is still supposed to describe the effective action. Of particular interest is the case when a complex metric (manifold) can be found such that there exist both a Lorentzian and an Euclidean real sections. In this case, one can relate the functional $Z[g]$ to the Euclidean partition function

$$Z_E[g] = \int d[\phi]\, e^{-S_E[\phi,g]}, \tag{1.2}$$

for which the methods of the theory of the elliptic operators can be applied [S.W. Hawking (1977)]. It should be noted that, in the general case, this is not a unique procedure. However, for a static metric, it is always possible to find the associated Euclidean section by simply looking at the form of the metric tensor. Thus, in general, it could not be possible to define consistently the partition function. A sufficient condition is to restrict the analysis to the so called "strongly elliptic metrics", namely complex metrics for which a function f exists such that $\mathrm{Re}\,(f g_{\mu\nu})$ is positive definite. In this case, the path integral is well defined [S.W. Hawking (1977)].

Since the dominant contributions to $Z_E[g]$ will come from fields which are near the classical solution ϕ_c, which extremizes the action, for our purposes it will be sufficient to consider classical actions which are quadratic in the quantum fluctuations $\tilde{\phi} = \phi - \phi_c$. In fact, a functional Taylor expansion of the action around ϕ_c gives

$$S[\phi, g] = S[\phi_c, g] + \left.\frac{\delta^2 S[\phi, g]}{\delta\phi^2}\right|_{\phi=\phi_c} \frac{\tilde{\phi}^2}{2} + \text{ higher order terms in } \tilde{\phi}, \quad (1.3)$$

where $S[\phi_c, g]$ is the classical action and so the one-loop approximated theory is determined by the "zero temperature partition function"

$$Z[\phi_c, g] \sim \exp(-S[\phi_c, g]) \int d[\tilde{\phi}] \exp\left(-\frac{1}{2}\int d^D x\, \tilde{\phi} A \tilde{\phi}\right). \quad (1.4)$$

To ensure diffeomorphism invariance of the functional measure, the functional integral dummy variables $\tilde{\phi}$ are chosen to be scalar densities of weight $-1/2$ [K. Fujikawa (1980); D.J. Toms (1987)]. In Eq. (1.4), the analogue of the Wick rotation of the time axis in the complex plane has to be understood. In this manner, the metric $g_{\mu\nu}$ becomes Euclidean and the small disturbance operator A self-adjoint and non-negative. For non self-interacting fields minimally coupled to gravity, A is equal to the field operator, that is, the Klein-Gordon operator for scalar fields and the Dirac operator for fermion fields.

With these assumptions, the partition function can be formally computed in terms of the real eigenvalues λ_n of the operator A. This can be easily done by observing that the functional measure $d[\tilde{\phi}]$ reads

$$d[\tilde{\phi}] = \prod_n \frac{dc_n}{\sqrt{2\pi\ell}}, \quad (1.5)$$

c_n being the expansion coefficients of $\tilde{\phi}$ in terms of the eigenvectors of A and ℓ an arbitrary parameter necessary to adjust dimensions. It will be fixed by renormalization. In this manner, for neutral scalar fields one has [S.W. Hawking (1977)]

$$Z[A]_\ell = \prod_n \frac{1}{\sqrt{2\pi\ell}} \int_{-\infty}^{\infty} dc_n e^{-\frac{1}{2}\lambda_n c_n^2} = \left[\det(\ell^2 A)\right]^{-1/2}. \quad (1.6)$$

The functional $Z[A]_\ell$ determines the one-loop quantum corrections to the classical theory. An expression very similar to the latter is valid for charged scalar fields and also for anticommuting spinor fields. To consider all the cases, we may write

$$\log Z[A]_\ell = -\nu \log\det(\ell^2 A), \quad (1.7)$$

where $\nu = -1, 1, 1/2$, according to whether one is dealing with Dirac spinor, charged or neutral scalar fields respectively. Equation (1.7) is valid also for a multiplet of matter fields. In such a case A is a matrix valued differential operator.

Looking at the equations above, we see that in order to define the (one-loop) quantum theory via a path integral, one needs a suitable regularization of the determinant of (elliptic) (pseudo)differential operators, which is formally infinity, since the naive definition of the product over the eigenvalues gives rise to a badly divergent quantity. This stems from the fact that the number $\mathcal{N}(\lambda)$ of eigenvalues whose value is less than λ admits an asymptotic expansion of the form [L. Hormander (1968); P.B. Gilkey (1975); S.W. Hawking (1978); G. Cognola et al. (1989a)] (see also Chapter 2)

$$\mathcal{N}(\lambda) \simeq \sum_{n<D} \frac{K_n(A)}{\Gamma(\frac{D-n}{2})+1} \lambda^{\frac{D-n}{2}} , \qquad \lambda \to \infty , \tag{1.8}$$

where $K_n(A)$ are the integrated spectral coefficients associated with the operator A, which we are going to introduce in the next subsection. As one can see, they control the degree of divergence of the determinant. For future use, we may introduce the spectral density defined by

$$\rho_\lambda(A) = \frac{d\mathcal{N}(\lambda)}{d\lambda} \simeq \sum_{n<D} \frac{K_n(A)}{\Gamma(\frac{D-n}{2})} \lambda^{\frac{D-n}{2}-1} , \qquad \lambda \to \infty . \tag{1.9}$$

1.2 Schwinger-Like Regularizations and Heat-Kernel Expansion

For the sake of simplicity, here we shall restrict ourselves to neutral scalar fields. In such case, the one-loop Euclidean effective action reads [S.W. Hawking (1977)]

$$W := W[\phi_c, g] = -\log Z = S + \frac{1}{2}\log(\det \ell^2 A) , \tag{1.10}$$

where, in order to have Euclidean quantities, rotation of the time axis in the complex plane has to be understood. As far as the regularization is concerned, we may consider a large class of regularizations of functional determinants. They are often used in the literature and can be called generalized proper-time or Schwinger regularizations, since the prototype among them was introduced by Schwinger [J. Schwinger (1951a)].

The class of regularization we shall be dealing with is defined by [R.D. Ball (1989); A.A. Bytsenko et al. (1996)]

$$\log\det(A)(\varepsilon) = -\int_0^\infty dt\, t^{-1} g_\varepsilon(t) \mathrm{Tr}\, e^{-tA}\,, \tag{1.11}$$

where $g_\varepsilon(t)$ is a suitable regularizing function, necessary in order to control the (ultraviolet) divergences of $\mathrm{Tr}\, e^{-tA}$ for small t in the integral. In the latter equation (and in the following), for simplicity, we have dropped the scale parameter ℓ^2. When necessary, it shall be indroduced again by simple dimensional arguments.

From Eq. (1.11), we see that a crucial role i played by the asymptotics for $t \to 0$ of the so-called heat operator trace $\mathrm{Tr}\, e^{-tA}$.

Let us start with the simpler cases, that is smooth manifolds (with smooth boundaries) and second order differential Laplace-type operators (with local boundary conditions). The heat-kernel asymptotics reads

$$\mathrm{Tr}\, e^{-tA} \simeq \sum_{n=0}^{\infty} K_n(A)\, t^{\frac{n-D}{2}}\,. \tag{1.12}$$

The coefficients $K_n(A)$ can be evaluated by means of several techniques (see for example [P.B. Gilkey (1995); K. Kirsten (2002); D.V. Vassilevich (2003)] and references quoted therein). In particular, for manifolds without boundary $K_n(A) = 0$ for any odd n.

Let us introduce the zeta function $\zeta(s|A) = \mathrm{Tr}\, A^{-s}$ related to the elliptic non-negative operator A, via the Mellin transform (see Appendix A, Section A.2),

$$\Gamma(s)\zeta(s|A) = \int_0^\infty dt\, t^{s-1} \mathrm{Tr}\, e^{-tA}\,. \tag{1.13}$$

The integral is convergent for $\mathrm{Re}\, s > D/2$. If the asymptotic expansion (1.12) holds, a standard argument leads to the following meromorphic extension of the zeta function

$$\Gamma(s)\zeta(s|A) = \sum_{n=0}^{\infty} \frac{K_n(A)}{s + \frac{n-D}{2}} + J(s)\,, \tag{1.14}$$

$J(s)$ being the analytic part. As a result, at $s = 0$ the zeta function is regular and its derivative exists. We also have $\zeta(0|A) = K_D(A)$. Such a quantity is vanishing in boundary-less manifolds with odd dimension since, as we said above, for manifolds without boundary all odd heat-kernel coefficients are vanishing.

When the zeta function is regular at the origin, the zeta-function regularization is defined by the choice for the regularizing function [D.B. Ray and I.M. Singer (1973); S.W. Hawking (1977)]

$$g_\varepsilon(t) = \frac{d}{d\varepsilon}\left(\frac{t^\varepsilon}{\Gamma(\varepsilon)}\right) . \tag{1.15}$$

This choice is motivated by the regularity at $s = 0$ of the zeta function, and yields the finite result

$$\begin{aligned} \log\det\left(\ell^2 A\right)(\varepsilon) &= -\frac{d}{d\varepsilon}\left(\int_0^\infty dt\, \frac{t^{\varepsilon-1}}{\Gamma(\varepsilon)} \mathrm{Tr}\, e^{-t\ell^2 A}\right) \\ &= -\frac{d\zeta(\varepsilon|\ell^2 A)}{d\varepsilon} = -\zeta'(0|A) + \zeta(0|A)\log \ell^2 + \mathcal{O}(\varepsilon) . \end{aligned} \tag{1.16}$$

Another simple choice is the analytic regularization defined by $g_\varepsilon(t) = t^\varepsilon$ [J.S. Dowker and R. Critchley (1976)]. This corresponds to

$$\begin{aligned} \log\det\left(\ell^2 A\right)(\varepsilon) &= -\Gamma(\varepsilon)\zeta(\varepsilon|\ell^2 A) = -\frac{1}{\varepsilon}\zeta(0|A) - \zeta'(0|A) \\ &\quad +(\gamma + \log \ell^2)\zeta(0|A) + \mathcal{O}(\varepsilon) , \end{aligned} \tag{1.17}$$

where γ is the Euler-Mascheroni constant. In this case there is a divergent contribution which depends on the heat-kernel coefficient $K_D(A)$, but the finite part contains the zeta-function regularization result. It is equivalent to the usual dimensional regularization employed in perturbation theory, since formally $D \to D + \varepsilon$.

In the physics literature another class of proper-time regularizations are often used with the following important property: the functions $g_\varepsilon(t)$ admit the Mellin transform

$$\widehat{g}_\varepsilon(s) = \int_0^\infty dt\, t^{s-1} g_\varepsilon(t) . \tag{1.18}$$

In this case, making use of the Mellin-Parseval identity, (Appendix A, Section A.2) we may rewrite Eq. (1.11) in terms of a complex integral involving $\hat{g}_\varepsilon(z)$ and the Mellin transform of $\mathrm{Tr}\, e^{-tA}$, that is $\Gamma(z)\zeta(z|A)$.

An important example of this kind is the ultraviolet cutoff regularization, $g_\varepsilon(t) = \theta(t-\varepsilon)$. In this case we have $\varepsilon(s) = -\varepsilon^s/s$ and so

$$\log\det\left(\ell^2 A\right)(\varepsilon) = -\frac{1}{2\pi i}\int_{\mathrm{Re}\, z > D/2} \frac{dz}{z}\,\Gamma(z)\zeta(z|\ell^2 A)\ell^{-2z}\varepsilon^{-z}\,. \tag{1.19}$$

The meromorphic properties of $\Gamma(z)\zeta(z|A)$ are given by Eq. (1.14). Then, the complex function in the latter integral has simple poles at $z = (D-r)/2$, $r < D$ and a double pole at $z = 0$. Shifting the vertical contour and using the residue theorem one easily gets

$$\begin{aligned}\log\det\left(\ell^2 A\right)(\varepsilon) &= -\zeta'(0|A) + \zeta(0|A)\left(\log(\ell^2\varepsilon) + \gamma\right) \\ &\quad + \sum_{n\neq D} K_n(A)\varepsilon^{\frac{n-D}{2}} + \mathcal{O}(\varepsilon)\,.\end{aligned} \tag{1.20}$$

As a consequence, the finite part coming from the double pole at $z = 0$ contains $\zeta'(0|A)$. Under the hypothesis made, this fact justifies the use of zeta-function regularization. The sum over n contains the ultraviolet divergences, controlled by ε. In the one-loop approximation, these divergent terms are removed by the corresponding counterterms.

1.3 Logarithmic Terms in the Heat-Kernel Expansion

In this Section we would like to investigate more general cases, since there exist situations such that the heat-kernel asymptotics admits also logarithmic terms. Typically such logarithmic terms are present in the heat trace associated with a pseudo-differential operator defined on a smooth manifold [J. Duistermaat and V. Guillemin (1975); J. Bruning (1982); G. Cognola et al. (1992)], a spinorial Laplacian defined on a manifold with boundary, subject to spectral boundary conditions [P.B. Gilkey and G. Grubb (1998)], and a Laplace-type operator with singular potential [C. Callias (1983); J. Bruning and R. Seeley (1987)]. A recent survey can be found in [G. Grubb (2002)].

One may have logarithmic terms in the heat-kernel trace also in the case of non-smooth manifolds. Examples are the heat-kernel asymptotics of a Laplace operator defined on higher dimensional cones [M. Bordag et al. (1996c); A.A. Bytsenko et al. (1997b)] and on 4-dimensional space-times with a 3-dimensional non-compact hyperbolic section of finite volume [A.A. Bytsenko et al. (1997b)]. More recently the presence of logarithmic terms in self-interacting scalar field theory defined on manifolds with non-commutative coordinates have also been observed [A.A. Bytsenko et al. (2001b); A.A. Bytsenko et al. (2001a)]. Some of these examples will be considered in detail in the rest of the book.

Here we shall investigate the consequences of the presence of logarithmic terms in the heat-kernel expansion, as far as the one-loop effective action is concerned. Let A be a non-negative (pseudo)-differential operator (see Appendix B), defined on a D-dimensional manifold and let us assume that the heat trace asymptotics reads

$$\operatorname{Tr} e^{-tA} \simeq \sum_{n=0}^{\infty} K_n(A) t^{\frac{n-D}{2}} + \sum_{n=0}^{\infty} B_n(A) t^{n-q} \log t\,, \tag{1.21}$$

where, in general, q is a given non-negative, integer or half-integer number, which depends on the problem one is dealing with. In order to obtain the meromorphic structure of the zeta function, we use the Mellin representation, Eq. (1.13) and we separate the integration over t into two parts, namely $0 < t < 1$ and $t > 1$. Then, making use of the latter heat-kernel asymptotics we easily gets

$$\Gamma(s)\zeta(s|A) = \sum_{n=0}^{\infty} \frac{K_n(A)}{s+\frac{n-D}{2}} - \sum_{n=0}^{\infty} \frac{B_n(A)}{(s+n-q)^2} + J(s)\,, \tag{1.22}$$

$J(s)$ being again the analytic part. We see that in this case the zeta function has also double poles and, if q is an integer, it is no longer regular at the origin, but it has a simple pole with residue equal to $B_q = 0$. This latter case is, of course, of crucial importance in the regularization of the one-loop effective action, since it depends on the zeta function at the origin. To investigate what are the consequences of the presence of such a pole, here we use the dimensional regularization.

Assuming q to be an integer, we have in this case

$$\begin{aligned} \log\det\left(\ell^2 A\right)(\varepsilon) &= -\Gamma(\varepsilon)\zeta(\varepsilon|\ell^2 A) = -\frac{\Gamma(1+\varepsilon)}{\varepsilon^2}\,\omega(\varepsilon|\ell^2 A) \\ &= -\frac{\Gamma(1+\varepsilon)}{\ell^{2\varepsilon}}\left[\frac{\omega(0|A)}{\varepsilon^2} + \frac{\omega'(0|A)}{\varepsilon} + \frac{\omega''(0|A)}{2} + \mathcal{O}(\varepsilon)\right], \end{aligned} \tag{1.23}$$

where we have introduce the "new zeta function"

$$\omega(z|A) = z\operatorname{Tr} A^{-z} = z\zeta(z|A)\,. \tag{1.24}$$

It has the properties $\omega(z|\ell^2 A) = \ell^{-2s}\omega(z|A)$ and

$$\omega(0|A) = -B_q(A)\,, \qquad \omega'(0|A) = K_D(A) + \gamma B_q(A)\,. \tag{1.25}$$

Thus, the theory under investigation is one-loop renormalizable if the coefficients $K_D(A)$ and $B_q(A)$, as functions of the background field, have the same formal structure of the classical action S.

For the sake of completeness, let us illustrate the use of the cutoff regularization. We start with

$$\log\det\left(\ell^2 A\right)(\varepsilon) = -\frac{1}{2\pi i}\int_{\mathrm{Re}\, z>D/2}\frac{dz}{z}\Gamma(z)\zeta(z|\ell^2 A)\,(\varepsilon)^{-z}\,. \tag{1.26}$$

We are interested in real non-negative poles which give rise to the divergences and the finite part. Now, one has simple real and non-negative poles at $z = D/2-n$, $n = 0, 1, ..., D/2-1$, a third order pole at $z = 0$, and double poles at $z = q-k$, $k = 0, 1, q-1$. Thus, the residue theorem gives

$$\begin{aligned}
\log\det\left(\ell^2 A\right)(\varepsilon) \;=\; & -\sum_{n=0}^{D-1}\frac{2K_n(A)}{D-n}\left(\ell^2\varepsilon\right)^{n-D} \\
& +\sum_{k=0}^{q-1} B_k(A)\left(\ell^2\varepsilon\right)^{k-q}\log\left(\ell^2\varepsilon\right) \\
& -\frac{1}{2}\omega''(0|A) + \omega'(0|A)\left[\log\left(\ell^2\varepsilon\right)+\gamma\right] \\
& -\omega(0|A)\left[(\log(\ell^2\varepsilon+\gamma)^2) + \psi'(1)\right] + \mathcal{O}(\varepsilon)\,,
\end{aligned}$$

where $\psi(z)$ is the derivative of the $\Gamma(z)$ function. It should be noted that, with respect to zeta function regularization, here more divergent terms appear. In the usual situations, $B_k = 0$ and the logarithmic terms are absent for every k. As a consequence

$$\zeta'(0|A) = \frac{1}{2}\omega''(0|A)\,, \qquad \zeta(0|A) = \omega'(0|A)\,.$$

The above result, with regard to the finite part of the effective action, suggests us to adopt the following generalized zeta-function regularization

$$\log Z = -\nu\log\det\left(\ell^2 A\right) = \frac{\nu}{2}\lim_{s\to 0}\frac{d^2}{ds^2}\left[s\,\mathrm{Tr}(\ell^2 A)^{-s}\right] = \frac{\nu}{2}\,\omega''(0|\ell^2 A)\,, \tag{1.27}$$

which reduces to the original one, Eq. (1.16), in the absence of logarithmic terms. In such a case it explicitly reads

$$\log Z = -\nu \log det\left(\ell^2 A\right) = \nu \lim_{s\to 0} \frac{d}{ds}\left[\mathrm{Tr}(\ell^2 A)^{-s}\right] = \nu\zeta'(0|\ell^2 A)\,. \quad (1.28)$$

1.4 One-Loop Renormalization Group Equations

In this Section, following the approach considered in [D. Hochberg et al. (1998)], we shall illustrate the crucial role played by the spectral coefficient $K_D := K_D(A)$ in the one-loop renormalization group equations. To this aim we focus our attention on a theory defined in a compact manifold without boundary. In such a case, all spectral coefficients K_n with odd n vanish and thus, only the even dimensional case is non-trivial. Then we set

$$A_n(A) = K_{2n}(A) = \int_{\mathcal{M}} dx\sqrt{g(x)}\, a_n(x,x|A)\,,$$

$a_n(x,x|A)$ being the diagonal heat-kernel coefficients related to the operator A (see Chapter 2, Section 2.2 for their rigorous definition).

We assume all bare coupling constants appearing in the one-loop effective action to depend on the scale parameter ℓ. For technical reasons, in this Section it is convenient to make use of a dimensionless scale parameter defined by $\ell = \bar{\ell}\exp\varrho$, $\bar{\ell}$ being a fixed scale introduced only for dimensional reasons and which disappears in all renormalization group equations. Thus, we write

$$W := W(\varrho) = S(\varrho) + \frac{1}{2}\log\det(e^{2\varrho}A)\,, \qquad \varrho = \log\frac{\ell}{\bar{\ell}}, \quad (1.29)$$

where it is understood that all coupling constants $\lambda_i = \lambda_i(\varrho)$ (mass included), present also in the classical action, are functions of ϱ. The next step consists in imposing the independence on the renormalization scale of the one-loop regularized effective action, namely

$$\frac{d}{d\varrho}W(\varrho) = 0\,. \quad (1.30)$$

First, let us consider the usual case in which the zeta function is regular at the origin. For this situation, zeta function regularization gives

$$W(\varrho) = S(\varrho) - \frac{1}{2}\zeta'(0|A) + \varrho A_{D/2}(\varrho)\,. \quad (1.31)$$

$\zeta(0|A) = A_{D/2}$ and $\zeta'(0|A)$ represent the one-loop contributions to the effective action. This means that

$$\frac{d\zeta(0|A)}{d\varrho} = \frac{d\zeta(0|A)}{d\lambda_k} \frac{d\lambda_k}{d\varrho}, \tag{1.32}$$

$$\frac{d\zeta'(0|A)}{d\varrho} = \frac{d\zeta'(0|A)}{d\lambda_k} \frac{d\lambda_k}{d\varrho}, \tag{1.33}$$

give higher order quantum corrections [D. Hochberg et al. (1998)] and can be neglected, since $d\lambda_k/d\varrho$ is of first order in $\hbar$. From (1.30), one gets

$$\frac{d}{d\varrho} S(\varrho) + A_{D/2}(\varrho) = 0\,. \tag{1.34}$$

This expression shows the significance and the physical importance of the Seeley-DeWitt coefficient $A_{D/2}$: the one-loop renormalization group equations are completely determined by the knowledge of $A_{D/2}$. It should be noted that this coefficient appears in the leading divergence of the dimensional regularization and in the logarithmic divergences of the other Schwinger regularizations we have discussed.

As a function of the background fields ϕ, in general one has

$$S(\varrho) = \int dx \sum_i \lambda_i(\varrho) F_i(\phi)\,, \tag{1.35}$$

where $F_i(\phi)$ are linear independent invariants of the fields. If the theory is one-loop renormalizable, one must have

$$A_{D/2} = \int dx \sum_i c_i(\lambda_k(\varrho)) F_i(\phi)\,, \tag{1.36}$$

where c_i are known expressions of the coupling constants λ_k. Now, the renormalization group equations read

$$\frac{d}{d\varrho} \lambda_i + c_i(\lambda_k) = 0\,. \tag{1.37}$$

Let us illustrate how the method works, by means of a very simple example. Consider a self-interacting massive scalar field in 4 dimensions. The classical action as a function of the background field and renormalization parameter ϱ is given by

$$S(\varrho) = \int dx \left(-Z(\varrho)\phi\Box\phi + \frac{m^2(\varrho)}{2}\phi^2 + \frac{\lambda(\varrho)}{4!}\phi^4 + C(\varrho) \right) , \tag{1.38}$$

where m^2 is the bare mass, λ the bare coupling constant and $C(\varrho)$ a "cosmological" trivial constant. As usual, $Z(\varrho)$ is associated with the wave function renormalization and shall not be confused with the partition function. The one-loop fluctuation operator is

$$A = -\Box + m^2(\varrho) + \frac{\lambda(\varrho)}{2}\phi^2 . \tag{1.39}$$

In 4 dimensions, the relevant Seeley-DeWitt coefficient A_2 related to A is (neglecting trivial surface terms)

$$A_2(\varrho) = \int dx \, \frac{1}{32\pi^2} \left(m^2(\varrho) + \frac{\lambda(\varrho)}{4}\phi^2 \right)^2 . \tag{1.40}$$

Thus, the condition (1.34) reduces to

$$\begin{aligned} \int dx \left(-\frac{d}{d\varrho} Z(\varrho)\phi\Box\phi + \frac{1}{2}\frac{dm^2(\varrho)}{d\varrho}\phi^2 + \frac{1}{24}\frac{d}{d\varrho}\lambda(\varrho)\phi^4 + \frac{d}{d\varrho}C(\varrho) \right) \\ = -\int dx \, \frac{1}{32\pi^2} \left(m^4(\varrho) + m^2(\varrho)\lambda(\varrho)\phi^2 + \frac{\lambda(\varrho)^2}{4}\phi^4 \right) . \end{aligned} \tag{1.41}$$

As a consequence, one obtains the well known one-loop renormalization group equations

$$\begin{aligned} \frac{d}{d\varrho} f &= 0 , \\ \frac{d}{d\varrho} m^2 &= -\frac{1}{16\pi^2} m^2 \lambda , \\ \frac{d}{d\varrho} \lambda &= -\frac{3}{16\pi^2} \lambda^2 , \\ \frac{d}{d\varrho} C(\varrho) &= -\frac{1}{16\pi^2} m^4 . \end{aligned}$$

The first equation says that there is no wave function renormalization for this model in 4 dimensions. The third one gives the so-called one-loop beta function of the model. It can be easily solved, the solution being

$$\lambda(\varrho) = \frac{\lambda_0}{1 - \frac{3}{16\pi^2}\lambda_0(\varrho - \varrho_0)} . \tag{1.42}$$

It turns out that there exists a singularity at $\varrho = \varrho_0 = -16\pi^2/(3\lambda_0)$, which corresponds to $\ell = \ell_0 \exp\left[-16\pi^2/(3\lambda_0)\right]$, the so-called Landau pole. With regard to this issue, one should notice that there is no reason to still believe in the one-loop approximation.

1.5 Static Spacetimes: Thermodynamic Effects

Finite-temperature effects in the canonical and grand canonical ensembles require thermal equilibrium and this is compatible only with static spacetimes. Among the globally static spacetimes, besides the Minkowski spacetime with cosmological constant $\Lambda = 0$, de Sitter ($\Lambda > 0$) and anti-de Sitter ($\Lambda < 0$) spacetimes are particularly interesting. The first one is simply connected, while the second has a simple connected universal covering (see for example [S.W. Hawking and G.F. Ellis (1973)]). These manifolds have constant curvature. It is an interesting fact that the Euclidean sections corresponding to Minkowski, de Sitter and anti-de Sitter spacetime are the three constant curvature space forms $\mathbb{R}^4$, $\mathbb{S}^4$ and $\mathbb{H}^4$ respectively. As a consequence, the Euclidean field theory on these manifolds can have some relevance. However, in this section, we shall deal with an arbitrary and globally static spacetime and we shall briefly discuss how it is possible, making use of conformal transformation techniques, to restrict ourselves to ultrastatic spacetimes.

1.5.1 *Static and ultrastatic spacetimes*

By definition, a static spacetime is such that the metric admits a global time-like Killing vector field orthogonal to a space-like hypersurface. More strongly, if the Killing vector has constant norm, the spacetime is said to be ultrastatic. In almost all physical applications of the present book, we shall deal with an ultrastatic spacetime $\mathcal{M}^D = \mathbb{R} \times \mathcal{M}^N$ ($D = N + 1$) which admits a global coordinate systems where, as a consequence of the definition, conditions $g_{00} = -1$ and $g_{0i} = 0$ hold true. The use of ultrastatic spacetime is not a strong restriction since, barring some physically interesting case that needs a special treatment, any static metric may be transformed in an ultrastatic one (optical metric) by means of a conformal transformation [G.W. Gibbons and M.J. Perry (1978)]. This fact permits us to compute all physical quantities in an ultrastatic manifold and, at the end of calculations, transform them back to a static one, with an arbitrary g_{00}. This has been done, for example, in [J.S. Dowker and G. Kennedy (1978); D. Page (1982); M.R. Brown and A.C. Ottewill (1985); M.R. Brown et al. (1986); V.P. Gusynin and V.V. Roman'kov (1987); J.S. Dowker and J.P. Schofield (1988); J.S. Dowker and J.P. Schofield (1989)] and in [K. Kirsten (1991)], where the existence of boundaries has also been taken into account. In particular, the approximation for the heat kernel on a static Einstein spacetime was first introduced in [D. Page (1982)], where the approximate propagator has been obtained in closed form by making use of the results contained in [J.D. Bekenstein and L. Parker (1981)]. Here we would like to review the techniques of those papers.

To start with, we consider a scalar field ϕ on a $D = (N+1)$-dimensional static spacetime defined by the metric

$$ds^2 = g_{00}(\mathbf{x})(dx^0)^2 + g_{ij}(\mathbf{x})dx^i dx^j \,, \qquad \mathbf{x} = x^j \,, \qquad i,j = 1,...,N \,, \tag{1.43}$$

The one-loop partition function is given by Eq. (1.4) where the operator A has the form

$$A = -g^{00}(\partial_\tau - \mu)^2 + L_N \,, \tag{1.44}$$

where

$$L_N = -\Delta_N + m^2 + \xi R \,, \tag{1.45}$$

Δ_N being the Laplace-Beltrami operator on the N-dimensional hypersurface $x^0 = \tau =$ const, m and ξ arbitrary parameters and R the scalar curvature of the manifold. For the sake of completeness, we consider the combination $\partial_\tau - \mu$, which is relevant in the finite-temperature theories with chemical potential and for the sake of simplicity we will limit ourselves to the simplest cases, namely when the zeta function is regular at the origin and the manifolds are without boundary.

The ultrastatic metric $g'_{\mu\nu}$ can be related to the static one by the conformal transformation

$$g'_{\mu\nu}(\mathbf{x}) = e^{2\sigma(\mathbf{x})} g_{\mu\nu}(\mathbf{x}) \,, \tag{1.46}$$

$\sigma := \sigma(\mathbf{x})$ being a scalar function. We have to choose $\sigma(\mathbf{x}) = -(1/2)\log g_{00}$. In this manner, $g'_{00} = 1$ and $g'_{ij} = g_{ij}/g_{00}$ (optic metric).

Recalling that, by a conformal transformation,

$$R' \;=\; e^{-2q\sigma}\left[R - 2(D-1)\Delta_D\sigma - (D-1)(D-2)g^{\mu\nu}\partial_\mu\sigma\partial_\nu\sigma\right] \,,$$

$$\tilde{\phi}' = e^{q\sigma}\tilde{\phi}\,,$$

$$\begin{aligned}\Delta'_D\tilde{\phi}' &= e^{-q\sigma}\left[\Delta_D - \frac{D-2}{2}\Delta_D\sigma - \frac{(D-2)^2}{4}g^{\mu\nu}\partial_\mu\sigma\partial_\nu\sigma\right]\tilde{\phi} \\ &= e^{-\sigma}\left[\Delta_D + \xi_D(e^{2\sigma}R' - R)\right]\tilde{\phi}\,,\end{aligned}$$

where $\xi_D = (D-2)/4(D-1)$ is the conformal factor, R' and Δ'_D the scalar curvature and Laplace operator associated with the metric g', one obtains

$$A\tilde{\phi} = e^{\sigma}\left\{-g'^{00}(\partial_\tau - \mu)^2 - \Delta'_N + \xi_D R' + e^{-2\sigma}\left[m^2 + (\xi - \xi_D)R\right]\right\}\tilde{\phi}'\,.$$

From the latter equation, we have $\tilde{\phi}A\tilde{\phi} = \tilde{\phi}'A'\tilde{\phi}'$, where, by definition

$$A' = e^{-\sigma}Ae^{-\sigma} = -g'^{00}(\partial_\tau - \mu)^2 - \Delta'_N + \xi_D R' + e^{-2\sigma}\left[m^2 + (\xi - \xi_D)R\right]\,.$$

This means that the action $S' = S$, by definition. Note that classical conformal invariance requires the action to be invariant in form and thus $A' = A$. As is well known, this happens only for conformally coupled massless fields ($\xi = \xi_D$).

For the one-loop partition function we have

$$Z' = J[g, g']\,Z\,, \tag{1.47}$$

where $J[g, g']$ is the Jacobian of the conformal transformation. Such a Jacobian can be computed for any infinitesimal conformal transformation (see for example [V.P. Gusynin and V.V. Roman'kov (1987)] and references cited therein). To this aim, it is convenient to consider a family of continuous conformal transformations

$$g^q_{\mu\nu} = e^{2q\sigma(\mathbf{x})}g_{\mu\nu} \tag{1.48}$$

in such a way that the metric is $g_{\mu\nu}$ or $g'_{\mu\nu}$ according to whether $q = 0$ or $q = 1$ respectively. In this manner one has

$$\log J[g_q, g_{q+\delta q}] = \log\left(\frac{Z_{q+\delta q}}{Z_q}\right) = \frac{\delta q}{(4\pi)^{D/2}}\int d^Dx\,\sqrt{g^q}a_{D/2}(x, x|A_q)\sigma(x)\,.$$

In the case of conformal invariant theories, $a_{D/2}(x, x|A_q)$ is proportional to the trace anomaly.

The Jacobian for a finite transformation can be obtained from the latter equation by an elementary integration in q [V.P. Gusynin and V.V. Roman'kov (1987)]. In particular, we have

$$\log J[g,g'] = \frac{1}{(4\pi)^{D/2}} \int_0^1 dq \int d^D x \sqrt{g^q} a_{D/2}(x,x|A_q)\sigma(x) \,, \tag{1.49}$$

and finally

$$\log Z = \log Z' - \log J[g,g'] \,. \tag{1.50}$$

Then we see that in principle the knowledge of the partition function Z' in the ultrastatic manifold, and the heat coefficient $a_{D/2}(x,x|A_q)$, are sufficient in order to get the partition function in the static manifold. We know that the heat coefficients depend on invariant quantities build up with curvature (field strength) and their derivatives. As a consequence, they do not depend on the parameter μ, since we may regard such a parameter as the temporal component of a (pure gauge) Abelian potential. Thus we can simply put $\mu = 0$ in the computation of the Jacobian determinant $J[g,g']$ using Eq. (1.49).

1.5.2 *Finite-temperature effects*

There are several reasons to consider finite-temperature effects in relativistic quantum field theories. A main one is based on important recent developments in cosmological models. According to the standard Big-Bang cosmology and the more recent inflationary models, the very early universe has passed through a phase of thermal equilibrium at very high temperature and density, where the symmetry was restored but with a large cosmological constant. As the universe has become cool, it has gone through several phase transitions (see for example [R.H. Brandenberger (1985)]). Although the usual thermodynamical concepts may be inappropriate in the presence of very strong gravitational interactions, the need for considering finite-temperature field theory in a curved background has arisen. Strictly speaking, it has been shown that thermal equilibrium can be maintained for conformally invariant field theories in conformally flat expanding space-times [B.L. Hu (1982)]. Otherwise the expansion must be nearly adiabatic [L. Parker and S.A. Fulling (1974); L. Parker et al. (1974); B.L. Hu (1981); B.L. Hu (1983)]. On the other hand, particle production in a hot Friedman universe is exponentially suppressed due to the increase of the mass by thermal effects at temperatures smaller than the Planck one [A.D. Linde (1983)] and the expansion should be nearly adiabatic. Phase transitions in de Sitter spacetime have been first considered in [B. Allen (1983)] with the important result that the critical behavior strongly depends on curvature. Very little is known about those in anti-de Sitter spacetime, mainly because de Sitter space is more relevant than anti-de Sitter in inflationary scenarios.

Now we would like to present a survey of finite-temperature quantum field theory and discuss some useful representations of thermodynamical quantities.

1.5.3 *The free energy*

Let us consider (scalar) fields in thermal equilibrium at finite temperature $T = 1/\beta$. It is well known that the corresponding partition function Z_β may be obtained, within the path-integral approach, simply by Wick rotation $\tau = ix^0$ and imposing a β periodicity in τ for the field $\phi(\tau, x^i)$ ($i = 1, ..., N$, $N = D-1$) [C. Bernard (1974); L. Dolan and R. Jackiw (1974); S. Weinberg (1974); J.I. Kapusta (1989)]. In this way the one-loop approximation reads

$$Z_\beta = e^{-S[\phi_c, g]} \int_{\tilde{\phi}(\tau,x^i)=\tilde{\phi}(\tau+\beta,x^i)} d[\tilde{\phi}] \exp\left(-\int_0^\beta d\tau \int d^N x \tilde{\phi} A \tilde{\phi}\right) . \quad (1.51)$$

If the spacetime is ultrastatic, then we are dealing with the manifold $\mathbb{S}^1 \times \mathcal{M}^N$. The relevant differential operator is $A = -\partial_\tau^2 + L_N$ and its eigenvalues $\lambda_{n,j}$ can be expressed in term of the eigenvalues ω_j^2 of L_N, by

$$\lambda_{n,j} = \left(\frac{2\pi n}{\beta}\right) + \omega_j^2 , \qquad n = 0, \pm 1, \pm 2, ... \quad (1.52)$$

Then, from Eq. (1.7), also using Eq. (A.1) in Appendix A, Section A.1, after some calculations one obtains for the one-loop partition function, the three equivalent representations [B. Allen (1986); A.A. Bytsenko et al. (1992)]

$$\begin{aligned} &\log Z_\beta^{(1)} + \nu\beta\zeta^{(r)}(-1/2|\ell^2 L_N) \\ &\quad = \frac{\nu\beta}{\pi} \sum_{n=1}^{\infty} \lim_{s\to 0} \int_{-\infty}^{\infty} dt\, e^{in\beta t}\, \zeta(s|L_N + t^2) && (1.53) \\ &\quad = \frac{\nu\beta}{\sqrt{\pi}} \sum_{n=1}^{\infty} \int_0^{\infty} dt\, t^{-3/2}\, e^{-n^2\beta^2/4t}\, \mathrm{Tr}\, e^{-tL_N} && (1.54) \end{aligned}$$

$$= \frac{\nu\beta}{\pi i}\int_{\mathrm{Re}\, z=c} dz\, \zeta_R(z)\Gamma(z-1)\zeta\left(\frac{z-1}{2}|L_N\right)\beta^{-z}\,, \quad (1.55)$$

where $\nu = 1$ ($\nu = 1/2$) for charged (neutral) scalar fields. The "regularized" function $\zeta^{(r)}(s_0|A)$ is defined by

$$\zeta^{(r)}(s_0|A) = \lim_{s\to s_0}\left[\zeta(s|A) - \frac{\mathrm{Res}\,\zeta(s_0|A)}{s-s_0}\right] + (2-2\log 2)\mathrm{Res}\,\zeta(s_0|A)\,. \quad (1.56)$$

Of course, when the zeta function is regular at s_0, $\zeta^{(r)}(s_0|A)$ coincide with $\zeta(s_0|A)$.

Equations (1.53)-(1.55) not only define the finite-temperature properties of quantum fields but, as we shall see in Sec. 1.5.5, they will be our starting point for the computation of the regularized vacuum energy.

The free energy is related to the canonical partition function by means of the equation

$$F(\beta) = -\frac{1}{\beta}\log Z_\beta = F_0 + F_\beta\,, \quad (1.57)$$

where F_β represents the temperature dependent part (statistical sum) and so Eqs. (1.53)-(1.55) give different representations of F_β.

1.5.4 *The thermodynamic potential*

The generalization to the more general case of a charged (scalar) field with a non-vanishing chemical potential μ, in thermal equilibrium by some unspecified process at finite temperature $T = \beta^{-1}$ is quite immediate. The grand canonical partition function has the path-integral representation [B. Freedman and L. McLerran (1977); J.I. Kapusta (1981); A. Actor (1985); A. Actor (1986)]

$$\begin{aligned} Z_{\beta,\mu} &= e^{-S[\phi_c,g]}\int_{\phi(\tau,x^i)=\phi(\tau+\beta,x^i)} d[\bar\phi]d[\phi]\, e^{-\int_0^\beta d\tau\int d^N x\, \bar\phi A(\mu)\phi} \\ &= \exp[-\beta\Omega(\beta,\mu)]\,, \end{aligned} \quad (1.58)$$

where now, the operator A depends on the chemical potential μ, i.e. $A(\mu) = -(\partial_\tau - \mu)^2 + L_N$ and the latter equation defines the thermodynamic potential $\Omega(\beta,\mu)$. The operator $A(\mu)$ is still elliptic but non-Hermitian, in fact it is normal and its eigenvalues are complex and read

$$\lambda_{n,j} = \left(\frac{2\pi n}{\beta} + i\mu\right)^2 + \omega_j^2, \qquad n = 0, \pm 1, \pm 2, \dots . \tag{1.59}$$

Nevertheless, one can still define the related zeta function. Thus, if one resorts to zeta-function regularization, one obtains

$$\Omega(\beta, \mu) = \frac{1}{\beta} S[\phi_c, g] - \frac{1}{\beta} \zeta'(0|\ell^2 A(\mu)) . \tag{1.60}$$

Using Eq. (A.1) in Appendix A and the Mellin representation for $\zeta(s|A)$, Eq. (1.13), after some algebra, one gets the three useful representations for the temperature dependent part $\Omega_\beta(\beta, \mu)$ of the thermodynamic potential (1.60) [A.A. Bytsenko et al. (1992); G. Cognola and L. Vanzo (1992)]

$$\begin{aligned}
\Omega_\beta(\beta, \mu) &= \Omega(\beta, \mu) - \Omega_0 \\
&= -\frac{1}{\pi} \sum_{n=1}^{\infty} \int_{-\infty}^{\infty} dt\, e^{in\beta t}\, \zeta'(0|L_N + [t + i\mu]^2) && (1.61) \\
&= -\frac{1}{\sqrt{\pi}} \sum_{n=1}^{\infty} \cosh n\beta\mu \int_0^{\infty} dt\, t^{-3/2} e^{-n^2\beta^2/4t}\, \mathrm{Tr}\, e^{-tL_N} && (1.62) \\
&= -\frac{1}{\pi i} \sum_{n=0}^{\infty} \frac{\mu^{2n}}{(2n)!} \int_{\mathrm{Re}\, s = c} \frac{ds}{\beta^s}\, \zeta_R(s) \Gamma(s + 2n - 1) \zeta\left(\frac{s + 2n - 1}{2} | L_N\right) . && (1.63)
\end{aligned}$$

Here $\zeta_R(s)$ is the Riemann zeta function and

$$\Omega_0 = \frac{1}{\beta} S[\phi_c, g] + \zeta^{(r)}(-1/2|\ell^2 L_N) , \tag{1.64}$$

is the zero-temperature contribution (classical and vacuum energy).

Equations (1.61)-(1.63) are valid for any $|\mu| < \omega_0$, ω_0 being the smallest eigenvalue of L_N. When $|\mu| = \omega_0$, a more careful treatment is needed. The physical reason for this behavior is the occurrence of Bose-Einstein condensation. Of course, in the limit $\mu \to 0$, Eqs. (1.61)-(1.63) reduces to Eqs. (1.53)-(1.55) which has been discussed in the previous section.

Since $\mathrm{Tr} \exp(-tL_N) = \sum_j \exp(-t\omega_j^2)$, Eq. (1.62) leads to the well known result

$$\Omega(\beta, \mu) = \Omega_0 + \frac{1}{\beta} \sum_j \log\left(1 - e^{-\beta(\omega_j + \mu)}\right) + \frac{1}{\beta} \sum_j \log\left(1 - e^{-\beta(\omega_j - \mu)}\right)$$

$$= \Omega_0 + \frac{1}{\beta}\,\mathrm{Tr}\,\log\left(1 - e^{-\beta Q^+}\right) + \frac{1}{\beta}\,\mathrm{Tr}\,\log\left(1 - e^{-\beta Q^-}\right) . \quad (1.65)$$

Above, we have introduced the two pseudo-differential operators $Q^{\pm} = \mathcal{L}_N^{1/2} \pm \mu$, whose eigenvalues are $\omega_j \pm \mu$. We see that $\Omega(\beta, \mu)$ is the sum of the vacuum energy (zero-temperature contribution) and two additional finite-temperature contributions, one for particles and one for anti-particles. We also see, from Eq. (1.65), that the temperature contribution, as a function of the complex parameter μ, has branch points when μ^2 is equal to an eigenvalue ω_j^2. Then $\Omega_\beta(\beta, \mu)$ is analytic in the μ complex plain with a cut from ω_0 to ∞ and from $-\omega_0$ to $-\infty$. Thus, the physical values of μ are given by $|\mu| \leq \omega_0$.

For the sake of completeness, we here report another integral representation of the thermodynamic potential, which can be obtained in a way similar to the one used for deriving Eq. (1.63) [G. Cognola and L. Vanzo (1992)]. To this aim, we observe that the poles of $\zeta(s|Q^{\pm})$ (related to the two pseudo-differential operators $Q^{\pm}$) are given by the Seeley theorem [R.T. Seeley (1967)] at the points $s = N - k$ $(k = 0, 1, \ldots)$ with residues which are polynomials in μ. The formula which generalizes the result of [H.E. Haber and H.A. Weldon (1982a)] to compact manifolds then reads

$$\Omega_\beta(\beta, \mu) = -\frac{1}{2\pi i}\int_{\mathrm{Re}\, s=c} ds \zeta_R(s)\Gamma(s-1)\left[\zeta(s-1|Q^+) + \zeta(s-1|Q^-)\right]\beta^{-s} . \quad (1.66)$$

This has to be used to discuss $\Omega_\beta(\beta, \mu)$ as a function of complex μ.

1.5.5 *Regularization of the vacuum energy*

The vacuum energy density (the first term on the right hand side in Eq. (1.65)) is the only source of divergences in the thermodynamic potential. Therefore, we shall now briefly discuss how to give it a mathematical meaning. We consider a general formula for the vacuum energy of a scalar field defined on an ultrastatic spacetime with compact spatial section consisting, in general, in a manifold with boundary.

To start with, we shall take a compact arbitrarily shaped manifold with boundary, which may be multi-connected. The main issue related to the vacuum energy E_V is the necessity of a regularization scheme. In fact, if one naively makes use of the canonical quantization of a scalar field, the result is (see for example [S.A. Fulling (1973)])

$$E_V = \nu \sum_j \omega_j \,, \tag{1.67}$$

ω_j^2 being the eigenvalues of L_N defined on $\mathcal{M}^N$, the eigenfunctions of which are the mode functions satisfying suitable conditions on the boundary of $\mathcal{M}^N$ and, as in all the present chapter, $\nu = 1$ ($\nu = 1/2$) for charged (neutral) scalar fields. Vcuum energy as defined by means of Eq. (1.67) is an ill defined object and one needs some prescription in order to extract from it a finite, observable quantity. A quite natural definition of the vacuum energy in terms of the zeta function, which is valid for a manifold with an arbitrary smooth boundary, can be obtained starting from the finite-temperature partition function Z_β. According to [G. Cognola et al. (1992)], we define

$$E_V = -\lim_{\beta\to\infty} \partial_\beta \log Z_\beta \,. \tag{1.68}$$

Using for $\log Z_\beta$ the regularized expression (1.54), one immediately gets (here $S[\phi_c, g] = 0$, because we are considering free fields)

$$E_V = \nu \zeta^{(r)}(-1/2|\ell^2 L_N) \,. \tag{1.69}$$

A similar prescription has appeared in many places in the literature (see for example [J.S. Dowker and R. Critchley (1976); J.S. Dowker and G. Kennedy (1978)]). Here we would like to stress the proposal contained in [S.K. Blau et al. (1988)], which is very close to our approach. In other words, the path integral zeta-function derivation presented here, leads directly to the "principal part prescription" of [S.K. Blau et al. (1988)]. It is interesting to observe that the term in the vacuum energy, which depends on ℓ (see Eq. (1.56)), is proportional to the integral of the conformal anomaly [N.D. Birrell and P.C.W. Davies (1982)].

1.5.6 *A generalized vacuum energy formula*

The prescription previously obtained is valid when the zeta function is regular at the origin. Thi is the usual situation, but, as we already said above, it is possible to deal with cases in which the relevant zeta function has a (simple) pole at the origin. One of these will be considered in Chapter 3.

Here we shall briefly discuss the generalization of the vacuum energy formula, obtained by means of the "generalized zeta-function regularization" [A.A. Bytsenko et al. (2001a)].

Our definition of vacuum energy in the path integral formalism is again given by Eq. (1.68), but now, the one-loop partition function has to be defined using the "generalized zeta function" $\omega(s|L_D)$, as in Eq. (1.27).

In the case of Bose fields one has periodic boundary conditions in the imaginary time and so, in ultrastatic manifolds, the eigenvalues of the Laplace-like operator L_D are given in terms of the eigenvalues ω_j^2 of L_N by Eq. (1.59). Using Eq. (1.13) and Jacobi identity (A.1), in Appendix A, Section A.1, we obtain

$$\begin{aligned}\omega(s|L_D) &= \frac{\beta\, s\,\Gamma(s-1/2)\zeta(s-1/2|L_N)}{\sqrt{4\pi}\,\Gamma(s)} \\ &\quad + \frac{\beta\, s}{\sqrt{\pi}\Gamma(s)} \sum_{n=1}^{\infty} \int_0^{\infty} dt\, t^{s-3/2}\, e^{-n^2\beta^2/4t}\, \mathrm{Tr}\, e^{-tL_N}. \quad (1.70)\end{aligned}$$

For the operators we are dealing with, one can see that, in the latter equation, the integral is exponentially vanishing in the limit $\beta \to \infty$ and so, as in the standard case, it does not give contributions to the vacuum energy. Then we obtain

$$\begin{aligned}E_V &= -\lim_{\beta\to 0} \partial_\beta \log Z_\beta = -\frac{\nu}{2} \lim_{\beta\to 0} \partial_\beta \omega''(0|\ell^2 L_D) \\ &= -\frac{\nu}{2\sqrt{4\pi}} \lim_{s\to 0} \frac{d^2}{ds^2} \left[\frac{s\Gamma(s-1/2)\zeta(s-1/2|L_N)}{\Gamma(s)}\right] \\ &= \nu \left[G(-1/2|L_N) - \frac{K_D(L_N)}{\sqrt{4\pi}}(\gamma + \log \ell^2) \right. \\ &\quad \left. + \frac{B_q}{\sqrt{\pi}} \left(\frac{\gamma^2}{4} - \frac{\pi^2}{24} - \gamma \log \ell + (\log \ell)^2\right)\right], \quad (1.71)\end{aligned}$$

where $G_N(-1/2|L_N)$ is the finite part, of $\zeta(s|L_N)$ at $s = -1/2$, that is

$$G_N(-1/2|L_N) = \lim_{s\to -1/2} \left[\zeta(s|L_N) + \frac{K_D + \psi(3/2)B_q}{\sqrt{4\pi}(s+1/2)} - \frac{B_q}{\sqrt{4\pi}(s+1/2)^2}\right].$$

Equation (1.71), when $B_q(L_N) = 0$, reduces to the usual one.

The explicit expression (1.71) can be obtained from the knowledge of the meromorphic structure of $\zeta(s|L_N)$. In fact, this can be directly read off from the ones of $\zeta(s|L_D)$, which are supposed to be of the form (1.21). The manifold $\mathcal{M}^D$ we are dealing with is the direct product of a flat manifold S^1 and N-dimensional manifold $\mathcal{M}^N$. Then, the heat-kernel asymptotics reads

$$\begin{aligned}\mathrm{Tr}\, e^{-tL_D} &\sim \sum_{n=0}^{\infty} K_n(L_D)\, t^{\frac{n-D}{2}} + \sum_{n=0}^{\infty} B_n(L_D, q)\, t^{n-q} \log t \\ &= \frac{\beta}{\sqrt{4\pi}} \sum_{n=0}^{\infty} K_n(L_N)\, t^{\frac{n-N}{2}} + \sum_{n=0}^{\infty} B_n(L_N)\, t^{n-p} \log t\,, \end{aligned} \tag{1.72}$$

where $p = q - 1/2$ and q is supposed to be a given integer. Then, according to Eq. (1.22), $\zeta(s|L_D)$ has a singularity at the origin, which corresponds to a singularity of $\zeta(s|L_N)$ at $s = -1/2$. From the latter equation one can read off the correspondence between $K_n(L_N)$ and $K_n(L_D)$ and something similar for $B_n(L_N)$ and $B_n(L_D)$. From Eq. (1.22) we have

$$\Gamma(z)\zeta(z|L_N) = \sum_{n=0}^{\infty} \frac{K_n(L_N)}{z + \frac{n-N}{2}} - \sum_{n=0}^{\infty} \frac{B_n(L_N)}{\left[z + n - \left(q - \frac{1}{2}\right)\right]^2} + J(z)\,, \tag{1.73}$$

from which, Eq. (1.71) easily follows.

Chapter 2

The Zeta-Function Regularization Method

2.1 Survey of the Chapter, Notation and Conventions

This Chapter contains a rigorous description of the main mathematical results on the heat-kernel theory and the subsequent local and global spectral zeta function techniques in quite a simple form, i.e., referred to an either real or complex scalar field propagating in a curved Euclidean space with an external potential. If the field is complex, an external $U(1)$ gauge field is also considered. We develop the kernel of the zeta function for compact manifolds, later we provide some comments on noncompact cases. The theory we are going to present for the heat kernel straightforwardly generalizes, with few changes, into vector valued field theory on fiber bundles and extensions exist also to those involving Dirac operators [N. Berline et al. (1996); J. Roe (1998)].

In this book, a *spacetime* is a Lorentzian manifold, i.e., a smooth D-dimensional manifold equipped with a smooth metric with Lorentzian signature $(-,+,\ldots,+)$. D-dimensional Minkowski spacetime is a manifold diffeomorphic to $\mathbb{R}^D$ equipped with a metric that takes constant form $\mathrm{diag}(-1,+1,\ldots,+1)$ in canonical coordinates of $\mathbb{R}^D$. A spacetime is said to be *curved* if it is not Minkowski spacetime. Quantum field theory can be implemented in a curved spacetime [N.D. Birrell and P.C.W. Davies (1982); S.A. Fulling (1991); R.M. Wald (1994)]. The standard procedure [R.M. Wald (1994)] requires that the D-dimensional spacetime is globally hyperbolic (roughly speaking this means that there must exist $(D-1)$-dimensional spacelike submanifolds Σ, called Cauchy surfaces, such that the assignment of Cauchy data on Σ determine the evolution of a field in the whole spacetime).

All the physical quantities we shall consider in this chapter arise from a Euclidean functional integral which plays a crucial rôle in quantum field theory in a curved spacetime. The functional integral works on fields defined in a Euclidean manifold $\mathcal{M}$ called the "Euclidean section" of the spacetime. A precise definition will be given shortly. Recall that for real scalar fields ϕ that functional reads:

$$Z_E[g] = \int d[\phi]\, e^{-\frac{1}{2}\int_{\mathcal{M}} d\mu_g\, \phi A \phi}\,, \tag{2.1}$$

$A\phi = 0$ is the equation of motion of the considered field where

$$A = -\Delta + V \tag{2.2}$$

with $\Delta = \nabla_a \nabla^a$. If the field is complex, in (2.1) $d[\phi]$ and $\phi A \phi$ have to be replaced respectively for $d[\phi]d[\phi^*]$ and $\phi^* A \phi$ with

$$A = -(\nabla_a - iX_a)(\nabla^a - iX^a) + V\,. \tag{2.3}$$

Here ∇ indicates the covariant derivative associated with the metric g (i.e. the gravitational field), X is a real vector field on $\mathcal{M}$ (for instance the electromagnetic field), V is any other real external potential, and $d\mu_g$ is the natural measure induced by the metric. Operators A are supposed to act on the space of compactly-supported smooth functions. Actually, as it will be clear later, the functional integral depends on the chosen self-adjoint extension (in a suitable Hilbert space) of A, $\mathcal{A}$ instead of A itself. In fact, $\mathcal{A}$ contains information about dynamics embodied in A, as well as the field configurations and $Z_E[g]$ is strongly dependent from boundary conditions.

2.1.1 *Feasibility of physical interpretation via Wick rotation*

As it stands, if some hypotheses are fulfilled the functional integrals above may be interpreted as a partition function of a particular quantum state in a static spacetime corresponding to either a *canonical* or *grand canonical ensemble* of particles associated to the field ϕ [S.W. Hawking (1977)]. The limit case of vanishing temperature (ground state) may also be studied. Further *external* fields, different from gravity/gauge fields, may be introduced in the functional integral above replacing A by a more complicated operator. Moreover ϕ may be replaced by a vector-valued field in a suitable fiber bundle. However, in the present part of the book we only deal with the simple case above specified since the basic aspects of the mathematical theory are not affected from the generalizations mentioned. The case of spinorial fields is more complicated.

This physical interpretation of $Z_E[g]$ in terms of a partition function makes sense provided some mathematical requirements, necessary to make contact with the Lorentzian quantum field theory from the Euclidean formalism, are accomplished. In particular, **(a)** one has to built up a spacetime and **(b)** associate to the operator A, or some Green function of it, a preferred quantum state within that spacetime. The latter can be done by specifying the definition of A, giving precise information about its domain and this, in fact, fixes a self-adjoint extension $\mathcal{A}$ of A. Let us examine these points in some detail.

(a) The procedure of building up a spacetime $\mathcal{M}_L$ —the *Lorentzian section* associated with the Euclidean manifold $\mathcal{M}$— is doable, whenever some hypothesis on $\mathcal{M}$ are fulfilled. Sufficient hypotheses are, for this matter, that *the initial Euclidean D-dimensional manifold $\mathcal{M}$ admits a global Killing field $K \neq 0$ which is orthogonal to a $(D-1)$-dimensional imbedded submanifold Σ that intersects each orbit of K once.*
Under these hypotheses, in a neighborhood of each point $p \in \mathcal{M}$, it is possible to define local coordinates $x^0, x^1, \ldots, x^{D-1}$, where $\tau \stackrel{\text{def}}{=} x^0$ is the parameter of the orbits of $\partial_\tau = K$, and $x^1, \ldots, x^{D-1}$ are local coordinates on Σ. Varying the point $p \in \mathcal{M}$, all the local coordinate frames considered form an atlas $\mathcal{B}$ of $\mathcal{M}$. Usually, the range of τ is fixed to be an interval $[-\beta/2, \beta/2]$, $\beta > 0$, with the identification $-\beta/2 \equiv \beta/2$, no matter which the considered orbit is. With that choice, the one-parameter group of isometries generated by ∂_t is $\mathbb{S}^1$. In local coordinate frames of $\mathcal{B}$, the metric g of $\mathcal{M}$ does not depend on x^0 and satisfies: $g_{00} > 0$, $g_{0i} = 0$ if $i = 1, \ldots, D_1$. The analytic continuation of $x^0 = \tau$ into imaginary values it gives rise to a *real* metric $g_{(L)}$ defined in local coordinates $y^0 \stackrel{\text{def}}{=} t, y^1 \stackrel{\text{def}}{=} x^1, \ldots, y^{D-1} \stackrel{\text{def}}{=} x^{D-1}$ where $g_{(L)00} = -g_{00}$, $g_{(L)0i} = 0$ if $i = 1, \ldots, D-1$ and $g_{(L)}$ does not depend on x^0. It is possible to show that the procedure, performed in all the coordinate patch of $\mathcal{B}$, produces a well-defined Lorentzian manifold $\mathcal{M}_L$. In such manifold, there is a globally defined timelike Killing field, ∂_t, which is orthogonal to a spacelike $(D-1)$-dimensional imbedded submanifold, Σ. In fact, Σ can be seen as imbedded both in $\mathcal{M}$ and $\mathcal{M}_L$. In other words, $\mathcal{M}_L$ is a *static* spacetime with respect to ∂_t and Σ. The only ambiguity in

defining $\mathcal{M}_L$ is the range of the global coordinate, t, which is usually fixed to be the whole space $\mathbb{R}$. $\mathcal{M}_L$ is called the *Lorentzian section* associated with $\mathcal{M}$ and $\mathcal{M}$ is called the *Euclidean section* associated with $\mathcal{M}_L$.

Assuming further hypotheses on K with respect to the Euclidean metric on Σ (see Chapter 6 in [S.A. Fulling (1991)]), it is possible to make Σ a Cauchy surface and, as a consequence, $\mathcal{M}_L$ globally hyperbolic. In particular, global hyperbolicity is assured by the *compactness* of Σ.

We remark that all the procedure can be reversed to build up a Euclidean section, when a static spacetime is given.

The presented machinery may be implemented, without guarantee for a well-behaved construction, in general, also to the case when K has some *fixed point* $x \in \mathcal{M}$, i.e., $K(x) = 0$. As fixed points may involve subtle problems, these situations must be considered case-by-case and with great care. This is also due to the appearance of conical singularities in the Euclidean section for values of the Killing period $\beta \neq \beta_0$, β_0 being a preferred value, fixed by requiring the smoothness of the manifold. Fixed points may be associated with the presence of Killing event horizons in the Lorentzian section. That is the case, for instance, of Schwarzschild spacetime where the preferred value β_0 determines the Hawking temperature of a black hole.

(b) A has to be interpreted as an operator in the Hilbert space $L^2(\mathcal{M}, d\mu_g)$ of the square-integrable functions defined on $\mathcal{M}$, where the measure μ_g is that induced from the metric. The usual domain of A is the space of smooth functions with compact support $C_0^\infty(\mathcal{M})$. With these definitions, A turns out to be symmetric and precise boundary conditions or a precise definition of its domain may make it self-adjoint, which gives rise to a self-adjoint extension $\mathcal{A}$. The positive self-adjoint extensions of A are those physically relevant in our context, because the (generalized) Wick rotation, i.e. the analytic continuation $\tau \to it$ deals with these as shown in [R.M. Wald (1979)]. Indeed, if a self-adjoint extension $\mathcal{A}$, is positive, it determines a unique Green function $\mathcal{A}^{-1}$, which, formally or in a proper sense in some backgrounds (e.g. if $\mathcal{M}$ is compact as we shall see later), may be represented by an integral kernel $G(x, y|\mathcal{A})$. Performing the analytic continuation $\tau \to it$ in the arguments of $G(x, y|\mathcal{A})$ (or following the more general prescription defined in [R.M. Wald (1979)] whenever an integral kernel for $\mathcal{A}^{-1}$ cannot be defined), the arising two-point function enjoys all relavant properties of a *Feynman propagator* of a quasifree quantum state defined on $\mathcal{M}_L$. Therefore, $\mathcal{A}$ determines a quantum state in the Lorentzian section via Wick rotation. More precisely, whenever the program above can be implemented, the pair of Wightman functions of the quantum state associated with $\mathcal{A}$ turns out to be periodic in the imaginary time with period β. Such periodicity property is formally equivalent to the *Kubo-Martin-Schwinger condition* which characterises a quasifree thermal quantum state [S.A. Fulling and S.N.M. Ruijsenaars (1987); B.S. Kay and R.M. Wald (1991)], i.e. a canonical ensemble, at statistical temperature* $1/\beta$.

Whenever the procedure based on points **(a)** and **(b)** can be implemented, $Z_E[g]$ is formally interpreted as the *partition function* $Z[g]$ of the given canonical ensemble at temperature $1/\beta$.

To conclude, suppose that the field ϕ is complex and a $U(1)$ external gauge field is also present. Suppose also that the gauge field does not depend on τ, and that the scalar product of X and ∂_τ attains the constant value μ on $\mathcal{M}$. Under further suitable hypotheses on the geometry of $\mathcal{M}$, involving the presence of a timelike Killing vector with orbits with period β [S.W. Hawking (1977)], a (generally stationary) Lorentzian section $\mathcal{M}_L$ can be defined and the analytic continuation of the Green function of $\mathcal{A}$ in $\mathcal{M}_L$ represents a quantum state associated with a grand canonical ensemble of charged particles with chemical potential μ and temperature $1/\beta$. In this case $Z_E[g]$ is interpreted as the partition function of the given grand canonical ensemble.

Once the physical meaning of the Euclidean functional integral (2.1) has been clarified, our subsequent goal is to give a mathematical interpretation of that functional integral which can be implemented in practical computations. This is done by generalizing the concept of determinant, as shown in Chapter 1. In order to implement, in a rigorous way, that interpretation, it is necessary to introduce several technical notions and results concerning the heat-kernel theory. They will be presented in Section 2.2. In fact there is a strong interplay between the spectral zeta function and the heat kernel, which will be examined in Section 2.3. The subsection below is devoted to specify the framework where we intend to develop all this mathematical theory.

*As is well known, the local thermodynamical temperature is given by Tolman's relation $T_\beta(x) = T_\beta/\sqrt{g(\partial_\tau, \partial_\tau)_x}$.

2.1.2 *Notation and general hypotheses*

In the following, $(\mathcal{M}, g)$, more often indicated simply by $\mathcal{M}$, is a smooth Riemannian D-dimensional manifold equipped with a smooth metric g. (Throughout this chapter "smooth" means C^∞.) The metric g in the point $x \in \mathcal{M}$ will be indicated by $g_{ab}(x)$ in components in local coordinates. $g_x(X, Y)$ denotes the scalar product $X_a Y^a = g_{ab}(x) X^a Y^b$, where X and Y are vectors defined in x. Sometimes we avoid to indicate the point x if this does not produce confusion. ∇ indicates the covariant derivative associated with the metric. $L^2(\mathcal{M}, \mu_g)$ denotes the Hilbert space of functions $f : \mathcal{M} \to \mathbb{C}$ such that $|f|^2$ is integrable with respect to the Borel measure μ_g naturally induced by the metric. That is $d\mu_g = \sqrt{\det(g)}\, dx^1 \wedge \ldots \wedge dx^D$ in every local coordinate patch, $\det(g)$ denoting the determinant of the metric tensor in those coordinates.

General hypotheses. *Throughout this chapter we assume the following General Hypotheses:*
(1) *$\mathcal{M}$ is second countable, Hausdorff, connected, oriented and compact (throughout the book, "compact" includes the absence of boundaries, unless otherwise stated).*
(2) *Concerning the operators, we consider real elliptic differential operators with a "generalized Laplace-Beltrami operator plus potential" form, either*

$$A \stackrel{def}{=} -\Delta^{(X)} + V \ : \ C^\infty(M) \to L^2(\mathcal{M}, \mu_g)\,, \tag{2.4}$$

where, locally,

$$\Delta^{(X)} \stackrel{def}{=} (\nabla_a - iX_a)(\nabla^a - iX^a)\,, \tag{2.5}$$

or

$$A \stackrel{def}{=} -\Delta + V \ : \ C^\infty(M) \to L^2(\mathcal{M}, \mu_g)\,. \tag{2.6}$$

Above , X is a real smooth vector field on $\mathcal{M}$, and V is a real *function of $C^\infty(\mathcal{M})$ class, with physical dimensions of a* squared mass.

In the first case, A is the Euclidean operator of motion of charged bosons, in the latter A is the Euclidean operator of motion of neutral bosons. A few comments on these hypotheses are in order. First of all, a countable base of the topology is required in order to endow the manifold with a partition of the unity and allow the use of Hilbert-Schmidt's theory (which is fundamental in our dealing with the heat-kernel theory [I. Chavel (1984)]). Indeed, the Hilbert space $\mathcal{M}$ $L^2(\mathcal{M}, \mu_g)$ of the complex valued functions on $\mathcal{M}$ results to be separable [P.R. Halamos (1969)] and this allows one to describe Hilbert-Schmidt operators in terms of integral kernels. Anyhow, the requirement of compactness implies completeness (equivalent to geodesic completeness), paracompactness (the manifold being Hausdorff) and thus the presence of a countable base of the topology being automatically assured by our general hypotheses.

With the given domain, trivially A turns out to be Hermitian. If S is a Hermitian operator in a Hilbert space with domain $\mathcal{D}(S)$ and $c \in \mathbb{R}$, by $S \geq c$, we mean $(f, Sf) \geq c(f, f)$ for all of $f \in \mathcal{D}(A)$. The operators A satisfy $A - V_0 I \geq 0$ where $V_0 = \inf_{x \in \mathcal{M}} V(x)$ and I is the identity operator, they are symmetric on $C^\infty(\mathcal{M})$ because $\overline{C^\infty(\mathcal{M})} = L^2(\mathcal{M}, \mu_g)$ and they admit self-adjoint extensions since they commutes with the anti-unitary complex-conjugation operator in $L^2(\mathcal{M}, \mu_g)$ [M. Reed and B. Simon (1980)]. In particular, if A is one of the operators given above, one may consider the self-adjoint extension given by $\mathcal{A} = B + V_0 I$, where B is the Friedrichs self-adjoint extension of the positive operator $B' = A - V_0 I$ [M. Reed and B. Simon (1980)] which, as is well known, is bounded from below by the same bound of B'. Actually, as we shall say, A is essentially self-adjoint and thus $\mathcal{A}$ is its unique self-adjoint extension.

We use the fact that the *injectivity radius* r of $\mathcal{M}$ is strictly positive in compact manifolds. We remind the reader that the *injectivity radius* can be defined, for an Euclidean manifold, as

$$r = \inf_{p \in \mathcal{M}} d(p, C_m(p)) \,, \tag{2.7}$$

where the *cut locus* $C_m(p)$ is the set of the union of the *cut points* of p along all of geodesics that start from p, d is the *geodesic distance*† [I. Chavel (1984); M.P. do Carmo (1992)]. The relevant fact for our considerations in this book is that if $r > 0$ and if $p, q \in \mathcal{M}$ and $d(p, q) < r$, there is a *normal coordinate system* centered in p (resp. q) which contains also the point q (resp. p). Within this neighborhood, the function $x \mapsto (1/2)d^2(p, x) \stackrel{\text{def}}{=} \sigma(p, x)$ (resp. $x \mapsto (1/2)d^2(x, q)$) is C^∞. Moreover, by this result, taking into account that $(x, y) \mapsto d(x, y)$ is everywhere continuous in $\mathcal{M} \times \mathcal{M}$ for

† $d(p, q)$ is defined as $\inf\{\ell(\gamma)\}$ for all $\gamma : [0, 1] \to \mathcal{M}$ differentiable such that $\gamma(0) = p$, $\gamma(1) = q$, $\ell(\gamma)$ denoting the length of γ.

Euclidean metrics and employing Sobolev's Lemma [W. Rudin (1997)], one finds that the function $(x, y) \mapsto \sigma(x, y)$ belongs to the space of smooth functions $C^\infty(\{(x, y) \in \mathcal{M} \times \mathcal{M} \mid d(x, y) < r\})$. Sufficient conditions for having $r > 0$ are stated in Chapter 13 of [M.P. do Carmo (1992)]. In particular, a strictly positive upper bound k_0 for the sectional curvature of a compact manifold is sufficient to have $r > 0$, indeed, in this case $r \geq \pi/\sqrt{k_0}$. Notice also that, for instance, a Riemannian manifold symmetric under a Lie group of isometries involves $r > 0$ trivially. Finally $r > 0$ if $\mathcal{M}$ is compact as one may easily show.

We also remind the reader that in either Lorentzian or Euclidean smooth manifolds (no matter the value of r in the latter case) the so-called *Synge world function* $(x, y) \mapsto \sigma(x, y)$ introduced above can be defined as follows (and this definition is equivalent to that given above in the Euclidean case). Let x, y belong to a geodesically convex ‡ neighborhood $\mathcal{N}$. $\sigma(x, y) \stackrel{\text{def}}{=} (1/2)g_x(X, X)$, where X is the tangent vector at $\lambda = 0$ to the unique geodesic segment $\gamma = \gamma(\lambda)$, with $\gamma(0) = x$, $\gamma(1) = y$, $\gamma(\lambda) \in \mathcal{N}$ if $\lambda \in [0, 1]$. On $\mathcal{N} \times \mathcal{N}$, σ turns out to be smooth, symmetric ($\sigma(x, y) = \sigma(y, x)$) and it holds $\nabla_x \sigma = X$ and $2\sigma(x, y) = g_x(\nabla_x \sigma, \nabla_x \sigma) = g_y(\nabla_y \sigma, \nabla_y \sigma)$ [B.S. DeWitt and R.W. Brehme (1960); F.G. Friedlander (1975); S.A. Fulling (1991)].

2.2 Heat-Kernel Expansion and Coefficients

2.2.1 *The heat-kernel expansion on compact manifolds*

The zeta function technique is based on several mathematical properties of the heat kernel associated with the operator $\mathcal{A}$. In the following we summarize relevant statements of the heat kernel on a compact manifold.

Let us start with the definition of *heat kernel* for operators A as defined above [R.M. Wald (1979); I. Chavel (1984); P.B. Gilkey (1995); E.B. Davies (1989)].

Definition 2.1 *Taking into account the general hypotheses on $\mathcal{M}$ and A,*

‡An open set $\mathcal{N}$ is geodesically convex if, for every pair $x, y \in \mathcal{N}$ there is a unique geodesic segment γ from x to y which is completely contained in $\mathcal{N}$ (but there can be others which are not completely contained in $\mathcal{N}$) [M.P. do Carmo (1992)].

consider a one-parameter class of operators

$$(K_t\psi)(x) \stackrel{def}{=} \int_{\mathcal{M}} d\mu_g(y)\, K(t,x,y|A)\psi(y) \quad t \in (0,+\infty) \tag{2.8}$$

where the integral kernel is $C^0((0,+\infty)\times\mathcal{M}\times\mathcal{M})$, $C^1((0,+\infty))$ *in the variable* t, $C^2(\mathcal{M})$ *in the variable* x *and satisfies the "heat equation"*

$$[\partial_t + A_x]\, K(t,x,y|A) = 0 \tag{2.9}$$

(the index x *in* A_x *means that* A *acts on the variable* x*) with initial condition,*

$$\lim_{t\to 0^+} \int_{\mathcal{M}} d\mu_g(y)\, K(t,x,y|A)\psi(y) = \psi(x) \quad \textit{for each } x \in \mathcal{M} \tag{2.10}$$

uniformly in x*, for all* $\psi \in C^0(\mathcal{M})$*. The kernel* $K(t,x,y|A)$*, if exists, is called heat kernel of* A*.*

If $\psi \in C^0(\mathcal{M})$, using usual theorems based on Lebesgue dominate convergence, it turns out that $(t,x) \mapsto (K_t\psi)(x)$ belongs to $C^0([0,+\infty)\times\mathcal{M})$, $C^1(0,+\infty)$ in t and $C^2(\mathcal{M})$ in x. (2.10) has two important consequences:

$$\lim_{t\to 0^+} \int_{\mathcal{M}} d\mu_g(y)\, K(t,x,y|A) = 1 \quad \text{for each } x \in \mathcal{M} \tag{2.11}$$

and, using (2.11) and the compactness of $\mathcal{M}$,

$$\lim_{t\to 0^+} \int_{\mathcal{M}} d\mu_g(y)\, K(t,x,y|A)\psi(t,y) = \psi(0,x) \quad \text{for each } x \in \mathcal{M} \tag{2.12}$$

for every $\psi \in C^0((-\varepsilon,\varepsilon)\times\mathcal{M})$ with $\varepsilon > 0$.

The existence of heat kernel for suitable operators A, also dropping the hypothesis of $\mathcal{M}$ compact, can be proven directly, by studying the integral kernel of the exponential of the Friedrichs self-adjoint extension of $\mathcal{A}$ as done in Chapter 5 of [E.B. Davies (1989)] whenever $A \geq 0$ (or more weakly $\mathcal{A} \geq c$ for some constant $c \in \mathbb{R}$).

Holding our general hypotheses on $\mathcal{M}$ and A, we have several important theorems.

Theorem 2.1 *Under the stated conditions on* $\mathcal{M}$ *and* A*, if an operator* K_t *defined in Definition 2.1 exists, it is uniquely determined and is*

Hermitian:

$$K(t,x,y|A) = K(t,y,x|A)^* , \tag{2.13}$$

for $t > 0, x, y \in \mathcal{M}$, where * denote the complex conjugation.

Proof. Define $L \stackrel{\text{def}}{=} -A - \partial/\partial t$. Let u, v be complex functions defined on $\mathcal{M} \times (0,t)$, $t > 0$ being any constant and such that they are C^2 in the first variable and C^1 in the second. Since $\mathcal{M}$ is compact without boundary, a direct computation shows that

$$\begin{aligned}&\int_{\mathcal{M}} d\mu_g(z)\,[(Lu)^*(z,t-\tau)v(z,\tau) - u^*(z,t-\tau)(Lv)(z,\tau)]\\ &= \int_{\mathcal{M}} d\mu_g(z)\frac{\partial}{\partial\tau}u^*(z,t-\tau)v(z,\tau)\,.\end{aligned} \tag{2.14}$$

Integrating with respect to τ we get, for every interval $[\alpha,\beta] \subset (0,t)$,

$$\begin{aligned}&\int_\alpha^\beta d\tau \int_{\mathcal{M}} d\mu_g(z)\,[(Lu)^*(z,t-\tau)v(z,\tau) - u^*(z,t-\tau)(Lv)(z,\tau)]\\ &= \int_{\mathcal{M}} d\mu_g(z)\,[u^*(z,t-\beta)v(z,\beta) - u^*(z,t-\alpha)v(z,\alpha)]\ .\end{aligned} \tag{2.15}$$

To conclude, fix $x, y \in \mathcal{M}$ and define $u(z,\tau) \stackrel{\text{def}}{=} K(\tau,x,z|A)$ and $v(z,\tau) \stackrel{\text{def}}{=} K(\tau,y,z|A)$ where $K(\tau,u,v|A)$ is a heat kernel of A in the sense of Definition 2.1. Inserting these functions in (2.15), taking the limits $\alpha \to 0^+$ and $\beta \to t^-$ and using both (2.9) and (2.12) one gets the hermiticity property $K(t,y,x|A)^* = K(t,x,y|A)$. Moreover, by the same procedure, if $u(z,\tau) \stackrel{\text{def}}{=} K(\tau,x,z|A)$ and $v(z,\tau) \stackrel{\text{def}}{=} K'(\tau,y,z|A)$ where $K'(\tau,y,z|A)$ is another heat kernel, one gets $K(t,x,y|A)^* = K'(t,y,x|A)$. Therefore $K(t,x,y|A) = K'(t,x,y|A)$ because of the hermiticity of $K(t,x,y|A)$. □

A relevant proposition concerning the existence and uniqueness of the heat-kernel equation, which we shall use in the theorem below, holds.

Proposition 2.1 *Under the general hypotheses on $\mathcal{M}$ and A, assume that a heat-kernel exists for A. For each $\psi \in C^0(\mathcal{M})$ there is a unique function $\phi = \phi(t,x)$ such that $\phi \in C^0([0,+\infty)\times\mathcal{M})$, ϕ is in $C^1(0,+\infty)$ in the former argument, and in $C^2(\mathcal{M})$ in the latter, and satisfies the heat equation $(\partial_t + A_x)\phi(t,x) = 0$ on $(0,+\infty) \times \mathcal{M}$ with initial condition $\phi(0,x) = \psi(x)$ for all $x \in \mathcal{M}$.*

Proof. Given any ψ as above, $\phi(t,x) \stackrel{\text{def}}{=} (K_t\psi)(x)$ is a solution of the heat equation with the requested regularity properties. By linearity, the uniqueness holds if it holds for the case $\psi \equiv 0$. Let ϕ be a solution of the heat equation with initial condition $\phi(0,x) = 0$, for all $x \in \mathcal{M}$, and with the regularity properties stated in the hypotheses. Let us show that $\phi \equiv 0$. To this end consider, for any smooth function h on $\mathcal{M}$, and $0 < \tau < t$, where $t > 0$ is any fixed constant,

$$f(\tau) = \int_{\mathcal{M}\times\mathcal{M}} d\mu_g(x) \otimes d\mu_g(y) K(t-\tau, x, y|A_0) h(y) \phi(\tau, x) .$$

Since $\mathcal{M}$ has no boundary, it holds $f'(\tau) = 0$, for $0 < \tau < t$, by the heat equation. Thus, f is constant on $(0,t)$. Taking the limit as $\tau \to 0^+$ and $\tau \to t^-$, one also gets that, for each $t > 0$,

$$\int_{\mathcal{M}} d\mu_g(x) \phi(t,x) h(x) = 0 .$$

Since h is smooth and arbitrary, this implies that $\phi(t,x) = 0$, for all $x \in \mathcal{M}$ and all $t > 0$. □

Theorem 2.2 *Taking into account the general hypotheses on $\mathcal{M}$ and the operator A let $V_0 \stackrel{def}{=} \inf\{V(x) \,|\, x \in \mathcal{M}\}$. If the heat kernel of A_0 exists and is jointly smooth for $t \in (0,+\infty)$, $x, y \in \mathcal{M}$, the following statements hold.*

(a) *$\{K_t\}_{t\geq 0}$ with $K_0 \stackrel{def}{=} I$ is a semigroup, i.e. $K_t K_{t'} = K_{t+t'}$, and the operators are bounded, self-adjoint, positive and Hilbert-Schmidt if $t > 0$.*

(b) *The following expansion holds for $t > 0$:*

$$K(t,x,y|A) = \sum_{j=0}^{\infty} e^{-t\lambda_j} \phi_j(x) \phi_j^*(y) \tag{2.16}$$

where the series converges on $[\alpha, +\infty) \times \mathcal{M} \times \mathcal{M}$ regularly (i.e., uniformly together with the series of the absolute values) for any fixed $\alpha > 0$.

(c) *The class $\{\phi_j | j = 0,1,2,...\}$ is a Hilbert base of $L^2(\mathcal{M}, \mu_g)$ made of smooth eigenvectors of A, the dimension of each eigenspace being finite, and the associated eigenvalues satisfy*

$$V_0 \leq \lambda_0 \leq \lambda_1 \leq \lambda_2 \leq \ldots \leq \lambda_j \leq \ldots \to +\infty . \tag{2.17}$$

One may have, in particular, for some j, $\lambda_{j-1} < \lambda_j = \lambda_{j+1} = \ldots = \lambda_{j+d-1} < \lambda_{j+d}$, where $d < +\infty$ is the dimension of the eigenspace of λ_j.

Proof. Notice that any $L^2(\mathcal{M} \times \mathcal{M}, \mu \otimes \mu)$ kernel on a general measure space $(\mathcal{M}, \mu)$ induces a Hilbert-Schmidt (and thus bounded and compact) operator on $L^2(\mathcal{M}, \mu)$ if μ is separable [F. Riesz and B.S. Nagy (1972)]. This is our case because, for $t > 0$, the heat kernel is $L^2(\mathcal{M} \times \mathcal{M}, \mu_g \otimes \mu_g)$, being smooth and μ_g being separable since $\mathcal{M}$ is compact. Self-adjointness of K_t is a direct consequence of (2.13). Proposition 2.1 entails the semi-group property: $K_0 = I$ by definition, moreover take $t' > 0$ and $\psi \in C^0(\mathcal{M})$, the map $t \mapsto K_{t+t'}\psi$ is a regular (in the sense of Proposition 2.1) solution of the heat equation with initial condition $K_{t'}\psi$. On the other hand, the map $t \mapsto K_t K_{t'}\psi$ is another regular solution of the same equation with the same initial condition: Thus, they must coincide. Finally, since $\psi \in C^0(\mathcal{M})$, which is dense in $L^2(\mathcal{M}, \mu_g)$, one has $K_0 = I$ and $K_t K_{t'} = K_{t+t'}$ when these operators work on $L^2(\mathcal{M}, \mu_g)$. Each K_t is positive because $(f, K_t f) = (f, K_{t/2}K_{t/2}f) = (K_{t/2}f, K_{t/2}f) \geq 0$ for all $f \in L^2(\mathcal{M}, \mu_g)$. As is well-known, bounded positive (and thus self-adjoint) compact operators have spectrum made of a discrete-spectrum part $\Lambda_0(t) \geq \Lambda_1(t) \geq \ldots \geq \Lambda_j(t) \geq \ldots \to 0$ and a possible continuous-spectrum part containing 0 at most, finally the set of eigenvectors $\phi_j(t)$ forms a Hilbert base. Notice that, in particular, the dimension of every eigenspace is finite and K_t is strictly positive. Since K_t is self-adjoint, $K_t\phi \to \phi$ as $t \to 0^+$ for $\phi \in C^0(\mathcal{M})$ and $C^0(\mathcal{M})$ is dense in $L^2(\mathcal{M}, \mu_g)$, we conclude that the one-parameter semigroup of bounded operators K_t on $L^2(\mathcal{M}, \mu_g)$ is weakly continuous and thus it is also strongly continuous (Theorem 1.4 in Chapter 2 of [A. Pazy (1983)]). Since the semigroup is strongly continuous, it must hold $K_t = e^{-tB}$ where B is the densely-defined, closed, normal operator obtained as the strong derivative of K_t at $t = 0$ (Theorem 13.38 in [W. Rudin (1997)]). Moreover as K_t is self-adjoint, it is simply proven that B must be self-adjoint too. The spectral theorem implies that $\text{Spec}(K_t) = \{e^{-t\lambda} \,|\, \lambda \in \text{Spec}(K_t)\}$ and thus the spectrum of B must be discrete (in particular each eigenspace has finite dimension) with elements $\lambda_0 \leq \lambda_1 \leq \lambda_2 \leq \ldots \leq \lambda_j \leq \ldots \to +\infty$. Moreover, for every fixed $t > 0$, if ϕ_j is an eigenvector of B with eigenvalue λ_j, it must be (up to a suitable re-arrangement of the eigenvectors of K_t) $\phi_j(t) = \phi_j$ and $\Lambda_j(t) = e^{-t\lambda_j}$. As $\phi_j(t) = \Lambda_j(t)^{-1}K_t\phi_j(t)$, and K_t is defined by a smooth kernel, ϕ_j must be as smooth as the heat kernel $K(t, x, y|A)$ in the variable x. As a consequence, the t-derivative of $K_t\phi_j$ can be computed using the heat equation, obtaining $(d/dt)K_t\phi_j = -K_t A\phi_j$. On the other hand, it is simply proven that the pointwise derivative must coincide with the analogous derivative of $K_t\phi_j = e^{-tB}\phi_j$ in the strong topology sense which reads $K_tB\phi_j$. We conclude that $K_tA\phi_j = K_tB\phi_j$. But K_t is injective (0 is not an eigenvalue), and so $A\phi_j = B\phi_j$ and thus $A\phi_j = \lambda_j\phi_j$ for $j = 0, 1, 2, \ldots$. Finally, the lower bound in (2.17) is a consequence of $(\phi, A\phi) \geq V_0(\phi, \phi)$ for each $\phi \in C^\infty(\mathcal{M})$. From the Hilbert-Schmidt theory, the expansion (2.16) holds in the $L^2(\mathcal{M} \times \mathcal{M})$ sense. Then the convergence properties of (2.16) follows from Mercer's theorem [F. Riesz and B.S. Nagy (1972)] by the positivity of the Hilbert-Schmidt operator K_t. □

Concerning the issue of the pointwise positivity of the heat kernel, see also [E.B. Davies (1989)]. Notice that the stronger requirement $A \geq 0$ entails $\lambda_j \geq 0$. This constraint is essential to give rise to an exponential decay of $K(t, x, y|A)$ (barring zero modes) as $t \to +\infty$. This decay allows the definition of the various zeta functions we shall see shortly.

2.2.2 *The self-adjoint extension*

The next theorem proves that A is essentially self-adjoint and thus, that it uniquely determines its self-adjoint operator $\mathcal{A}$.

Theorem 2.3 *Under the same hypotheses as in Theorem 2.2, we have*

(a) *A and its positive powers A^n, $n = 1, 2, \ldots$ are essentially self-adjoint, hence they admit unique self-adjoint extensions, $\mathcal{A}^n$, which coincide with $\overline{A^n}$ (and with Friedrichs' extension of A^n if A is positive) moreover $\mathcal{A}^n$ is bounded from below by the same bound of A^n;*

(b) $\mathrm{Spec}(\mathcal{A}) = \mathrm{Spec}_p(\mathcal{A}) = \mathrm{Spec}_d(\mathcal{A}) = \{\lambda_n | n = 0, 1, 2 \ldots\}$*;*

(c) *in the usual spectral-theory sense, for $t \in (0, +\infty)$, $K_t = e^{-t\mathcal{A}}$;*

(d) *$\{K_t | t \in (0, +\infty)\}$ is a strongly continuous one-parameter semigroup of bounded operators such that, for each $\psi \in L^2(\mathcal{M}, \mu_g)$,*

(i) *$K_t\psi \xrightarrow{L^2} \psi$ as $t \to 0^+$,*

(ii) *if $A \geq 0$, $K_t\psi \xrightarrow{L^2} P\psi$ as $t \to +\infty$ and this limit also holds pointwise and uniformly on $\mathcal{M}$, P being the projector onto* $\mathrm{Ker}(\mathcal{A})$ *and* $\mathrm{Ker}(\mathcal{A}) = \mathrm{Ker}(A)$.

Above $\mathrm{Spec}(\mathcal{A})$, $\mathrm{Spec}_\mathrm{p}(\mathcal{A})$ and $\mathrm{Spec}_\mathrm{d}(\mathcal{A})$ denote the spectrum, the *point* spectrum and *discrete* spectrum of $\mathcal{A}$ respectively.

Proof. The essentially self-adjointness of A^n is a consequence of Nelson's theorem [B. Simon (1972)], using the class of all linear combinations of vectors ϕ_j as a dense set of analytic vectors. The remaining part of **(a)** is standard spectral theory. **(b)** is a direct consequence of part **(c)** of Theorem 2.2, and the uniqueness of self-adjoint extensions of A using the fact that $\mathcal{A}f \stackrel{\text{def}}{=} \sum_j \lambda_j \phi_j^*(\phi_j, f)$ (in the strong limit sense and with obvious domain) extends A and is self-adjoint. The proof of **(c)** is trivial if one notices that from **(b)** and **(c)** of Theorem 2.2, K_t and $e^{-t\mathcal{A}}$ coincide on the dense set of linear combinations of eigenvectors ϕ_j and both operators are bounded. The proof of the first part of **(d)** and **(i)** are direct consequences of **(c)** via standard spectral theory. The existence of the L^2 limit stated in **(ii)** can be proven using the spectral decomposition of $e^{-t\mathcal{A}}$, taking the presence of a discrete spectrum into account. Afterwards, the statement concerning the uniform convergence can be proven as follows. Consider the identity $|K_{t+T}\psi - K_{\tau+T}\psi|(x) = |K_T(K_t\psi - K_\tau\psi)|(x)$. The right-hand side can be rewritten $|\int_{\mathcal{M}} d\mu_g(y) K(T,x,y|A)(K_t\psi - K_\tau\psi)(y)| \leq C||K_t\psi - K_\tau\psi||$ by the Cauchy-Schwartz inequality and the continuity of the heat kernel, where C is some positive constant not depending on T and x. Therefore we have obtained that $\sup_{\mathcal{M}} |K_{t+T}\psi - K_{\tau+T}\psi| \leq C||K_t\psi - K_\tau\psi||$. As $K_t\psi$ converges in the sense of L^2 for $t \to +\infty$, $K_t\psi$ must converge uniformly to a continuous function on $\mathcal{M}$. The two limits must coincide, since uniform convergence implies that in the L^2 sense. □

Remark. The point **(c)** states that the heat kernel of A coincides with the integral kernel of $e^{-t\mathcal{A}}$. In our hypotheses, A completely determines $\mathcal{A}$ since the former is essentially self adjoint. However there are cases (e.g. manifolds with boundary) where different self-adjoint extensions of A take place. In those cases, there are various heat kernels of A respectively associated with the various self-adjoint extensions of A itself. For this reason, from now on we indicate the heat kernel by $K(t,x,y|\mathcal{A})$ instead of $K(t,x,y|A)$.

2.2.3 *Existence of the (differentiated) heat-kernel expansion*

A constructive proof of existence of the heat kernel can be obtained by the straightforward generalization of $-\Delta$, contained in Sections 3 and 4 of Chapter VI of [I. Chavel (1984)], to operators $-\Delta^{(X)} + V$. The next theorem presents a small-t off-diagonal expansion of the differentiated heat kernel, which generalizes the well-known *on-diagonals* asymptotic expansion of the heat kernel [S. Minakshisundaram (1949); S. Minakshisundaram and A. Pleijel (1949); I. Chavel (1984); P.B. Gilkey (1995); R. Camporesi (1990); S.A. Fulling (1991)].

To state and prove the theorem, we have to introduce some relevant ingredients and specify the notation used. From now on, $\mathbb{N} := \{0, 1, 2, \ldots\}$, $\gamma = (\gamma_1, \ldots, \gamma_D) \in \mathbb{N}^D$ is a multiindex, $|\gamma| = \sum_i \gamma_i$ and D_z^α denotes $\partial^{|\alpha|}/(\partial z^{1\alpha_1} \cdots \partial z^{D\alpha_D})$ referred to a local coordinate patch. The smoothing function $\chi \in C^\infty([0, +\infty))$ is defined as follows. $\chi(u) = 1$ if $|u| < r^2/16$ and $\chi(u) = 0$ when $|u| \geq r^2/4$, where r is the injectivity radius of the manifold $\mathcal{M}$ (see Section 2.1.2). The precise cutoff values $r^2/4$ and $r^2/16$ are not so important and could be changed, anyway in the following we assume those values.

The *off-diagonal heat-kernel coefficients*, $a_j(x, y|A)$, ($j = 0, 1, 2...$), which appear in the theorem below are smooth functions defined in $\mathcal{M} \times \mathcal{M}$ when x and y belong to a common geodesically convex neighborhood, in particular, if $d(x, y) < r$. However they can be trivially defined to the whole manifold $\mathcal{M} \times \mathcal{M}$, making use of the smoothing function: $\chi(\sigma)a_j \in C^\infty(\mathcal{M} \times \mathcal{M})$. $\sigma = \sigma(x, y)$ is one half the squared geodesic distance of x and y (see Section 2.1.2). We shall give the definition of these heat-kernel coefficients and discuss the interplay of these and other geometric objects, as σ itself in Section 2.2.4. For any fixed $N = 0, 1, 2, ...$, the function of class $C^\infty((0, +\infty) \times \mathcal{M} \times \mathcal{M})$

$$F_N(t, x, y) \stackrel{\text{def}}{=} \frac{e^{-\sigma(x,y)/2t}}{(4\pi t)^{D/2}} \chi(\sigma(x, y)) \sum_{j=0}^{N} a_j(x, y|A) t^j$$

is called the *heat-kernel parametrix*. It is useful to build up a heat kernel, taking, in a certain sense, the limit as $N \to \infty$ [I. Chavel (1984)]. (Notice that in a proper sense, the limit of F_N as $N \to +\infty$ generally does *not* converge.) Reducing to a normal coordinate system centered on y, and using the fact that $a_0(y, y|A) = 1$ and $(4\pi t)^{-D/2} e^{-|x-y|^2/4t} \to \delta(x - y)$ in the weak sense as $t \to 0^+$ in $\mathbb{R}^D$, one finds that

$$F_N(t, x, y) \to \delta(x, y) \text{ as } t \to 0^+ . \tag{2.18}$$

We can now state our theorem on the differentiated expansion of the heat kernel.

Theorem 2.4 *Under the general hypotheses on $\mathcal{M}$ and A, a smooth heat* *kernel $K(t,x,y|\mathcal{A})$ of A exists. Let I_u be a sufficiently small neighborhood of any $u \in \mathcal{M}$, equipped with a coordinate system. If $x, y \in I_u$, for any pair of multiindices α, β, an integer $N > D/2 + 2|\alpha| + 2|\beta|$ and a real $\eta \in (0,1)$, then*

$$\begin{aligned} D_x^\alpha D_y^\beta K(t,x,y|\mathcal{A}) &= D_x^\alpha D_y^\beta \left\{ \frac{e^{-\sigma(x,y)/2t}}{(4\pi t)^{D/2}} \chi(\sigma) \sum_{j=0}^{N} a_j(x,y|A) t^j \right\} \\ &+ \frac{e^{-\eta\sigma(x,y)/2t}}{(4\pi t)^{D/2}} t^{N-|\alpha|-|\beta|} O_{\eta,N}^{(\alpha,\beta)}(t;x,y) \,. \end{aligned} \tag{2.19}$$

Above, the derivative D_x^α and D_x^β refer to the coordinates on I_u, and it turns out that $O_{\eta,N}^{(\alpha,\beta)}$ is continuous in $[0,+\infty) \times I_u \times I_u$ (or in $[0,+\infty) \times \mathcal{M} \times \mathcal{M}$ for $\alpha = \beta = 0$) and (x,y)-uniformly bounded by $M_{\eta,N}^{(\alpha,\beta)} t$ in $[0,T) \times I_u \times I_u$ for some constants $T, M_{\eta,N}^{(\alpha,\beta)} > 0$.

Proof. The proof of the existence of the heat kernel given in [I. Chavel (1984)], Section 4 of Chapter VI, for the operator $A = -\Delta$ in a compact Riemannian manifold, holds true for an operator $A = -\Delta^{(X)} + V$ as it stands. Therefore, using the mathematical machinery presented in that proof, we see that (2.19) only. By (45) in Section 4 of Chapter VI of [I. Chavel (1984)], one then has that the heat kernel can be expanded as

$$K(t,x,y|\mathcal{A}) = F_N(t,x,y) + (F_N * F)(t,x,y) \,, \tag{2.20}$$

where

$$F_N(t,x,y) = \frac{e^{-\sigma(x,y)/2t}}{(4\pi t)^{D/2}} \chi(\sigma(x,y)) \sum_{j=0}^{N} a_j(x,y|A) t^j \,, \tag{2.21}$$

provided $N > D/2 + 2$ is a fixed integer [§]. The convolution $*$ is defined as

$$(G * H)(t,x,y) \stackrel{\text{def}}{=} \int_0^t \int_{\mathcal{M}} d\tau \, d\mu_g(z) G(\tau,x,z) H(t-\tau,z,y) \,, \tag{2.22}$$

[§] Actually, there is a missprint in Eq. (45) in Section 4 of Chapter VI of [I. Chavel (1984)], because of the unnecessary presence of the operator L_x on the right hand side of the first line of (45) in Section 4 of Chapter VI of [I. Chavel (1984)]. Owing to the absence of this operator in the correct formula, we cannot get the second line of (45) in a direct way. Some other parts of Section 4 of Chapter VI in [I. Chavel (1984)] contain additional missprints, like the requirement $F \in C^0(\mathcal{M} \times \mathcal{M} \times [0,+\infty))$ in Lemma 2, which has to be corrected into $F \in C^1(\mathcal{M} \times \mathcal{M} \times [0,+\infty))$.

whenever the right-hand side makes sense. Finally, the function F which appears in (2.20) is defined by a uniformly convergent series in $[0,T]\times\mathcal{M}\times\mathcal{M}$ for any $T>0$

$$F(t,x,y) \stackrel{\text{def}}{=} \sum_{l=1}^{\infty}[(A_x - \partial/\partial t)F_N]^{*l}(t,x,y) \tag{2.23}$$

(B^{*l} means $B*B*\cdots*B$, l times). This function belongs to $C^L([0,+\infty)\times\mathcal{M}\times\mathcal{M})$ provided $M > D/2+2L$ (see Section 4 of Chapter VI in [I. Chavel (1984)]).

The latter term on the right-hand side of (2.20), computed up to the coefficient $a_N(x,y|A)$ using the given definition of convolution, can be explicitly written as

$$(F_N * F)(t,x,y) = \int_0^t d\tau\tau^{-D/2}(t-\tau)^{N-D/2}\int_{\mathcal{M}} d\mu_g(z)\mathcal{F}_N(\tau,x,z)$$
$$\times\mathcal{F}(t-\tau,z,y)e^{-\sigma(x,z)/2\tau}e^{-\sigma(z,y)/2(t-\tau)}\,, \tag{2.24}$$

where

$$\mathcal{F}_N(t,x,z) = \chi(\sigma(x,y))\sum_{j=0}^{N} a_j(x,y|A)t^j\,.$$

This function is of class $C^\infty([0,+\infty)\times\mathcal{M}\times\mathcal{M})$ and vanishes smoothly whenever the geodesic distance between x and z is sufficiently large, that is $d(x,z)\geq r/2$, with our definition of χ. $\mathcal{F}(t,z,y)$ happens to enjoy similar properties, but it is defined only when the distance of z and y is sufficiently small, for instance if $d(x,y)<r$ and $t>0$. In that case $\mathcal{F}$ is of class C^L provided $N > D/2+2L$. Then, pick out a geodesically spherical open neighborhood of any fixed $u\in\mathcal{M}$, J_u, with a geodesic radius $r_0 < r/8$. By the triangular inequality $d(x,y) = \sqrt{2\sigma(x,y)} < r/4$ whenever $x,y\in J_u$, and thus, by the definition of the function χ, it holds $\chi(\sigma(x,y)) = 1$, whenever $x,y\in J_u$. We conclude that the coefficient χ can be omitted in the heat-kernel expansion working with any coordinate system defined in a neighborhood of J_u (e.g. normal Riemannian coordinates). From now on, the points x and y are assumed to belong to J_u. Moreover, we take $N > D/2+2|\alpha|+2|\beta|$, so that $\mathcal{F}$ is of class $C^{|\alpha|+|\beta|}$.

We can apply operators D_x^α and D_y^β to both sides of (2.20). In this way, the former term in the right-hand side of (2.19) arises by the first term on the right-hand side of (2.20). Let us focus on the derivatives of the latter term in (2.20). Our question concerns the possibility to pass these under the integration symbol in (2.24). The action of the derivatives can be carried out under integration symbols (obtaining also x, y-continuous final function if the derivatives of the integrand are continuous) provided, for any fixed choice of a couple of multiindices α, β, that the derivatives of the integrand are locally x, y-uniformly bounded by an integrable function (dependent on the multiindices in general). We shall see that this is here the case after a suitable manipulation of the integral. Notice that the derivatives (with respect to the manifold variables) of the function $\mathcal{F}$ do exist, because the second integral in the right-hand side of (2.24) takes contribution only from points z such that both $d(y,z) < r$ and $d(x,z) < r$ are fulfilled, as required above. Indeed, it must be $d(x,z) < r/2$ otherwise $\mathcal{F}_N(\tau, x, z)$ smoothly vanishes as pointed out above, and, taking account of $d(x,y) < r/4$, the triangular inequality entails also $d(y,z) \leq d(x,z) + d(x,y) < r/2 + r/4 = 3r/4$. Then, fix a new open neighborhood of u, I_u, such that its closure $\overline{I_u}$ is contained in J_u, and fix $T > 0$. Barring $\tau \mapsto \tau^{-D/2}$, all functions of τ, x, y, z and all their x- ,y- ,z- derivatives we shall consider are bounded in the compact $[0,T] \times \overline{I_u} \times \overline{I_u} \times \mathcal{M}$ where we are working because these are continuous therein. We can rearrange the expression (2.24) into

$$
\begin{aligned}
&(F_N * F)(t,x,y) \\
&= \int_0^t d\tau \tau^{-D/2}(t-\tau)^{N-D/2} \int_{\mathbb{S}^{D-1}} d\mathbf{v} \int_0^{+\infty} d\rho \rho^{D-1} J(x,\mathbf{v},\rho) \\
&\times e^{-\rho^2/2\tau} \mathcal{F}_M(\tau, x, z(x,\rho,\mathbf{v})) \mathcal{F}(t-\tau, z(x,\rho,\mathbf{v}), y) \\
&\times e^{-\sigma(z(x,\rho,\mathbf{v}),y)/2(t-\tau)}\,, \qquad (2.25)
\end{aligned}
$$

where, to determine the position of z, we have employed a spherical system of normal coordinates $\rho, \mathbf{v}$ centered in any x. ρ is the distance of z from x, its range is maximized in the integrals above because the integrand vanishes smoothly for $\rho > r/2$, and thus all the functions contained in the integrand are well-defined for $\rho \geq 0$. $\mathbf{v}$ is a unitary $(D-1)$-dimensional vector and $d\mu_g(z) = d\rho d\mathbf{v} \rho^{D-1} J(x,\mathbf{v},\rho)$, $d\mathbf{v}$ is the natural measure in $\mathbb{S}^{D-1}$. The function J is continuous and bounded in $\overline{I_u} \times \mathbb{S}^{D-1} \times \{\rho \in [0, r/2]\}$ together with all derivatives. Changing variables $\rho \to \rho' \stackrel{\text{def}}{=} \rho/\sqrt{\tau}$ we have

$$
\begin{aligned}
&(F_N * F)(t,x,y) \\
&= \int_0^t d\tau (t-\tau)^{N-D/2} \int_{\mathbb{S}^{D-1}} d\mathbf{v} \int_0^{+\infty} d\rho' \rho'^{D-1} J(x,\mathbf{v},\tau^{1/2}\rho')
\end{aligned}
$$

$$\times e^{-\rho'^2/2}\mathcal{F}_M(\tau, x, z(x, \tau^{1/2}\rho', \mathbf{v}))\mathcal{F}(t-\tau, z(x, \tau^{1/2}\rho', \mathbf{v}), y)$$
$$\times e^{-\sigma(z(x,\tau^{1/2}\rho',\mathbf{v}),y)/2(t-\tau)} . \tag{2.26}$$

The formal action of the operators D_x^α and D_y^β under the integration produces a sum of continuous and bounded functions (now the function $\tau \mapsto \tau^{-D/2}$ has disappeared and the remaining functions and their x-, y-, z- derivatives are bounded since they are products of bounded functions). Also, it changes $(t-\tau)^{N-D/2}$ into several terms of the form $(t-\tau)^{N-D/2-L_i}$ (where each $L_i \leq |\alpha| + |\beta|$), because of the derivatives of the second exponential. These functions of τ are continuous and bounded being $N > D/2 + |\alpha| + |\beta| \geq L_i$ in our hypotheses. We can bound the absolute value of these functions by $Ce^{-\rho'^2/2}$, where C is a sufficiently large constant. This function is trivially integrable in the measure we are considering. This (x, y, t)-uniform bound assures that, concerning the x-,y-derivatives of $F_N * F$, one can interchange the symbols of derivatives with those of integrals and also that the derivative of $(t, x, y) \mapsto (F_N * F)(t, x, y)$ are continuous functions in $(0, +\infty) \times I_u \times I_u$.

In order to finish this proof, let us consider a finer estimate of $O_{\eta,N}^{(\alpha,\beta)}(x, y)$. We have the inequality [I. Chavel (1984)] for $\tau \in [0, t]$:

$$\frac{d^2(x,y)}{t} \leq \frac{d^2(x,z)}{\tau} + \frac{d^2(z,y)}{t-\tau} , \tag{2.27}$$

and thus, picking out any $\eta \in (0,1)$ and posing $\delta \stackrel{\text{def}}{=} 1 - \eta \in (0,1)$ we get (notice that $t - \tau \geq 0$)

$$\begin{aligned} e^{-\sigma(x,z)/2\tau}e^{-\sigma(z,y)/2(t-\tau)} &\leq e^{-\eta\sigma(x,y)/2t}\left(e^{-\delta\sigma(x,z)/2\tau}e^{-\delta\sigma(z,y)/2(t-\tau)}\right) \\ &\leq e^{-\eta\sigma(x,y)/2t}e^{-\delta\sigma(x,z)/2\tau} . \end{aligned} \tag{2.28}$$

We can use this relation in the x- and y-derivatives of (2.26) obtaining

$$\begin{aligned} |D_x^\alpha D_y^\beta(F_N * F)(t,x,y)| &\leq \sum_i e^{-\eta\sigma(x,y)/2t}\int_0^t d\tau(t-\tau)^{N-D/2-L_i} \\ &\times \int_{\mathbb{S}^{D-1}} d\mathbf{v} \int_0^{+\infty} d\rho' \rho'^{D-1} e^{-\delta\rho'^2/2} C_i , \end{aligned} \tag{2.29}$$

where the coefficients C_i are upper bounds of the absolute values of the continuous functions missed in the integrand. We can execute the integral in τ obtaining, for $0 < t \leq T$ and $x, y \in I_u$ (recall that $N > D/2 + |\alpha| + |\beta|$)

$$\begin{aligned} |D_x^\alpha D_y^\beta (F_N * F)(t,x,y)| &\leq e^{-\eta\sigma(x,y)/2t} \sum_i C'_{i,\delta} t^{N+1-L_i-D/2} \\ &\leq \frac{C_\delta}{(4\pi)^{D/2}} e^{-\eta\sigma(x,y)/2t} t^{N+1-D/2-|\alpha|-|\beta|} . \end{aligned} \tag{2.30}$$

Above C_δ is a positive sufficiently large constant which depends on T, α, β in general. This proves the remaining part of the thesis posing $M_{\eta,N}^{(\alpha,\beta)} \stackrel{\text{def}}{=} C_\delta$. Indeed, the remaining $O_{\eta,N}^{(\alpha,\beta)}$ we wanted to compute coincides with $D_x^\alpha D_y^\beta (F_N * F)$ just up to the factor $(4\pi t)^{-D/2} t^{N-|\alpha|-|\beta|} \exp(-\eta\sigma/2t)$. $O_{\eta,N}^{(\alpha,\beta)}$ can be defined in $t = 0$ as $O_{\eta,N}^{(\alpha,\beta)}(0; x, y) = 0$, obtaining a continuous function in $[0, +\infty) \times I_u \times I_u$. □

The function $O_{\eta,N}^{(\alpha,\beta)}$ above depends on χ while the coefficients a_j are independent. Moreover, the map

$$R_{\eta,N}^{(\alpha,\beta)}(t,x,y)) \stackrel{\text{def}}{=} \frac{e^{-\eta\sigma(x,y)/2t}}{(4\pi t)^{D/2}} O_{\eta,N}^{(\alpha,\beta)}(t,x,y) \tag{2.31}$$

belongs to either $C^\infty((0,+\infty) \times I_u \times I_u)$ or $C^\infty((0,+\infty) \times \mathcal{M} \times \mathcal{M})$ for $\alpha = \beta = 0$.

An important point is that the expansion (2.19) may *not* be asymptotic if $x \neq y$ and $t > 0$ as $N \to \infty$. Indeed, due to the presence of $\eta > 0$, there is no guarantee that $R_{\eta,N}$ which has the factor $e^{-\eta\sigma(x,y)/2t}$ is more infinitesimal than the last term in the summation in (2.19) where the exponential $e^{-\sigma(x,y)/2t}$ appears. Conversely, if $x = y$ the expansion turns out to be asymptotic because the exponential terms disappear everywhere in the expansion (2.19).

Concerning similar expansions to (2.19) employed by physicists (sometimes they call WKB expansion any expansion similar to (2.19)), we remark that the expansion we have presented here (see also [S.A. Fulling (1991); R. Camporesi (1990); A.A. Bytsenko et al. (1996)]) is slightly different from the more usual Schwinger-De Witt one [N.D. Birrell and P.C.W. Davies (1982); R.M. Wald (1979)], where a further overall exponential $\exp(-tm^2)$ appears on the right-hand side of the expansion of $K(t, x, y|\mathcal{A})$, m being the mass of the field. Moreover the coefficients $a_j(x, y|A)$ obtained employing the Schwinger-De Witt expansion are related to those in (2.19) by trivial relations, i.e., taking the usual $t = 0$ Taylor expansion of $\exp(-tm^2)$ into account and regrouping coefficients with the same power of t [R. Camporesi (1990)].

As a final comment let us consider the Lorentzian case where the picture changes dramatically. Generally, $A = -\Delta^{(X)} + V : C^\infty(\mathcal{M}) \to C^\infty(\mathcal{M})$, where $\Delta^{(X)}$ is referred to a *Lorentzian* metric on a (compact or not) manifold $\mathcal{M}$, is not bounded below. The drawback, in general, is preserved by self-adjoint extensions. They exist because A, thought as an operator on $L^2(\mathcal{M}, \mu_g)$, is symmetric (e.g., on $C^\infty(\mathcal{M})$) and it commutes with the antiunitary operator given by complex conjugation [M. Reed and B. Simon (1980)]. This introduces several pathologies when dealing with the heat equation and the associated formal *semigroup* of exponentials generated by self-adjoint extensions of A, $\exp(t\mathcal{A})$, which, as a consequence, contains unbounded operators. Therefore one is addressed to consider a different "heat-kernel" $H(s, x, y|\mathcal{A})$ solution of the "Schrödinger" equation

$$-i\frac{\partial}{\partial s}H(s,x,y|\mathcal{A}) + A_x H(s,x,y|\mathcal{A}) = 0\,, \tag{2.32}$$

with initial condition (holding on locally-integrable smooth test functions)

$$\lim_{s\to 0} H(s,x,y|\mathcal{A}) = \delta(x,y)\,. \tag{2.33}$$

(See [M.A. Shubin (1987); R. Camporesi (1990); S.A. Fulling (1991)] and [I.G. Avramidi (1991); I.G. Avramidi (1999)] for details.)

The kernel $H(s, x, y|\mathcal{A})$ formally related to the one-parameter group of the *imaginary* exponentials operators $\exp(is\mathcal{A})$ which are bounded. Then one expects that some local "asymptotic" expansion of the form ¶

$$H(t,x,y|\mathcal{A}) \sim \frac{e^{i\sigma(x,y)/2s}}{(4\pi i s)^{D/2}}\sum_{j=0}^{\infty} a_j(x,y|A)(is)^j \tag{2.34}$$

should hold (where the coefficients a_j are those defined by (2.41) and (2.42) below). The actual picture is much more complicate [S.A. Fulling (1991); S.A. Fulling (1999); R. Estrada and S.A. Fulling (1999)] and we shall not address this issue here.

¶Notice that $\sigma(x,y)$ can also be negative in Lorentzian manifolds.

2.2.4 *The heat-kernel coefficients*

Let us consider the heat-kernel coefficients. The precise form of the off-diagonal heat-kernel coefficients $a_j(x,y|A)$, which does not depend on χ, is obtained as follows [I. Chavel (1984); R. Camporesi (1990)]. One formally inserts the expansion (2.19) for $N=+\infty$ and $\alpha,\beta=0$, omitting χ and the remaining $O^{(\alpha,\beta)}_{\eta,N}$:

$$\frac{e^{-\sigma(x,y)/2t}}{(4\pi t)^{D/2}}\sum_{j=0}^{N} a_j(x,y|A)t^j \tag{2.35}$$

into the heat-kernel equation (2.9), and requires that the coefficients of each power t^j vanish separately. That procedure gives rise to a set of recurrent differential equations which make sense in any geodesically convex neighborhood $\mathcal{N}$ containing x and y (where $\sigma(x,y)$ is smooth). To give the precise form of these equations we introduce the *van Vleck-Morette determinant* [B.S. DeWitt (1965)].

In a geodesically convex neighborhood of either a Lorentzian or a Euclidean (also *non*compact) manifold (or for x,y with $d(x,y)<r$ in the latter case), the van Vleck-Morette determinant is the positive biscalar defined as, in a local coordinate patch

$$\Delta_{VVM}(x,y)\stackrel{\text{def}}{=}(-1)^D\frac{\text{sgn}(g(x))}{\sqrt{\det g(x)\det g(y)}}\det\left[\frac{\partial^2\sigma(x,y)}{\partial x^a\partial y^b}\right]. \tag{2.36}$$

Δ_{VVM} gives rise to the natural measure in normal Riemannian coordinates centered on y, $x^1,\ldots,x^D$, in fact it holds

$$\Delta_{VVM}(x,y)^{-1}=\sqrt{\frac{\det g(x)}{\det g(y)}} \tag{2.37}$$

(and thus $\Delta_{VVM}(y,y)=1$). Δ_{VVM} satisfies the identity [B.S. DeWitt and R.W. Brehme (1960)]

$$g_x(\nabla_x\sigma(x,y),\nabla_x\log\Delta_{VVM}(x,y))=D-\Delta_x\sigma(x,y)\,. \tag{2.38}$$

Using (2.38), the properties of σ previously stated, and following the procedure outlined above to get the equations for the heat-kernel coefficients in normal coordinates about y, one finds that these coefficients must satisfy:

$$\frac{d}{d\lambda}\left(a_0(x(\lambda),y|A)\Delta^{-1/2}_{VVM}(x(\lambda),y)\right)=0\,, \tag{2.39}$$

$$-\lambda^j \Delta_{VVM}^{-1/2}(x(\lambda), y)\, A_{x(\lambda)} a_j(x(\lambda), y|A) =$$
$$\frac{d}{d\lambda}\left(\lambda^{j+1} a_{j+1}(x(\lambda), y|A)\, \Delta_{VVM}^{-1/2}(x(\lambda), y)\right) . \tag{2.40}$$

Here $x = x(\lambda)$ is the unique geodesic segment in $\mathcal{N}$ with $x(0) = y$, $x(1) = x$. These equations uniquely determine all of coefficients $a_j(x, y|A)$ if $a_0(x, y|A)$ is known and it is required that each $a_j(x, y|A)$ is bounded as $x \to y$.

From (2.37), to assure the validity of (2.18), it is sufficient to require that $a_0(x, y|A)$ which satisfies (2.39) is smooth and uniformly $a_0(x, y|A) \to 1$ as $x \to y$. Then (2.39) and (2.40) can be solved into:

$$a_0(x, y|A) = \Delta_{VVM}^{1/2}(x, y) , \tag{2.41}$$
$$\frac{a_{j+1}(x, y|A)}{\Delta_{VVM}^{1/2}(x, y)} = -\int_0^1 d\lambda\, \lambda^j \left[\Delta_{VVM}^{-1/2} A_x a_j\right](x(\lambda), y) . \tag{2.42}$$

These equations are the *definitions* of the heat-kernel coefficients, for x, y which belong to a common geodesically convex neighborhood of either a Euclidean or Lorentzian smooth (noncompact) manifold.

A straightforward consequence of (2.19) and (2.13) is

$$D_x^\alpha D_y^\beta a_j(x, y|A)|_{x=y} = \left(D_x^\alpha D_y^\beta a_j(x, y|A)|_{x=y}\right)^* . \tag{2.43}$$

A related interesting question concerns the hermiticity of $a_j(x, y|A)$ under interchange of the arguments,

$$a_j(x, y|A) = a_j(y, x|A)^* .$$

In the real case for $A = -\Delta + V$ it is possible to show that, in fact,

$$a_j(x, y|A) = a_j(y, x|A) .$$

This question is relevant because, in the Lorentzian case, the heat-kernel coefficients are related to the coefficients of the Hadamard parametrix which is involved in the point-splitting renormalization procedure for the stress-energy tensor and the symmetry plays a rôle in the conservation of the point-splitting renormalized stress-energy tensor. Comments and the proof (also for Lorentzian manifolds) of the symmetry property in the real case can be found in [V. Moretti (1999c); V. Moretti (2000)]. In the complex case, a proof of hermiticity should be constructed similarly.

We stress that, whenever A admits different self-adjoint extensions (that is the case if $\mathcal{M}$ is not compact or boundary conditions are imposed), the heat-kernel coefficients do not depend on the chosen self-adjoint extension $\mathcal{A}$, namely, *they do not depend on the particular physical quantum state considered* in computing the various physical quantities related to the path integral.

On the physical ground, we can say that the length dimensions of the coefficients $a_j(x, y|A)$ and the other relevant quantities are trivially obtained from $[\sigma(x, y)] = L^{-2}$. Therefore, we have $[t] = L^2$, $[a_j(x, y|A)] = L^{-2j}$, $[A] = L^{-2}$ and $[K(t, x, y|\mathcal{A})] = L^{-D}$.

The explicit expression of some of the *on-diagonal heat-kernel coefficients* $a_j(x|A) \stackrel{\text{def}}{=} a_j(x, x|A)$ can be found in Appendix B of [A.A. Bytsenko et al. (1996)]. In general, apart from terms depending on the particular form of V, they are polynomials in the curvature tensors of the manifold. It is finally worth stressing that many efforts have been invested in computing explicitly the coefficients $a_j(x|A)$ for $j \geq 4$ in several cases [K. Kirsten (1998)] and finding automatic techniques for computing these coefficients spending a relatively short amount of time. Approximated expressions for $x \neq y$, for the coefficients of the Schwinger-De Witt expansion but in the case of Lorentzian signature of the metric (so that one has to formally continue the results in the Euclidean metric) the can be found in [N.D. Birrell and P.C.W. Davies (1982)].

2.3 Local and Global Spectral Zeta Functions on Compact Manifolds

The formalism and the results presented in the first part of Section 2.2 allow us to implement a procedure which interprets the partition function (see Section 2.1) $Z = \left[\det\left(\ell^2 A\right)\right]^{-1/2}$ as $\exp\{\zeta'(0|A/\mu)/2\}$, $\zeta(s|A/\mu)$ being the analytic continuation of $\sum_j \left(\ell^2 \lambda_j\right)^{-s}$. It is obvious that, in order to give a non-ambiguous meaning to the series above, since $s \in \mathbb{C}$, it must be Spec($\mathcal{A}$) ≥ 0. This is equivalent to $A \geq 0$. A sufficient condition to have $A \geq 0$ is $V \geq 0$. However different conditions, which do not assume $V \geq 0$, can be found in our general hypotheses. For instance, in the case $A = -\Delta + V$, by Theorem 4.2.1 in [E.B. Davies (1989)], one has that, not depending on the sign of V, if a strictly positive function $p \in C^2(\mathcal{M})$ exists such that $p(x)V(x) \geq \Delta p(x)$ everywhere on $\mathcal{M}$, then

$$A \geq \inf_{x\in\mathcal{M}} \left\{ V(x) - \frac{\Delta p(x)}{p(x)} \right\} \geq 0 . \tag{2.44}$$

2.3.1 *Weyl's asymptotic formulae*

Within our initial hypotheses and assuming that $A \geq 0$, a very famous estimation of the eigenvalues of A, λ_j for large j arises. This estimation is very useful for dealing with the techniques of the zeta function.

Theorem 2.5 *Under the general hypotheses, assuming $A \geq 0$ and denoting the eigenvalues of A as in Theorem 2.2, we have:*

(a) *K_t is the trace-class for all $t > 0$ and its trace $\mathrm{Tr}\, K_t$ equals*

$$\int_{\mathcal{M}} d\mu_g(x) K(t,x,x|\mathcal{A}) = \sum_{j=0}^{+\infty} e^{-\lambda_j t} < \infty ; \tag{2.45}$$

(b) *Weyl's asymptotic formulae hold*

$$\lim_{\lambda\to+\infty} \frac{N(\lambda)}{\lambda^{D/2}} = \frac{\omega_D \mathrm{Vol}(\mathcal{M})}{(2\pi)^D} , \tag{2.46}$$

$$\lim_{j\to+\infty} j^{-1}\lambda_j^{D/2} = \frac{(2\pi)^D}{\omega_D \mathrm{Vol}(\mathcal{M})} , \tag{2.47}$$

where $\omega_D = \pi^{D/2}/\Gamma(1+D/2)$ is the volume of the unit disk in $\mathbb{R}^D$, $N(\lambda) \stackrel{def}{=} \sum_{\lambda_j\leq\lambda} 1$ is the function which gives the number of eigenvalues λ_j (taking the degeneration into account) in $[0,\lambda]$ and $\mathrm{Vol}(\mathcal{M}) = \int_{\mathcal{M}} d\mu_g$ is the volume of the manifold $\mathcal{M}$.

Proof. Taking advantage of Theorem 2.2, integrating both sides of (2.16) in the variable x when $x = y$ and using the uniform convergence of the series, as well as the normalization of the function ϕ_j, one gets (2.45) and this also shows that K_t is trace class since it admits a trace on a Hilbert basis. Karamata's theorem on the Laplace transform of (probability) measures, Vol. 1, p. 446 of [W. Feller (1971)], shows that if m is a positive Borel measure on $[0,+\infty)$ such that $\int_0^{+\infty} dm(x)\exp(-tx) < \infty$ for all $t > 0$ and there are positive constants a, C such that $t^a \int_0^{+\infty} dm(x)\exp(-tx) \to C$ as

$t \to 0^+$, then

$$\lim_{t\to 0^+} \int_0^{+\infty} dm(x) f(e^{-tx}) e^{-tx} = \frac{C}{\Gamma(a)} \int_0^{+\infty} dt\, f(e^{-tx}) t^{a-1} e^{-t},$$

for every continuous function $f : [0,1] \to \mathbb{R}$. If $\{f_n\}$ is a decreasing sequence of continuous functions such that, for $n \to \infty$, $f_n(x) \to 1/x$ for $x \in [1/\delta, 1]$ and $f_n(x) \to 0$ for $x \in [0, 1/\delta)$, $\delta \in (0,1)$ being fixed, the use of Karamata's theorem straightforwardly implies that

$$\lim_{t\to 0^+} t^a \int_0^{1/t} dm(x) = \frac{C}{\Gamma(a)} \int_0^1 t^{a-1} dt\,.$$

As a consequence we have that, in the hypotheses of Karamata's theorem on the measure m and constants a, C:

$$\lim_{t\to 0^+} t^a \int_0^{1/t} dm(x) = \frac{C}{\Gamma(a)}\,. \tag{2.48}$$

Employing the expansion in (2.19) with $\alpha = \beta = 0$ and using $a_0(x, x|A) = 1$ we get (with obvious notation)

$$\sum_{j=0}^{+\infty} e^{-\lambda_j t} = \frac{\mathrm{Vol}(\mathcal{M})}{(4\pi t)^{D/2}} + t^{-D/2} \mathcal{O}(t)\,. \tag{2.49}$$

We can consider the point measure on $[0, +\infty)$ induced by the density $m(x) \stackrel{\text{def}}{=} \sum_j \delta(\lambda_j - x)$, where λ_j are the eigenvalues of A, which are positive by hypotheses. Defining $a \stackrel{\text{def}}{=} D/2$ and $C \stackrel{\text{def}}{=} (4\pi)^{-D/2}\mathrm{Vol}(\mathcal{M})$ we can apply (2.48) obtaining the former equation in (**b**). (2.47) is a straightforward consequence of the former. □

2.3.2 *Spectral zeta functions*

Let us give the definition of the various spectral zeta functions [S.W. Hawking (1977); R.M. Wald (1979); N.D. Birrell and P.C.W. Davies (1982); R. Camporesi (1990); E. Elizalde et al. (1994); E. Elizalde (1995)] starting from the local ones. From now on $\ell > 0$ is an arbitrary length scale.

Definition 2.2 *Following the general hypotheses on $\mathcal{M}$ and $A \geq 0$, we define the off-diagonal local spectral zeta function of A for* $\mathrm{Re}\, s > D/2$ *and*

$x, y \in \mathcal{M}$ as

$$\zeta(s, x, y|\ell^2 \mathcal{A}) \stackrel{\text{def}}{=} \int_0^{+\infty} d\left(\frac{t}{\ell^2}\right)\left(\frac{t}{\ell^2}\right)^{s-1} \frac{[K(t, x, y|\mathcal{A}) - P(x, y|\mathcal{A})]}{\Gamma(s)} . \tag{2.50}$$

The scale ℓ (almost always omitted at this step in the literature) is actually necessary for dimensional reasons and it is not fixed by the theory: it is an ambiguity which affects all the physical quantities regularized by means of the zeta function technique. By the way, and conversely, the heat-kernel coefficients $a_j(x, y|A)$ do not depend on any scale ℓ. $P(x, y|\mathcal{A})$ is the integral kernel of the projector onto the kernel of $\mathcal{A}$, $\text{Ker}(\mathcal{A})$.

As we shall see in the proof of Theorem 2.7 below, the given definition is well-posed since the integral above converges absolutely for $\text{Re}\, s > D/2$, because of $\text{Spec}(\mathcal{A}) \geq 0$, and it is the subsequent exponential decay of $K - P$ at large t, and the expansion (2.19) as $t \to 0^+$ which fixes the bound $\text{Re}\, s > D/2$.

A first explanation of the term "spectral" in the definition above is given by noticing the appearance of the eigenvalues of A on the right-hand side of the statement (**c**) of Theorem 2.6 below based on results by Minakshisundaram and Pleijel [S. Minakshisundaram (1949); S. Minakshisundaram and A. Pleijel (1949)].

Theorem 2.6 *In the general hypotheses on $\mathcal{M}$ and $A \geq 0$, and for $\ell > 0$:*

(**a**) *the integral in (2.50) converges absolutely for* $\text{Re}\, s > D/2$ *and the function $\zeta(\cdot, \cdot, \cdot|\ell^2 A)$ is continuous s-analytic in* $\{s \in \mathbb{C} \mid \text{Re}\, s > D/2\} \times \mathcal{M} \times \mathcal{M}$;

(**b**) *if* $\text{Re}\, s > D/2$, $(\ell^2 \mathcal{A})^{-s}$ *defined by the spectral theory in* $\text{Ker}(\mathcal{A})$ *is compact, Hilbert-Schmidt and trace-class, then for* $\text{Re}\, s > 3D/2$, *the map* $(x, y) \mapsto \zeta(s, x, y|\ell^2 \mathcal{A})$ *is its integral kernel;*

(**c**) *for $\beta > \alpha > 3D/2$, it holds*

$$\zeta(s, x, y|\ell^2 \mathcal{A}) = \sum_{j \in \mathbb{N}}{}' \left(\ell^2 \lambda_j\right)^{-s} \phi_j(x) \phi_j^*(y) , \tag{2.51}$$

the convergence being regular jointly in s, x, y if $\beta \geq \text{Re}\, s \geq \alpha$, *and the prime in $\sum_j' f(\lambda_j)\{\ldots\}$ indicates that the indices j corresponding to vanishing eigenvalues are omitted in the sum.*

Proof. Throughout the first part of the proof we omit ℓ^2, i.e. $\ell = 1$, without loss of generality. (**a**) is a direct consequence of the heat-kernel expansion (2.19), and the behavior of the heat kernel at large t, as shown in the first part of the proof of Theorem 2.7 below. Let us show (**b**) and (**c**). Consider the operator $Q_s \stackrel{\text{def}}{=} A^{-s}$ given by spectral theory, $Q_s = \sum_j' \lambda_j^{-s}\phi_j(\phi_j, \cdot)$ where the used convergence is strong uniform operator convergence. If $\mathrm{Re}\, s > D/2$, by (2.47) taking the finite dimension of every eigenspace into account, one has that the convergence holds in the uniform sense and Q_s is compact because of the uniform limit of compact operators [M. Reed and B. Simon (1980)]. Notice that ϕ_j and λ_j are eigenvectors and eigenvalues of Q_s, by direct inspection or employing spectral theory. Since $\sum_j' |\lambda_j^{-s}|^2 < \infty$ as $\mathrm{Re}\, s > D/2$ by (2.47), Q_s must be Hilbert-Schmidt [M. Reed and B. Simon (1980)], and thus it can be defined by a $L^2(\mathcal{M}\times\mathcal{M}, \mu_g \otimes \mu_g)$ kernel given by the right-hand side of (2.51) where the convergence is that of $L^2(\mathcal{M}\times\mathcal{M}, \mu_g \otimes \mu_g)$. Under the same hypotheses, $\sum_j |\lambda_j^{-s}| < \infty$ holds true too, and thus Q_s is trace class [M. Reed and B. Simon (1980)]. If $D/2 < s \in \mathbb{R}$, Q_s is positive and thus the series in (2.51) converges regularly to the kernel of the operator, by Mercer theorem [F. Riesz and B.S. Nagy (1972)]. If $\mathrm{Re}\, s \in [\alpha, \beta]$ with $D/2 < \alpha < \beta$, $|\lambda_j^{-s}\phi_j(x)\phi_j^*(y)| \leq |\lambda_j^{-\alpha}\phi_j(x)\phi_j^*(y)|$ and thus the convergence holds regularly and jointly in s, x, y also if $\mathrm{Re}\, s \in [\alpha, \beta]$, $x, y \in \mathcal{M}$. To finish, we have to show that the kernel of Q_s is the local zeta function, i.e., for $\mathrm{Re}\, s > 3D/2$

$$\int_0^{+\infty} d(\ell^{-2}t)(\ell^{-2}t)^{s-1}\left[K(t,x,y|\mathcal{A}) - P(x,y|\mathcal{A})\right]$$
$$= \sum_{j\in\mathbb{N}}{}' \left(\ell^2\lambda_j\right)^{-s} \phi_j(x)\phi_j^*(y)\,. \tag{2.52}$$

This arises from the well-known identity, $a > 0$, $s \in \mathbb{C}$, $\mathrm{Re}\, s > 0$,

$$a^{-s} = \frac{1}{\Gamma(s)}\int_0^{+\infty} dt t^{s-1}\, e^{-ta}\,, \tag{2.53}$$

with $a = \ell^2\lambda_j$. Details are supplied in the proof of Theorem 2.15 where a more general case is considered. □

As a consequence of the proven theorem, $\zeta(s,x,y|\ell^2\mathcal{A})$ could be equivalently defined by (2.51) for $\mathrm{Re}\, s > 3D/2$ without reference to the heat kernel, obtaining the same results.

Definition 2.3 *Using the general hypotheses on $\mathcal{M}$ and $A_0 \geq 0$, we define for $\ell > 0$ the local spectral zeta function of the operator $\mathcal{A}$ as the function of $x \in \mathcal{M}$ and $s \in \mathbb{C}$:*

$$\zeta(s, x|\ell^2 \mathcal{A}) \stackrel{def}{=} \zeta(s, x, x|\ell^2 \mathcal{A}) , \tag{2.54}$$

where $\text{Re } s > D/2$. *The spectral zeta function $\zeta(s|\ell^2 \mathcal{A})$ is given by*

$$\zeta(s|\ell^2 \mathcal{A}) \stackrel{def}{=} \text{Tr}\left[\left(\ell^2 \mathcal{A}\right)^{-s}\right] = \sum_{j \in \mathbb{N}}{}' \left(\ell^2 \lambda_j\right)^{-s} , \tag{2.55}$$

$s \in \mathbb{C}$ and $\text{Re}\, s > D/2$.

The operator $Q_s = \mathcal{A}^{-s}$ is defined via spectral theory (dropping the part corresponding to the kernel of $\mathcal{A}$ whenever the kernel is not trivial). The given definition is well-posed by (**b**) of Theorem 2.6. On the other hand by (2.47), the series on the right-hand side of (2.55) converges absolutely for $\text{Re}\, s > D/2$ and diverges for $\text{Re}\, s < D/2$.

The uniform convergence of the series in (2.51) for $\text{Re}\, s > 3D/2$ allows the integration on $\mathcal{M}$ of both sides giving

$$\begin{aligned} \text{Tr}\left[\left(\ell^2 \mathcal{A}\right)^{-s}\right] &= \int_{\mathcal{M}} d\mu_g(x)\, \zeta(s, x, x|\ell^2 \mathcal{A}) , \\ \zeta(s|\ell^2 \mathcal{A}) &= \int_{\mathcal{M}} d\mu_g(x)\, \zeta(s, x|\ell^2 \mathcal{A}) . \end{aligned} \tag{2.56}$$

Remark. Throughout the remaining part of the chapter, we shall make use of the following definitions

$$\mathcal{D} \stackrel{\text{def}}{=} \{(x, y) \in \mathcal{M} \times \mathcal{M} \mid x = y\} , \tag{2.57}$$

$$\delta_D \stackrel{\text{def}}{=} \begin{cases} 1, & \text{when } D \text{ is even} \\ 0 & \text{when } D \text{ is odd} \end{cases} , \quad \delta_{x,y} \stackrel{\text{def}}{=} \begin{cases} 1, & \text{when } x = y \\ 0 & \text{when } x \neq y \end{cases} .$$

We are now able to state and prove the crucial theorem on the spectral zeta function that summarizes all the relevant properties and the interplay between the various zeta functions introduced above.

Theorem 2.7 *Let $\mathcal{M}$, $A \geq 0$ satisfy the general hypotheses and $\ell > 0$, then*

(a) *in the case of the off-diagonal local zeta function,*

(i) *for $x \neq y$ in $\mathcal{M}$, the map $s \mapsto \zeta(s,x,y|\ell^2\mathcal{A})$ can be analytically continued, to the whole s-complex plane, into an entire function which satisfies (2.50) for* $\mathrm{Re}\, s > 0$ *(or continued everywhere in $s \in \mathbb{C}$ if $P \equiv 0$);*

(ii) *the analytically continued map $(s,x,y) \mapsto \zeta(s,x,y|\ell^2\mathcal{A})$, and all of the s-derivatives, are continuous in $\mathbb{C} \times (\mathcal{M} \times \mathcal{M} \setminus \mathcal{D})$.*

(b) *In the case of the local zeta function,*

(i) *for $x \in \mathcal{M}$, the map $s \mapsto \zeta(s,x|\ell^2\mathcal{A})$ can be analytically continued into a meromorphic function on $\mathbb{C}$ whose possible simple poles are only*

$$\begin{aligned} s_j &= D/2 - j, \quad j = 0,1,2,\ldots \qquad \text{if } D \text{ is odd}\,, & (2.58) \\ s_j &= D/2 - j, \quad j = 0,1,2,\ldots D/2-1 \quad \text{if } D \text{ is even}\,, & (2.59) \end{aligned}$$

with residues

$$\mathrm{Res}(s_j, x) = \frac{a_j(x|A)}{(4\pi)^{D/2}\Gamma(D/2-j)\ell^{D-2j}}\,; \qquad (2.60)$$

(ii) *the analytically continued map $(s,x) \mapsto \zeta(s,x|\ell^2\mathcal{A})$, and all of the s-derivatives, are continuous in $(\mathbb{C} \setminus \mathcal{P}) \times \mathcal{M}$, $\mathcal{P}$ being the set of the actual poles (each for some x).*

(c) *If x,y are fixed in $\mathcal{M}$, the s-continued functions above satisfy*

$$\zeta(0,x,y|\ell^2\mathcal{A}) + P(x,y|\mathcal{A}) = \frac{a_{D/2}(x,x|A)}{(4\pi)^{D/2}}\,\delta_{x,y}\,\delta_D\,. \qquad (2.61)$$

For $x \neq y$ the zero at $s = 0$ of the right-hand side of (2.61) is of order ≥ 1.

Proof. Take a cutoff of arbitrary length $\ell_0 > 0$. The idea is to break the integration in (2.50), for Re $s > D/2$, as

$$\begin{aligned} \zeta(s,x,y|\ell^2\mathcal{A}) &= \int_0^{+\infty} \frac{dt\, t^{s-1}}{\ell^{2s}\Gamma(s)} \left[K(t,x,y|\mathcal{A}) - P(x,y|\mathcal{A})\right] \\ &= \frac{\mu^{2s}}{\Gamma(s)} \int_0^{\ell_0^2} dt\,\{\ldots\} + \frac{1}{\ell^{2s}\Gamma(s)} \int_{\ell_0^2}^{+\infty} dt\,\{\ldots\}\,. \qquad (2.62) \end{aligned}$$

We now study the properties of these integrals separately. Let us first focus on the second integral in the right-hand side of (2.62), considered as a function of $s \in \mathbb{C}$, $x,y \in \mathcal{M}$. From (**b**) of Theorem 2.2, and using the Cauchy-Schwartz inequality, one finds

$$\begin{aligned} &|K(t,x,y|\mathcal{A}) - P(x,y|\mathcal{A})|^2 \leq [K(t,x,x|\mathcal{A}) - P(x,x|\mathcal{A})] \\ &\times [K(t,y,y|\mathcal{A}) - P(y,y|\mathcal{A})]\,. \qquad (2.63) \end{aligned}$$

Define

$$p(t,x) \stackrel{\text{def}}{=} e^{\lambda t}[K(t,x,x|\mathcal{A}) - P(x,x|\mathcal{A})] \geq 0\,, \qquad (2.64)$$

λ being the first strictly positive eigenvalue of $\mathcal{A}$. From the expansion (2.16) it is obvious that, since $\lambda_j - \lambda \geq 0$ whenever $\lambda_j \neq 0$, $p(t,x) \geq p(t',x)$ for $t' \geq t$, and thus, $t \in [\ell_0^2, +\infty)$, $p(t,x) \leq p(\ell_0^2, x)$. The last function is continuous in $\mathcal{M}$ by construction. $\max_{x\in\mathcal{M}} p(\ell_0^2, x)$ does exist because the compactness of $\mathcal{M}$. Putting all together, from (2.64) one has the following (x,y)-uniform upper bound of the heat kernel in $t \in [\ell_0^2, +\infty)$,

$$\begin{aligned} &|t^s||K(t,x,y|\mathcal{A}) - P(x,y|\mathcal{A})| \\ &\leq \max_{z\in\mathcal{M}}[K(\ell_0^2,z,z|\mathcal{A}) - P(z,z|\mathcal{A})]\ |t^s|e^{-\lambda(t-\ell_0^2)}\,. \end{aligned} \qquad (2.65)$$

This result proves that the second integral in (2.62) converges absolutely, not depending on $s \in \mathbb{C}$, and that the final function is in $C^0(\mathcal{M}\times\mathcal{M})$, for any fixed $s \in \mathbb{C}$, because the upper bound in (2.65) does not depend on x, y. Actually, studying the function $t \mapsto t^s \exp(-\lambda t)$, one finds that, for $\ell_0 > 0, \lambda > 0, \epsilon \in (0,1)$, there is a positive constant $B = B(\ell_0, \lambda, \epsilon)$ such that, for $s \in \mathbb{C}$, $|t^s|\exp(-\lambda t) \leq \exp[B\ (\text{Re}\ s)^2]\exp(-\epsilon\lambda t)$, whenever $t \in [\ell_0^2, +\infty)$. Therefore $|t^s e^{(-\lambda t)}| \leq C_{s_0} e^{-\epsilon\lambda t}$ for some $C_{s_0} > 0$ when s belongs to a sufficiently small neighborhood of any $s_0 \in \mathbb{C}$. This implies that the continuous function in (2.65) is locally uniformly bounded in s, x, y jointly by a t-integrable function, and thus the integral considered defines a jointly-continuous function of s, x, y by Lebesgue's dominate convergence theorem. The same bound (the s derivatives produce integrable factors $\log t$ in the integrand) proves that the integral considered is also an analytic function of $s \in \mathbb{C}$, for x, y fixed in $\mathcal{M}$. This is because, by Lebesgue's dominate convergence theorem, one can interchange the symbol of the t integration with the derivative in $\text{Re}\, s$ and $\text{Im}\, s$ proving Cauchy-Riemann's identities. Moreover, in the same way, it is trivially proven that all of s derivatives of the relevant integral belong to $C^0(\mathbb{C}\times\mathcal{M}\times\mathcal{M})$.

The former integral in the last line of (2.62) needs further manipulations. Theorem 2.4 and (2.19) for $N > D/2 + 2$ suggest to consider the function

$$\begin{aligned} \zeta(N,s,x,y|\ell^2\mathcal{A},\ell_0^2) &\stackrel{\text{def}}{=} \int_{\ell_0^2}^{+\infty} \frac{dt\, t^{s-1}\mu^{2s}}{\Gamma(s)}[K(t,x,y|\mathcal{A}) - P(x,y|\mathcal{A})] \\ &+ \frac{1}{\ell^{2s}\Gamma(s)}\int_0^{\ell_0^2} dt\, t^{s-1+N}\frac{e^{-\eta\sigma(x,y)/2t}}{(4\pi t)^{D/2}}O_{\eta,N}(t;x,y)\,, \end{aligned} \qquad (2.66)$$

where by definition $O_{\eta,N}(t;x,y) \stackrel{\text{def}}{=} O^{(0,0)}_{\eta,N}(t;x,y)$. Let us study the second integral in the right-hand side for $N > D/2+2$. Take $\epsilon > 0$ and $\beta > 0$ to be sufficiently large. For $\text{Re}\, s \in [D/2 - N + \epsilon, \beta]$ and $x, y \in \mathcal{M}$, the integrand is jointly-continuous and vanishes at $t = 0$, no matter if $x \neq y$. The same result arises concerning s-derivatives. Therefore, we can s-, x-, y-uniformly bound the integrand by a constant and use the same argument employed above, based on Lebesgue's theorem to find that the second integral in (2.66) and all of its s-derivatives converge absolutely and, considered as functions of s, x, y, belong to $C^0(\{s \in \mathbb{C} \mid \text{Re}\ s > D/2 - N\} \times \mathcal{M} \times \mathcal{M})$. Moreover, for x, y fixed, these functions are analytic in $\{s \in \mathbb{C} \mid \text{Re}\ s > D/2 - N\}$.

Notice that $\zeta(N, 0, x, y|\ell^2 \mathcal{A}, \ell_0^2) = 0$ for any integer $N > D/2 + 2$ and $x, y \in \mathcal{M}$. Putting everything together and coming back to the local zeta function, we have that, for $N > D/2 + 2$,

$$\zeta(s,x,y|\ell^2\mathcal{A}) = \zeta(N,s,x,y|\ell^2\mathcal{A},\ell_0^2) - \frac{\mu^{2s}P(x,y)}{\Gamma(s)}\int_0^{\ell_0^2} dt\, t^{s-1}$$
$$+\frac{1}{\ell^{2s}\Gamma(s)}\int_0^{\ell_0^2} dt\, t^{s-1}\frac{e^{-\sigma(x,y)/2t}}{(4\pi t)^{D/2}}\chi(\sigma(x,y))\sum_{j=0}^{N} a_j(x,y|A)t^j\,. \quad (2.67)$$

Up to now we have proven that, for x, y fixed, the first term on the right-hand side can be analytically continued, at least in the set $\mathcal{D}_N \stackrel{\text{def}}{=} \{s \in \mathbb{C} \mid \text{Re}\, s > D/2 - N\}$. Moreover, varying also x and y, one gets (at least) a (s, x, y)-continuous function also considering the s-derivatives. This function vanishes for $s = 0$.

The second term on the right-hand side of (2.67) can be computed for $\text{Re}\, s > 0$ and the result can be continued in the whole complex s-plane, defining an s-analytic function $C^\infty(\mathbb{C} \times \mathcal{M} \times \mathcal{M})$. This function attains the value $-P(x, y)$ for $s = 0$. We can rearrange (2.67) after the analytic continuation of the term containing P as

$$\zeta(s,x,y|\ell^2\mathcal{A}) = \zeta(N,s,x,y|\ell^2\mathcal{A},\ell_0^2) - \left(\frac{\ell_0}{\ell}\right)^{2s}\frac{P(x,y)}{s\Gamma(s)}$$
$$+\sum_{j=0}^{N}\frac{a_j(x,y|A)}{(4\pi)^{D/2}\ell^{2s}\Gamma(s)}\int_0^{\ell_0^2} dt\, t^{s-1+j-D/2}\chi(\sigma(x,y))e^{-\sigma/2t}\,. \quad (2.68)$$

Consider the last term of the right-hand side of (2.68) for $x \neq y$. Following procedures similar to those above, it is quite easy to prove that the last term in the right-hand side defines an everywhere s-analytic function $C^\infty(\mathbb{C} \times ((\mathcal{M} \times \mathcal{M}) \setminus \mathcal{D}))$, as it stands. This result is achieved because of the sharp decay of the exponential as $t \to 0^+$. Notice also that the considered term vanishes for $s = 0$. Moreover, the left-hand side of (2.67) defines an s-analytic function in each $\mathcal{D}_N$, $N > D/2+2$ which, at least, belongs also to $C^0(\mathcal{D}_N \times ((\mathcal{M} \times \mathcal{M}) \setminus \mathcal{D}))$, together with all of its s-derivatives. Moreover, it vanishes for $s = 0$, giving rise to (2.61) in the case $x \neq y$. The order of the zero at $s = 0$ in the right-hand side of (2.61) is at least 1, because of the overall factor $1/\Gamma(s)$ in (2.68). Up to now, we have proven statement (**a**) in domains $\mathcal{D}_N$ concerning the variable s and (**c**) in the case $x \neq y$.

Now, consider the last term on the right-hand side of (2.68) in the case $x = y$. We cannot take advantage of the sharp decay of the exponential. However, we can perform the integration for Re $s > D/2$ and then continue the result, as far as it is possible, to the remaining part of the complex s-plane. Notice that, away from the poles, the function obtained is in C^∞ in s, x, y trivially. We have, finally,

$$\begin{aligned}\zeta(s,x|\ell^2 A) &= \zeta(N,s,x,x|\ell^2 A,\ell_0^2) - \frac{(\ell_0/\ell)^{2s}P(x,x)}{s\Gamma(s)} \\ &\quad + \frac{1}{(4\pi)^{D/2}\ell^{2s}}\sum_{j=0}^{N}\frac{a_j(x,x|A)(\mu_0^{-2})^{(s+j-D/2)}}{\Gamma(s)(s+j-D/2)}\,. \end{aligned} \tag{2.69}$$

This identity defines an analytic continuation $\zeta(s,x|\ell^2 A)$ at least in each $\mathcal{D}_N$ for each integer $N > D/2+2$; indeed, therein both functions (and all of their s-derivatives) on the right-hand side are defined and continuous in (x,s), away from the possible poles. Summarizing, the left-hand side of the equation above is decomposed into a function analytic in $\mathcal{D}_N$ and a function which is meromorphic in the same set, both functions and their s derivatives being at least continuous in (s,x), away from possible poles.

Noticing that $\mathcal{D}_N \subset \mathcal{D}_{N+1}$ and $\bigcup_{N=1}^{+\infty}\mathcal{D}_N = \mathbb{C}$, the properties found out for the function $\zeta(s,x,y|\ell^2 A)$, either if $x \neq y$ or $x = y$, can be extended in the whole complex s-plane. In particular, the continued local zeta function and all of its s-derivatives belong to $C^0((\mathbb{C}\setminus\mathcal{P})\times\mathcal{M})$ at least, where $\mathcal{P}$ is the set of the actual poles of the last term in the right-hand side of (2.69). Notice that, for $s = 0$, (2.69) gives (2.61) in the case $x = y$. This proves (**b**) and complete the proof of (**c**). □

Corollary 2.1 *The spectral zeta function $s \mapsto \zeta(s|\ell^2 \mathcal{A})$ can be analytically continued into a meromorphic function on $\mathbb{C}$ which coincides with the integral over $\mathcal{M}$ of the analytic continuation of the local zeta function. The possible simple poles are those in $\mathcal{P}$ above with residues given by integrating of the residues (2.60).*

Proof.

Consider the local zeta function $\zeta(s, x|\ell^2 \mathcal{A})$ analytically continued in the variable s in the complex plane. Integrate it in the variable x over the manifold $\mathcal{M}$. The function thus obtained is analytic in the variable s, since Cauchy-Riemann's conditions can be proven to hold true, by interchanging the derivative in s with the symbol of integration. As a consequence of the uniqueness of the analytic continuation, the function found must be the analytic continuation of $\zeta(s|\ell^2 \mathcal{A})$. This procedure shows also that the possible (simple) poles of the analytically continued spectral zeta function arise from the corresponding poles of the local zeta function $\zeta(s, x|\ell^2 \mathcal{A})$ pointed out in (**b**) of Theorem 2.7. Finally, the associated residua are those obtained by integrating the corresponding residues of the local zeta function. $\square$

Several comments on this theorem and its corollary are in order.

(1) Assume that either Ker($\mathcal{A}$) is trivial, (i.e., $P \equiv 0$) and $x \neq y$ or, equivalently, $\text{Re}\, s > 0$ and $x \neq y$. In these cases the relation (2.50) is preserved also if the left-hand side is understood in the sense of analytic continuation and the right-hand side is computed as a proper integral. This property will be very useful for studying the Green function of $\mathcal{A}$.

(2) There is no guarantee that, actually, poles appear in the points (2.58) and (2.59), because the corresponding coefficients $a_j(x, x|A)$ (or the integrated ones) may vanish. Anyhow, *if* poles appear, they must belong to the listed set.

(3) (2.61) entails that the continued function $\zeta(s, x, y|\ell^2 \mathcal{A})$ is not continuous on the diagonal $x = y$, at least for $s = 0$. So the s-continuation procedure and the limit $x \to y$ *generally do not commute.*

Remark. From now on, barring different specification, the symbols of the various zeta function as $\zeta(s, x, y|\ell^2 \mathcal{A})$ indicate the *meromorphic functions continued* from the initial domain of definition $\text{Re}\, s > D/2$ as far as possible in the complex s-plane.

To conclude, within our general hypotheses on $\mathcal{M}$ and $A \geq 0$, we are able to define, with a mathematical rigorous meaning within the framework of the zeta function theory, the determinant of $\mathcal{A}$.

Definition 2.4 *Under the general hypotheses on $\mathcal{M}$ and $A \geq 0$, for $\ell > 0$ the determinant of the operator $\ell^2\mathcal{A}$ is defined as*

$$\det\left(\ell^2\mathcal{A}\right) \stackrel{\text{def}}{=} \exp\left(-\frac{d}{ds}\,|_{s=0}\,\zeta(s|\ell^2\mathcal{A})\right) . \tag{2.70}$$

2.4 Effective Action, Effective Lagrangian and Green Functions

Now we come back to physics and consider how the theory above may be employed in quantum field theory. In this Section, we prove some rigorous theorem concerning the spectral zeta function approach to define and regularize the effective action and the effective Lagrangian $\mathcal{L}(x|A)$. We also give some comments on formal definitions, often assumed by physicists. Moreover we analyze the relation between the local zeta function approach and the point-splitting procedure, in terms of the expansion (2.19) instead of the Schwinger-DeWitt one (hence, differently from several papers on the same subject, the obtained formulae do not distinguish between the cases of a finite or vanishing mass of the field) and quite a general scalar operator is considered here. However, in favour of Schwinger-DeWitt's expansion, it has to be noticed that that expansion, at least formally and for m strictly positive, should work also in noncompact manifolds, due to the sharp decay of the exponential $\exp(-m^2t)$ in the heat kernel (see the discussion in [R.M. Wald (1979)]). Finally we elaborate on some issues concerning the relation between off-diagonal zeta functions and two-point functions, in order to define the Feynman propagator and the associated Lorentzian quantum field theory. The technique presented will be developed further, towards the end of this chapter, to regularize $\langle\phi^2(x|\mathcal{A})\rangle$ and $\langle T_{ab}(x|\mathcal{A})\rangle$.

The results and the analysis presented in this Section were essentially given in [V. Moretti (1999a)]. Formal results, without rigorous proof are contained in [N.D. Birrell and P.C.W. Davies (1982)] (see also [S.A. Fulling (1991)]). In the (more rigorous) earlier work [R.M. Wald (1979)] similar results and proofs appeared for the particular case $V \equiv m^2$, starting from different hypotheses. There, the Schwinger-DeWitt expansion and an explicit use of large t-behavior of (t-derivatives of) $K(t,x,y|\mathcal{A})$ were used together with, in part, hypotheses on the essentially self-adjointness of A^n for any $n \in \mathbb{N}$. However, also the case of a noncompact manifold is considered.

We state the following crucial definition.

Definition 2.5 *Under the general hypotheses on $\mathcal{M}$ and $A \geq 0$, for $\ell > 0$, the effective action associated with the operator $\mathcal{A}$ is defined within the zeta function approach as*

$$W[\mathcal{A}]_\ell \stackrel{def}{=} -\log Z[\mathcal{A}]_\ell \,, \tag{2.71}$$

where the partition function $Z[\mathcal{A}]_\ell$ in the right-hand side is defined as

$$Z[\mathcal{A}]_\ell \stackrel{\text{def}}{=} \left[\det\left(\ell^2 \mathcal{A}\right)\right]^{-1/2} = \exp\left(\frac{1}{2}\frac{d}{ds}\bigg|_{s=0} \zeta(s|\ell^2\mathcal{A})\right) . \tag{2.72}$$

Therefore, we have defined $Z[\mathcal{A}]_\ell$ by (2.72), the right-hand side being rigorously interpreted as in (2.70) in the framework of the zeta function.

The definition of the effective Lagrangian is similar. The most natural choice is the following definition (where, from now on, the prime on a zeta function means s-derivative).

Definition 2.6 *Under the general hypotheses on $\mathcal{M}$ and $A \geq 0$, if $\ell^2 > 0$, then the effective Lagrangian associated with the operator $\mathcal{A}$ is defined, within the zeta function approach, by*

$$\mathcal{L}(x|\mathcal{A})_\ell \stackrel{\text{def}}{=} -\frac{1}{2}\zeta'(0,x|\ell^2\mathcal{A}) \,. \tag{2.73}$$

Notice that

$$W[\mathcal{A}]_\ell = -\log Z[\mathcal{A}]_\ell = \int_{\mathcal{M}} d\mu_g(x)\, \mathcal{L}(x|\mathcal{A})_\ell \tag{2.74}$$

is fulfilled by definition of integrated zeta function (2.56). Definition (2.73) is well-posed because Theorem 2.7 states that no singularity can appear at $s = 0$ in the local zeta function.

We comment on this, to point out what such definition actually "regularizes" also giving a justification for (2.73). Following [R.M. Wald (1979)] and starting from (2.71), using (2.72) and (2.70), one has (we omit ℓ for the sake of simplicity of the notations)

$$W[\mathcal{A}] = -\log Z[\mathcal{A}] = -\log\,[\det A]^{-1/2}\,.$$

At this step one may try to assume the validity of the matrix identity

$$\text{Trlog}\mathcal{A} = \log\det\mathcal{A}\,. \tag{2.75}$$

For instance, one may define to this end (in the strong operatorial topology)

$$\log\mathcal{A} = \lim_{\epsilon\to 0+}\left\{\int_\epsilon^{+\infty}\frac{dt}{t}e^{-t\mathcal{A}} + (\gamma - \log\epsilon)I\right\}\,. \tag{2.76}$$

The result is the same as that obtained from spectral theory. However the operator obtained, $\log A$, is not of the trace class because it is not bounded, as a direct consequence of Theorem 2.5. We conclude that (2.75) makes no proper sense and that it has to be interpreted in some formal sense, which gives a justification for the definition (2.73).

The way is just to drop the annoying term in (2.76), as well as the regulator ϵ, and use (2.75), and dropping $P(x,x)$ if any, interchanging several times the order of the trace symbol and integrals and the s-derivative:

$$\begin{aligned}
-2W[\mathcal{A}] &= \text{Trlog}\mathcal{A} = \int_0^{+\infty}\frac{dt}{t}\,\text{Tr}\,e^{-t\mathcal{A}} = \int_0^{+\infty}\frac{dt}{t}\,\text{Tr}\,K_t \\
&= \int_{\mathcal{M}}\int_0^{+\infty} dt\,t^{-1}\,d\mu_g(x)K(t,x,x|A) \\
&= \int_{\mathcal{M}}\int_0^{+\infty} dt\,d\mu_g(x)\left.\frac{d}{ds}\right|_{s=0}\left(\frac{1}{\Gamma(s)}t^{s-1}\right)K(t,x,x|A) \\
&= \int_{\mathcal{M}} d\mu_g(x)\,\zeta'(0,x|\mathcal{A})\,.
\end{aligned} \tag{2.77}$$

Notice that, above, also the t-integration of the heat kernel times t^{-1} evaluated for $x = x$ trivially diverges, and thus also the first passages above are incorrect. Anyhow, it is natural, from the formal identities above, to get the definition (2.73). $\mathcal{L}(x|\mathcal{A})$ may hence be considered as the "formal" integral kernel of the operator $\log\mathcal{A}$ evaluated on the diagonal.

As a final comment on (2.75) we remark that, nevertheless, one could use this (literally wrong) identity to *define* an extension of the concept of the trace of $\log\mathcal{A}$, this is because thc right-hand side is however well defined. However, this way leads to a generally loss of linearity due to the possible presence of the so-called *multiplicative anomaly* of the determinant defined in terms of zeta function. We will consider in detail this issue in Chapter 7.

2.4.1 *Comparison with the point-splitting regularization procedure*

Let us consider the relation between the technique and definitions above and the point-splitting renormalization procedure. The idea of the point-splitting method [R.M. Wald (1979); N.D. Birrell and P.C.W. Davies (1982)] consists, following the formal passages developed in (2.77), of the formal definition for the effective Lagrangian

$$\mathcal{L}(y|\mathcal{A}) \stackrel{\text{def}}{=} \lim_{x\to y}\left[-\int_0^{+\infty}\frac{dt}{2t}\,K(t,x,y|\mathcal{A}) + \text{“divergences”}\right]. \tag{2.78}$$

The divergences are those which appear evaluating by brute force the limit in the integral on the right-hand side [R.M. Wald (1979); N.D. Birrell and P.C.W. Davies (1982)]. We shall see the precise form of these by our local zeta function approach. We notice that the term “divergences” is quite ambiguous, because a divergent term plus a finite term is always a divergent term. In fact, one expects some ambiguity because of the dependence on ℓ, which must be preserved in the final expression of the renormalized effective action, for several reasons [R.M. Wald (1979); S.W. Hawking (1977); N.D. Birrell and P.C.W. Davies (1982)]. The actual value of ℓ might be fixed by experimental results (at least for $D = 4$).

As a further general comment, which is not so often remarked, we stress that the point-splitting procedure *works only in the case* $P(x, y) \equiv 0$. Indeed, whenever $\lambda = 0$ is an eigenvalue of $\mathcal{A}$, the integral for $x \neq y$ in (2.78) diverges due to the large t behavior of the integrand. To avoid this shortcoming, one could try to replace the integrand by $K(t, x, y|\mathcal{A}) - P(x, y)$. Nevertheless, this is not enough, indeed, direct inspection for $t \to 0^+$ using Theorem 2.4, proves that the replaced integrand, when inserted in (2.78), produces a divergent integral as $t \to 0^+$ just because of the presence of P.

Let us show that, within our general hypotheses for $A \geq 0$, the outlined point-splitting procedure is actually equivalent to the definition given in the local zeta function framework (2.73). For $N > D/2 + 2$ integer, $s \in \mathbb{C}$ and $x, y \in \mathcal{M}$ whenever the right-hand side makes sense, we consider the function introduced in (2.66) which can also be written as

$$\zeta(N,s,x,y|\ell^2 \mathcal{A},\ell_0^2) = \zeta(s,x,y|\ell^2\mathcal{A}) + \left(\frac{\ell_0}{\ell}\right)^{2s} \frac{P(x,y)}{s\Gamma(s)}$$

$$-\frac{\chi(\sigma(x,y))}{(4\pi)^{D/2}\ell^{2s}\Gamma(s)} \sum_{j=0}^{N} a_j(x,y|A) \int_0^{\ell_0^2} dt\, t^{s+j-D/2-1} e^{-\sigma(x,y)/2t}, \qquad (2.79)$$

where $\ell, \ell_0 > 0$ are fixed. The local zeta function in the right-hand side is that continued as far as is possible in the complex s-plane. Several properties of the function defined above are analyzed in the proof of Theorem 2.7.

Lemma 2.1 *Under the general hypotheses on $\mathcal{M}$ and $A \geq 0$, if $\ell, \ell_0 > 0$ and $N > D/2+2$, the function $(s,x,y) \mapsto \zeta(N,s,x,y|\ell^2\mathcal{A},\ell_0^2)$ is s-analytic and s-, x-, y-jointly continuous for* $\mathrm{Re}\, s > D/2 - N$, *$x, y \in \mathcal{M}$ with all s-derivatives. Moreover,*

$$\zeta'(N,0,x,y|\ell^2\mathcal{A},\ell_0^2) = \Gamma(s)\zeta(N,s,x,y|\ell^2\mathcal{A},\ell_0^2)|_{s=0}\,, \qquad (2.80)$$

$$\zeta(s,x|\ell^2\mathcal{A}) = \zeta(N,s,x,x|\ell^2\mathcal{A},\ell_0^2) - \frac{(\ell_0/\ell)^{2s}P(x,x)}{s\Gamma(s)}$$

$$+\frac{1}{(4\pi)^{D/2}\ell^{2s}} \sum_{j=0}^{N} \frac{a_j(x,x|A)(\ell_0^2)^{(s+j-D/2)}}{\Gamma(s)(s+j-D/2)}\,. \qquad (2.81)$$

Proof. The first part and the last identity have been proven within Theorem 2.7. (2.80) is a trivial consequence of (2.66) in the proof of Theorem 2.7, noticing that, there, the derivative at $s=0$ of the analytic continuation of $1/\Gamma(s)$ is equal to 1 and $1/\Gamma(s) \to 0$ as $s \to 0$. □

Now, let us consider the identity (2.81) ($N > D/2+2$) in the case $P \equiv 0$ and evaluate the s-derivative for $s=0$ necessary to get $\mathcal{L}(x|\mathcal{A})$ by (2.73),

$$\zeta'(0,y|\ell^2\mathcal{A}) = \zeta'(N,0,y,y|\ell^2\mathcal{A},\ell_0^2) - \sum_{j=0, j\neq D/2}^{N} \frac{a_j(y,y|A)}{(4\pi)^{D/2}\ell^{(D-2j)}(D/2-j)}$$

$$+\delta_D \left[\gamma + \log\left(\frac{\ell_0}{\ell}\right)^2\right] \frac{a_{D/2}(y,y|A)}{(4\pi)^{D/2}}\,. \qquad (2.82)$$

From Lemma 2.1, we obtain

$$\zeta'(N, 0, y, y|\ell^2 \mathcal{A}, \ell_0^2) = \lim_{x \to y} \left[\Gamma(s) \zeta(N, s, x, y|\ell^2 \mathcal{A}, \ell_0^2)|_{s=0} \right] . \tag{2.83}$$

By (2.79), taking also $P \equiv 0$ into account in (2.50), the right-hand side of the equation above reads

$$\lim_{x \to y} \left[\Gamma(s) \zeta(s, x, y|\ell^2 \mathcal{A})|_{s=0} - \sum_{j=0}^{N} \frac{a_j(x, y|A)}{(4\pi)^{D/2}} \int_0^{\ell_0^2} dt\, t^{j-D/2-1} e^{-\sigma(x,y)/2t} \right] ,$$

that is

$$\lim_{x \to y} \left[\int_0^{+\infty} \frac{dt}{t} K(t, x, y|\mathcal{A})\, dt - \sum_{j=0}^{N} \frac{a_j(x, y|A)}{(4\pi)^{D/2}} \int_0^{\ell_0^2} dt\, t^{j-D/2-1} e^{-\sigma(x,y)/2t} \right] .$$

Notice that the integral of the heat kernel times t^{-1} converges away from the diagonal, as follows from the proof of Theorem 2.7. To conclude, we just need to put the result obtained into the right-hand side of (2.82), using definition (2.73). By this way $\mathcal{L}(y|\mathcal{A})$ equals

$$\lim_{x \to y} \left\{ - \int_0^{+\infty} \frac{dt}{2t} K(t, x, y|\mathcal{A}) + \sum_{j=0}^{N} \frac{(\sigma/2)^{j-D/2} a_j(x, y|A)}{2(4\pi)^{D/2}} \right.$$

$$\times \int_{\sigma \ell_0^2/2}^{+\infty} du\, u^{D/2-j-1} e^{-u} + \sum_{j=0, j \neq D/2}^{N} \frac{a_j(y, y|A)}{2(4\pi)^{D/2} \ell_0^{(D-2j)} (D/2 - j)}$$

$$\left. -\delta_D \left[\gamma + \log \left(\frac{\ell_0}{\ell} \right)^2 \right] \frac{a_{D/2}(y, y|A)}{2(4\pi)^{D/2}} \right\} . \tag{2.84}$$

We have omitted the function $\chi(\sigma(x, y))$, which disappears in the considered limit, and we have changed variables in the integration.

It is possible to compute more explicitly the integrals above giving a closed form of the divergent terms in (2.84). This can be done by expanding the integrals in powers/logarithm of σ and keeping both the dominant divergent terms and those constant only. In the following, $j = 0, 1, 2, \ldots D/2 + 1, \ldots N$ ($N \geq D/2 + 2$) and $D > 1$. Let us define, with a little misuse of notation, since the right-hand side is not only a function of

σ,

$$H_j(\sigma) \stackrel{\text{def}}{=} \frac{(\sigma/2)^{j-D/2} a_j(x,y|A)}{2(4\pi)^{D/2}} \int_{\sigma/(2\ell_0^2)}^{+\infty} du\, u^{D/2-j-1} e^{-u}\,. \tag{2.85}$$

Therefore, one gets after some computations, $O_k(\sigma)$ being functions which vanish as $x \to y$, in the case of D even,

$$H_{j\geq D/2+1}(\sigma) = -\frac{1}{(D/2-j)\ell_0^{(D-2j)}} \frac{a_j(y,y|A)}{2(4\pi)^{D/2}} + O_j(\sigma)\,; \tag{2.86}$$

$$H_{D/2}(\sigma) = -\frac{a_{D/2}(x,y|A)}{2(4\pi)^{D/2}} \log\left(\frac{\sigma}{2\ell_0^2}\right) - \gamma\frac{a_{D/2}(y,y|A)}{2(4\pi)^{D/2}} + O_{D/2}(\sigma)\,; \tag{2.87}$$

$$H_{j<D/2}(\sigma) = (D/2-j-1)!\frac{a_j(x,y|A)}{2(4\pi)^{D/2}}\left(\frac{2}{\sigma}\right)^{D/2-j} - \frac{1}{(D/2-j)\ell_0^{(D-2j)}} \frac{a_j(y,y|A)}{2(4\pi)^{D/2}} + O_j(\sigma)\,, \tag{2.88}$$

and, if D is odd,

$$H_{j>(D+1)/2}(\sigma) = -\frac{1}{(D/2-j)\ell_0^{(D-2j)}} \frac{a_j(y,y|A)}{2(4\pi)^{D/2}} + O_j(\sigma)\,; \tag{2.89}$$

$$H_{(D+1)/2}(\sigma) = \frac{\ell_0 a_{(D+1)/2}(y,y|A)}{(4\pi)^{D/2}} + O_{(D+1)/2}(\sigma)\,; \tag{2.90}$$

$$H_{(D-1)/2}(\sigma) = \frac{a_{(D-1)/2}(x,y|A)}{2(4\pi)^{D/2}}\sqrt{\frac{2\pi}{\sigma}} - \frac{a_{(D-1)/2}(y,y|A)}{\ell_0(4\pi)^{D/2}} + O_{(D-1)/2}(\sigma)\,; \tag{2.91}$$

$$H_{j<(D-1)/2}(\sigma) = \left(\frac{2}{\sigma}\right)^{D/2-j} \frac{(D-2-2j)!!\sqrt{\pi}}{2^{(D+1)/2-j}} \frac{a_j(x,y|A)}{(4\pi)^{D/2}} - \frac{\ell_0^{(D-2j)}}{(D/2-j)} \frac{a_j(y,y|A)}{2(4\pi)^{D/2}} + O_j(\sigma)\,. \tag{2.92}$$

Above, if the natural n is odd, $n!! \stackrel{\text{def}}{=} 1\cdot 3\cdot 5\cdots n$. The point-splitting formula for the effective Lagrangian arises by substituting the results above into (2.84).

Theorem 2.8 *Under the general hypotheses on $\mathcal{M}$ and $A \geq 0$, if $\ell > 0$ and $D > 1$, then the effective Lagrangian computed by (2.73) may also be computed by a point-splitting procedure, provided $P \equiv 0$. In fact, $\mathcal{L}(y|\mathcal{A})_\ell$ respectively equals*

$$\lim_{x\to y}\left\{-\int_0^{+\infty}\frac{dt}{2t}K(t,x,y|\mathcal{A}) - \frac{a_{D/2}(x,y|A)}{2(4\pi)^{D/2}}\log\left(\frac{e^{2\gamma}\sigma}{2\ell^2}\right)\right.$$
$$\left. + \sum_{j=0}^{D/2-1}(D/2-j-1)!\frac{a_j(x,y|A)}{2(4\pi)^{D/2}}\left(\frac{2}{\sigma}\right)^{D/2-j}\right\}, \tag{2.93}$$

for D even, and

$$\lim_{x\to y}\left\{-\int_0^{+\infty}\frac{dt}{2t}K(t,x,y|\mathcal{A}) + \sqrt{\frac{2\pi}{\sigma}}\frac{a_{(D-1)/2}(x,y|A)}{2(4\pi)^{D/2}}\right.$$
$$\left. + \sum_{j=0}^{(D-3)/2}\frac{(D-2j-2)!!\sqrt{\pi}}{2^{(D+1)/2-j}}\frac{a_j(x,y|A)}{(4\pi)^{D/2}}\left(\frac{2}{\sigma}\right)^{D/2-j}\right\}, \tag{2.94}$$

for D odd.

Notice the absence of ℓ in the final result on the effective Lagrangian, whenever D is odd. Concerning the physical case $D = 4$ and the general case D even, the arbitrary scale ℓ is necessary, due to the logarithmic divergence in (2.93): ℓ^2 has to combine with σ in order to give a non-dimensional argument in the logarithm. Therefore, the effective Lagrangian turns out to be affected by ambiguous terms of the form, where α is any strictly positive real,

$$\delta\mathcal{L}(x,\alpha|\mathcal{A}) \stackrel{\text{def}}{=} -\delta_D\frac{a_{D/2}(x,x|A)}{2(4\pi)^{D/2}}\log\alpha\,, \tag{2.95}$$

generated by rescaling $\ell^2 \to \alpha\ell^2$. These terms cannot be determined within the theory and represent a remaining finite part of the renormalization procedure. The ambiguity here found is a subcase of a known ambiguity which arises also in the presence of P when the point-splitting approach cannot be implemented. This can be extracted directly from (2.73). In fact, already from the definitions (2.54) and (2.50), we have

$$\zeta(s,x|(\alpha\ell^2)\mathcal{A}) = \alpha^s\zeta(s,x|\ell^2\mathcal{A}) \tag{2.96}$$

and thus

$$\zeta'(0,x|(\alpha\ell^2)\mathcal{A}) = \zeta'(0,x|\ell^2\mathcal{A}) - \zeta(0,x|\mu^2\mathcal{A})\log\alpha\,. \tag{2.97}$$

Recalling (2.61) and (2.73) we get

$$\mathcal{L}(x|\mathcal{A})_{\sqrt{\alpha}\ell} = \mathcal{L}(x|\mathcal{A})_\ell - \left[\delta_D \frac{a_{D/2}(x,x|A)}{2(4\pi)^{D/2}} + \frac{1}{2}P(x,x)\right]\log\alpha\,. \tag{2.98}$$

All the results should be preserved also in the case of a noncompact manifold, because all proofs are based on Theorem 2.7, which should hold true also by dropping the hypothesis of compactness (by assuming some further hypotheses as completenesss).

2.4.2 *Green functions and zeta functions*

We want to focus on the relationship between spectral zeta function and Green function of the operator $\mathcal{A}$. A few preliminary comments on the inverse operators of positive self-adjoint operators are necessary. Consider a Hilbert space $\mathcal{H}$ and a self-adjoint operator $\mathcal{A} : \mathcal{D}(\mathcal{A}) \to \mathcal{H}$, $\overline{\mathcal{D}(\mathcal{A})} = \mathcal{H}$ (for instance, under our general hypotheses, $\mathcal{A}$ is the unique self-adjoint extension of A given in (2.4)). For the moment, we explicitly assume that $\text{Spec}(\mathcal{A}) \subset (0,+\infty)$. In this case, since the spectrum is closed, $\mathcal{A}$ has a well-defined unique inverse operator $\mathcal{A}^{-1} : \text{Ran}(\mathcal{A}) \to \mathcal{D}(\mathcal{A})$, with $\mathcal{A}^{-1}\mathcal{A} = I_{\mathcal{D}(\mathcal{A})}$ and $\mathcal{A}\mathcal{A}^{-1} = I_{\text{Ran}(\mathcal{A})}$.

Notice that $||\mathcal{A}^{-1}||$ is bounded by $\sup\{1/\lambda|\lambda \in \text{Spec}(\mathcal{A})\} < +\infty$ and $\text{Ran}(\mathcal{A})$ is dense in $\mathcal{H}$, because $\text{Ran}(\mathcal{A}) = \text{Ran}(\mathcal{A} - 0I)$ which is dense, $\mathcal{A}$ being self-adjoint and 0 belongs to the resolvent set $\rho(\mathcal{A})$. Therefore $\mathcal{A}^{-1}$ can be uniquely extended into a bounded operator defined on the whole $\mathcal{H}$, which we shall denote with the same symbol $\mathcal{A}^{-1}$.

With such an extended definition of $\mathcal{A}^{-1}$, $\mathcal{A}\mathcal{A}^{-1} = I$ holds true on the whole space $\mathcal{H}$ and not only in $\text{Ran}(\mathcal{A})$. Indeed, $\mathcal{A} = \mathcal{A}^\dagger$ is a closed operator and $\mathcal{A}^{-1}$ is bounded on $\text{Ran}(\mathcal{A})$. Consider $f \in \mathcal{H}$ and $\{f_n\} \subset \text{Ran}(\mathcal{A})$ with $f_n \to f$. Then $g_n = \mathcal{A}^{-1}f_n \in \mathcal{D}(\mathcal{A})$ and $g_n \to g = \mathcal{A}^{-1}f$ because $\mathcal{A}^{-1}$ is bounded but, we also have $\mathcal{A}g_n (= \mathcal{A}\mathcal{A}^{-1}f_n = f_n) \to f$, by the definition of closed operator $\mathcal{A}$ it must be $f \in \mathcal{D}(\mathcal{A})$ and $\mathcal{A}g = f$. In other words, $\mathcal{A}^{-1}(\mathcal{H}) \subset \mathcal{D}(\mathcal{A})$ and $\mathcal{A}\mathcal{A}^{-1} = I$.

We are now interested in integral representations of $\mathcal{A}^{-1}$ if $\mathcal{A}$ is given by the unique self-adjoint extension of $-\Delta^{(X)} + V$ and our general hypotheses hold. In the case $D < 4$, one gets from Theorem 2.5 that the series of elements $||\mathcal{A}^{-1}\phi_j||^2 = |\lambda_j|^{-2} = \lambda_j^{-2}$ converges and, since $\{\phi_j | j = 0, 1, 2, \ldots\}$ is a Hilbertian base, this means that [M. Reed and B. Simon (1980)] $\mathcal{A}^{-1}$ is a Hilbert-Schmidt operator. Thus, because of our hypotheses of a countable topology which entails a separable measure, $\mathcal{A}^{-1}$ is represented by an $L^2(\mathcal{M} \times \mathcal{M}, \mu_g \otimes \mu_g)$ integral kernel $G(x, y|\mathcal{A})$. As is well-known, such a *function* is called a *Green function* of $\mathcal{A}$, and satisfies almost everywhere, for any $\psi \in \mathcal{D}(\mathcal{A})$,

$$\int_{\mathcal{M}} d\mu_g(y)\, G(x, y|\mathcal{A})(\mathcal{A}\psi)(y) = \psi(x) \qquad (\text{namely } \mathcal{A}^{-1}\mathcal{A} = I_{\mathcal{D}(\mathcal{A})}) \quad (2.99)$$

and, for any $\psi \in L^2(\mathcal{M}, \mu_g)$, almost everywhere,

$$\mathcal{A}_x \int_{\mathcal{M}} d\mu_g(y)\, G(x, y|\mathcal{A})\psi(y) = \psi(x) \qquad (\text{namely } \mathcal{A}\mathcal{A}^{-1} = I)\,. \quad (2.100)$$

These relationships are often written under the synthetic form

$$\mathcal{A}_x G(x, y|\mathcal{A}) = \delta(x, y)\,. \quad (2.101)$$

In the case $D \geq 4$, $\mathcal{A}^{-1}$ is not Hilbert-Schmidt since the series of elements $|\lambda_j|^{-2} = \lambda_j^{-2}$ diverges as follows from Theorem 2.5. Anyhow, these facts do not forbid the existence of integral kernels, which are *not* $L^2(\mathcal{M} \times \mathcal{M}, \mu_g \otimes \mu_g)$, which represent $\mathcal{A}^{-1}$ and satisfy either (2.99) and/or (2.100) on suitable domains.

In the next theorem, we prove that, if $n = 1, 2, \ldots$, within our general hypotheses and $P \equiv 0$,

$$(x, y) \mapsto \ell^{2n}\zeta(n, x, y|\ell^2\mathcal{A})$$

defines an integral kernel of $\mathcal{A}^{-n}$, that is the Green function of $\mathcal{A}^n$, when one deals with functions of a suitable domain.

To this end, notice that, for $n = 1, 2, \ldots$, $A^n : C^\infty(\mathcal{M}) \to L^2(\mathcal{M}, \mu_g)$ is essentially self adjoint and its unique self-adjoint extension is $\mathcal{A}^n$ obtained via spectral theorem by $\mathcal{A}$. Indeed A^n is essentially self-adjoint on $C^\infty(\mathcal{M})$ by Nelson's theorem [B. Simon (1972)] because the eigenvectors of A are eigenvectors of A^n self and they give rise to a dense set of analytic vectors for A^n. Moreover $A(C^\infty(\mathcal{M})) \subset C^\infty(\mathcal{M})$ implies $\mathcal{A}^n f = A^n f$ for all $f \in C^\infty(\mathcal{M})$. By consequence $\mathcal{A}^n$ is a, and thus the unique, self-adjoint extension of A^n.

Theorem 2.9 *Assume the general hypotheses on $A \geq 0$ and $\mathcal{M}$ together and take $n > 0$ integer.*

(a) *If $P \equiv 0$, consider the continuous map for $(x, y) \in (\mathcal{M} \times \mathcal{M}) \setminus \mathcal{D}$*

$$(x, y) \mapsto G(x, y|\mathcal{A}^n) \stackrel{def}{=} \ell^{2n}\zeta(n, x, y|\ell^2 \mathcal{A}), \qquad (2.102)$$

does not depend on $\ell > 0$ and $G(x, y|\mathcal{A}^n) = G(y, x|\mathcal{A}^n)$ and

(i) *it defines a Green function of $\mathcal{A}^n$ on $C^\infty(\mathcal{M})$, in the sense that $G(\cdot, \cdot|\mathcal{A})$, for any $\psi \in C^\infty(\mathcal{M})$, and any $\phi \in \mathrm{Ran}(A^n)$ satisfies both*

$$\int_{\mathcal{M}} d\mu_g(y)\, G(x, y|\mathcal{A}^n)(\mathcal{A}^n\psi)(y) \;=\; \psi(x)\,, \qquad (2.103)$$

$$A^n_x \int_{\mathcal{M}} d\mu_g(y)\, G(x, y|\mathcal{A}^n)\phi(y) \;=\; \phi(x) \qquad (2.104)$$

and thus $G(\cdot, \cdot|\mathcal{A})$ defines the integral kernel of $\mathcal{A}^{-n}$ on $\mathrm{Ran}(A^n)$.

(ii) *In particular, if $n \geq (D + 1)/4$, (2.99) and (2.100) hold in the indicated extended domains for $\mathcal{A}^n$ in place of $\mathcal{A}$.*

(iii) *Any other map continuous on $(x, y) \in (\mathcal{M} \times \mathcal{M}) \setminus \mathcal{D}$ which satisfies either (2.103) or (2.104) coincides with $G(\,.,\, .|\mathcal{A}^n)$.*

(b) *Dropping the hypothesis $P \equiv 0$, for $(x, y) \in (\mathcal{M} \times \mathcal{M}) \setminus \mathcal{D}$*

$$(x, y) \mapsto G(x, y|\mathcal{A}^n) \stackrel{def}{=} \ell^{2n}\zeta(n, x, y|\ell^2 \mathcal{A})$$

is symmetric, it does not depend on $\ell > 0$, and

(i) *it satisfies (2.103) anymore for $\psi \in C^\infty(\mathcal{M}) \cap \{\mathrm{Ker}(\mathcal{A})\}^\perp$ and produces a vanishing right-hand side of* (2.103) *for $\psi \in C^\infty(\mathcal{M}) \cap \mathrm{Ker}(\mathcal{A})$;*

(ii) *it satisfies (2.104) for any $\phi \in \mathrm{Ran}(A^n)$.*

Proof. First notice that, from the definition (2.50) and the analytic extension given in (**a**) of Theorem 2.7, the $C^0((\mathcal{M} \times \mathcal{M}) \setminus \mathcal{D})$ function $\ell^{2n}\zeta(n, x, y|\ell^2\mathcal{A})$ does not depend on ℓ so, from now on, we shall omit ℓ. By the same definition, one gets the symmetry since the heat-kernel is symmetric. Let us consider the case (**a**) $P \equiv 0$. In this case $\mathcal{A}^{-n}$ exist (and is uniquely defined). $\mathrm{Spec}(\mathcal{A}^{-n}) = \{\lambda^{-n} \mid \lambda \in \mathrm{Spec}(\mathcal{A})\}$ and thus $\mathcal{A}^{-n}$ is Hilbert-Schmidt or trace class if, respectively, $n \geq (D + 1)/4$ or $n > D/2$.

We want to study the singularities of $G(x,y|\mathcal{A}^n)$. To this end, consider the function

$$y \mapsto \int_0^{+\infty} dt\, t^{n-1}|K(t,x,y|\mathcal{A})|\,. \tag{2.105}$$

Employing the same analysis of the proof of Theorem 2.7, by (2.79) and Lemma 2.1, one finds straightforwardly that the only possible singularities which may forbid the y-integrability of the function above come from the points at $t = 0$ and $y = x$. We have analyzed such divergences proving Theorem 2.8. These arise by taking the absolute value of integrals of the form (2.85) and, more precisely, of the form ($J = n, n+1, n+2, \ldots, N$, where N is a fixed sufficiently large integer)

$$G_J(\sigma) = \frac{(\sigma/2)^{J-D/2} a_{J-n}(x,y|A)}{2(4\pi)^{D/2}} \int_{\sigma/(2\ell_0^2)}^{+\infty} du\, u^{D/2-J-1} e^{-u}\,, \tag{2.106}$$

where we have taken into account that zeta function is evaluated at $s = n$, in the present case, rather than $s = 0$. As we have seen in (2.86) – (2.92), the dominant divergences arise for $J = n$ and they diverge as $d(x,y)^{(2n-D)}$ ($\log d(x,y)$ for $D = 2n$) whenever $D > 1$ is even, and once again, as $d(x,y)^{(2n-D)}$ whenever $D > 1$ is odd. Therefore, no divergence arises as $y \to x$ for $n > D/2$ and the integral in (2.113) can be defined, more weakly for $n \geq (D+1)/4$ divergences arise but $G(x,y|\mathcal{A}^n)$ turns out to be $L^2(\mathcal{M} \times \mathcal{M})$, i.e. it define a Hilbert-Schmidt operator. This result implies that if, $P \equiv 0$, $n \geq (D+1)/4$ and (2.103) and (2.104) are valid then (2.99) and (2.100) must hold by continuity. Indeed, with these hypotheses $G(x,y|\mathcal{A})$ defines a Hilbert-Schmidt operator which coincides with the Hilbert-Schmidt operator $\mathcal{A}^{-n}$ on a dense domain and the kernel of $\mathcal{A}^{-n}$, which satisfies (2.103) and (2.104) by definition, has to coincide with $G(x,y|\mathcal{A})$ given via zeta function, by uniqueness of the kernels of Hilbert-Schmidt operators. It remains to prove (2.103), (2.104), (**a, iii**) and (**b**). Notice that (2.103) and (2.104) are equivalent in the indicated domains, by the uniqueness of the inverse function of A^n. Therefore, let us prove (2.103). Take $\phi \in C^\infty(\mathcal{M})$ and consider

$$y \mapsto \int_0^{+\infty} dt\, t^{n-1}|K(t,x,y|\mathcal{A})\phi(y)|\,. \tag{2.107}$$

From the analysis of singularities given above, it turns out that the function in (2.107) is y-integrable and also the integrand in the right-hand side of

(2.107) belongs to $L^1(\mathcal{M} \times [0, +\infty), \mu_g \otimes dt)$.

First, consider the case $n = 1$. By Fubini's theorem, definition (2.50), the point (**a**) of Theorem 2.7 and using the second Green identity, noticing that $\partial\mathcal{M} = \emptyset$, we have for any $\psi \in C^\infty(\mathcal{M})$:

$$
\begin{aligned}
&\int_{\mathcal{M}} d\mu_g(y)\, \zeta(1, x, y|\mathcal{A})(\mathcal{A}\psi)(y) \\
&= \int_0^{+\infty} \int dt\, d\mu_g(y) K(t, x, y|\mathcal{A})(\mathcal{A}\psi)(y) \\
&= \lim_{\epsilon \to 0+} \int_\epsilon^{1/\epsilon} \int dt\, d\mu_g(y) K(t, x, y|\mathcal{A})(\mathcal{A}\psi)(y) \\
&= \lim_{\epsilon \to 0+} \int_\epsilon^{1/\epsilon} \int dt\, d\mu_g(y)(A_y K(t, x, y|\mathcal{A}))\psi(y) \\
&= -\lim_{\epsilon \to 0+} \int_\epsilon^{1/\epsilon} \int dt\, d\mu_g(y)\, dK(t, x, y|\mathcal{A})_t \psi(y) \\
&= -\lim_{\epsilon \to 0+} \int_\epsilon^{1/\epsilon} dt \left(\int d\mu_g(y)\, K(t, x, y|\mathcal{A})\psi(y) d\mu_{\mathbf{g}}(y) \right)_t \\
&= -\lim_{\epsilon \to 0+} (e^{-(1/\epsilon)\mathcal{A}}\psi)(x) + \lim_{\epsilon \to 0+} (e^{-\epsilon\mathcal{A}}\psi)(x) = \psi(x)\,, \qquad (2.108)
\end{aligned}
$$

where subscript index t indicates t-derivative. At the end, we have employed the point (**d**) of Theorem 2.3. The case $n > 1$ can be treated very similarly. Now, the first line of (2.108) has to be replaced by

$$
\begin{aligned}
&\int_{\mathcal{M}} d\mu_g(y)\zeta(n, x, y|\mathcal{A})(\mathcal{A}^n\psi)(y) \\
&= \frac{1}{\Gamma(n)} \int_0^{+\infty} \int_{\mathcal{M}} dt\, d\mu_{\mathbf{g}}(y) t^{n-1} K(t, x, y|\mathcal{A})(\mathcal{A}\psi)(y)\,. \qquad (2.109)
\end{aligned}
$$

Using manipulations similar to those for $n = 1$, decomposing $\mathcal{A}^n$ in $\mathcal{A}\mathcal{A}^{n-1}$ and employing integration by parts recursively, we get

$$
\int_{\mathcal{M}} d\mu_g(y)\, \zeta(n, x, y|\mathcal{A})(\mathcal{A}^n\psi)(y) = \frac{(n-1)!}{\Gamma(n)}\psi(x) = \psi(x)\,. \qquad (2.110)
$$

The proof of the point (**b**) of the thesis is very similar. It is sufficient to notice that, because of the presence of the further term $-P$ in the integrand of the right-hand side of (2.50), the procedure in (2.108) gives the final result 0 instead of $\psi(x)$ whenever $\psi \in \mathrm{Ker}(\mathcal{A})$. Conversely, we reduce to the case (**a**) if $P\psi = 0$ namely $\psi \in \mathcal{D}(\mathcal{A}^n) \cap \{\mathrm{Ker}(\mathcal{A})\}^\perp$. Let us prove the uniqueness property in (**a, iii**). If H is continuous on $\mathcal{M} \times \mathcal{M} \setminus \mathcal{D}$ and satisfies (2.103) (which is equivalent to (2.104)), then $K(\cdot,\cdot|\mathcal{A}) \stackrel{\text{def}}{=} G(\cdot,\cdot|\mathcal{A}^n) - H(\cdot,\cdot|\mathcal{A})$ is continuous on the same set. In particular if $x \in \mathcal{M}$ is taken and $x \notin E \subset \mathcal{M}$ with E measurable (with finite measure), for any $f \in \text{Ran}(A^n)$

$$\int_E d\mu_g(y) K(x,y|\mathcal{A}) f(y) = (f^* \restriction_E, K(x|\mathcal{A}) \restriction_E)_{L^2(E)} = 0\,.$$

$y \mapsto K(x,y|\mathcal{A}) \restriction_E$ is $L^2(E) \cap L^1(E)$ and since $\text{Ran}(A^n)$ is dense we may approximate in the norm L^2 the characteristic function of E obtaining

$$\int_E d\mu_g(y)\, K(x,y|\mathcal{A}) = 0\,.$$

Since E is arbitrary and $K(x,y|\mathcal{A})$ is continuous for $y \neq x$, we have $K(x,y|\mathcal{A}) = 0$ for each $y \neq x$. □

Corollary 2.2 *Referring to Theorem 2.9, in both cases* (**a**) *and* (**b**), $(x,y) \mapsto G(x,y|\mathcal{A})$ *turns out to be of class* $C^\infty(\mathcal{M} \times \mathcal{M} \setminus \mathcal{D})$.

Proof. Take $x \in \mathcal{M}$ and consider any open set $U' \subset \mathcal{M}$ equipped with a coordinate patch and such that $x \notin \overline{U} \subset U'$, U being any other relatively compact open set identified with a neighborhood of $\mathbb{R}^D$ by means of the coordinate system. Define $G_x(y) \stackrel{\text{def}}{=} G(x,y|\mathcal{A})$ for $y \in U$. (2.103) says that, viewing G_x as a distribution in $\mathcal{D}'(U)$, it holds in the weak sense

$$-g(y)^{1/2}\Delta_y G_x = -g(y)^{1/2} V(y) G_x\,. \tag{2.111}$$

However $G_x \in L^2(U) = W_0(U)$ because G is continuous on $\overline{U}$. $W_k(U)$ is the k-th Sobolev space. Notice that

$$-g(y)^{1/2}\Delta_y^{(X)} = -\partial_a g^{ab}(y) g(y)^{1/2}\partial_b + (\text{order} < 2)$$

is *strongly elliptic* on U because $y \mapsto [g_{ab}(y)]$ is continuous and everywhere positive defined, moreover gV is smooth and thus the right-hand side of (2.111) is of class $W_n(U)$, with $n = 0$ precisely. Hence, by Friedrichs' elliptic regularity theorem ([M. Reed and B. Simon (1980)] vol. II, IX p.112), G_x must be of class W_{2+n}. Iterating the procedure, one get $G_x \in W_n(U)$ for all $n \in \mathbb{N}$. This entails that $G_x \in C^\infty(U)$ by Sobolev's lemma [W. Rudin (1997)] because G_x is continuous. Since $G(x,y|\mathcal{A}) = G(y,x|\mathcal{A})^*$ by construction and U is arbitrary, we have that $G(\cdot,\cdot|A)$ is smooth in either arguments separately on $(\mathcal{M} \times \mathcal{M}) \setminus \mathcal{D}$. Using Sobolev's lemma again, the jointly continuity of G entails that $G(\cdot,\cdot|\mathcal{A}) \in C^\infty((\mathcal{M} \times \mathcal{M}) \setminus \mathcal{D})$. □

The fact that A^n is essentially self adjoint and $\mathcal{A}^n$ is the unique self-adjoint extension yields that $\text{Ran}(\mathcal{A}^n) \subset \overline{\text{Ran}(A^n)}$, because $\mathcal{A}^n$ is the closure of A^n. As a consequence, since the kernel of $\mathcal{A}^n$ coincides with the kernel of A^n and the kernel of A, if $P \equiv 0$, 0 is in the resolvent of $\mathcal{A}^n$ and we get $\overline{\text{Ran}(A^n)} = \overline{\text{Ran}(\mathcal{A}^n - 0I)} = L^2(\mathcal{M}, \mu_g)$.

We leave to the reader to prove that, if $P \equiv 0$ and $n = 1, 2, \ldots$, for any $\psi \in L^2(\mathcal{M}, \mu_g)$ and any $\psi \in \mathcal{D}(\mathcal{A}^n)$, almost everywhere,

$$\begin{aligned} \mathcal{A}_x \int_{\mathcal{M}} d\mu_g(y)\, G(x, y|\mathcal{A}^n)\psi(y) &= \psi(x)\,, \\ \int_{\mathcal{M}} d\mu_g(y) G(x, y|\mathcal{A}^n)(\mathcal{A}^n\psi)(y) &= \psi(x)\,, \end{aligned} \tag{2.112}$$

provided the integrals are understood in the improper sense if necessary. That is, if $f \in L^2(\mathcal{M}, \mu_g)$

$$\int_{\mathcal{M}} d\mu_g(y)\, G(x, y|\mathcal{A}^n) f(y) \stackrel{\text{def}}{=} \lim_{m\to\infty} \int_{\mathcal{M}} d\mu_g(y)\, G(x, y|\mathcal{A}^n) f_m(y)\,,$$

with $\{f_m\} \subset \text{Ran}(A^n)$, $f_m \to f$.

Finally, it is an easy task to show that if $n > D/2$, in both cases (**a**) and (**b**) discussed in Theorem 2.9, $G(x, y|\mathcal{A}^n)$ is continuous on $\mathcal{M} \times \mathcal{M}$ and

$$\int_{\mathcal{M}} d\mu_g(x)\, G(x, x|\mathcal{A}^n) = \text{Tr}\mathcal{A}^{-n}\,. \tag{2.113}$$

In the case (b) the operator $\mathcal{A}^{-n}$ above is supposed to be defined via the spectral theorem, dropping the part of the spectral measure associated with the kernel of $\mathcal{A}$.

2.4.3 *Differential calculus of the heat kernel and local zeta functions*

This part is very technical and concerns the smoothness of the local zeta functions obtained and some other related topics. We collect here several results which will be used towards the end of this chapter, when discussing the zeta-function regularization of the stress-energy tensor and field fluctuations [V. Moretti (1997b); V. Moretti (1999a); V. Moretti (1999b)]. The reader may skip this part now, to come back to it (if necessary) while reading the end of this chapter.

First of all, we want to investigate the possibility of passing the symbol of derivatives under the symbol of summation in the eigenvector expansion in (**b**) of Theorem 2.2.

Remark. Throughout this part, if $u \in \mathcal{M}$, I_u denotes a sufficiently small neighborhood of u, whose closure is compact and is contained in a coordinate patch domain. Derivative operators D^γ are referred to these coordinates.

Theorem 2.10 *Under the general hypotheses on M and A, we have*

$$D_x^\alpha D_y^\beta K(t,x,y|\mathcal{A}) = \sum_{j=0}^{\infty} e^{-\lambda_j t} D_x^\alpha \phi_j(x) D_y^\beta \phi_j^*(y)\,, \tag{2.114}$$

for $x, y \in I_u$, $u \in \mathcal{M}$, $t \in (0,+\infty)$ and any pair of multiindices α, β. The right-hand side of this formula is regular in $[\gamma,+\infty) \times I_u \times I_u$, for $\gamma > 0$. Moreover, if $T > 0$, $t > 2T$, $x, y \in I_u$, then

$$|D_x^\alpha D_y^\beta (K(t,x,y|\mathcal{A}) - P(x,y|\mathcal{A}))| \leq P_T^{(\alpha,\beta)} {\sum_{j\in\mathbb{N}}}' e^{-\lambda_j(t-2T)}, \tag{2.115}$$

$$\leq Q_T^{(\alpha,\beta)}\, e^{-\lambda(t-2T)}\,, \tag{2.116}$$

where λ is the first strictly positive eigenvalue of $\mathcal{A}$ and the constants $P_T^{(\alpha,\beta)}, Q_T^{(\alpha,\beta)} > 0$ do not depend on t, x, y.

Proof. We need a useful lemma.

Lemma 2.2 *For any $T > 0$ we have the following result*

$$|e^{-\lambda_j t} D_x^\alpha \phi_j(x) D_y^\beta \phi_j^*(y)| \leq P_T^{(\alpha,\beta)} e^{-\lambda_j(t-2T)}\,, \tag{2.117}$$

where $x, y \in I_u$, $t > 2T$, and

$$P_T^{(\alpha,\beta)} = \left[\sup_{x\in\overline{I_u}} ||D_x^\alpha K(T,x,\,.\,|\mathcal{A})||_{L^2(\mathcal{M},\mu_g)}\right] \times \left[\sup_{y\in\overline{I_u}} ||D_y^\beta K(T,\,.\,,y|\mathcal{A})||_{L^2(\mathcal{M},\mu_g)}\right]. \tag{2.118}$$

Let us now prove the lemma and the theorem. Consider an eigenvector ϕ_j and fix $T \in (0,+\infty)$, and consider a neighborhood of $u \in \mathcal{M}$, J_u endowed with a coordinate system. $x, y \in I_u$, I_u being another neighborhood of u

with $\overline{I_u} \subset J_u$. The derivative operators are referred to these coordinates. From the definition of the heat kernel, we have

$$e^{-T\lambda_j}\phi_j(x) = \int_{\mathcal{M}} d\mu_g(z)\, K(T,x,z|\mathcal{A})\phi_j(z)\,. \tag{2.119}$$

We can derive both sides of the equation above employing operators D^α_x. Since, for fixed y, T the derivatives of K are bounded ($K(T,\cdot,y|\mathcal{A})$ is in C^∞ on the compact $\overline{I_u}$), we can pass the derivative operator under the integral symbol, thus obtaining

$$\begin{aligned} |D^\alpha_x\phi_j(x)| = \quad & |e^{\lambda_j T}\int_{\mathcal{M}} d\mu_g(z) D^\alpha_x K(T,x,z|\mathcal{A})\phi_j(z)| \\ & \leq e^{\lambda_j T}||D^\alpha_x K(T,x,\,.\,|\mathcal{A})||\,. \end{aligned} \tag{2.120}$$

where we have made use of the Cauchy-Schwarz inequality, $||\phi_j|| = 1$ and the norms are referred to $L^2(\mathcal{M}, d\mu_g)$. It is easy to show that, under our hypotheses, $x \mapsto ||D^\alpha_x K(T,x,\,.\,|\mathcal{A})||$ for $x \in \overline{I_u}$ is a continuous map by Lebesgue's dominate convergence. The same results hold if one interchanges the rôle of y and x. Therefore, defining $P^{(\alpha,\beta)}_T$ as in (2.118), we have, for any $x, y \in I_u$, the λ_j-uniform upper bound (2.117):

$$|e^{-\lambda_j t} D^\alpha_x\phi_j(x) D^\beta_y\phi^*_j(y)| \leq P^{(\alpha,\beta)}_T e^{-\lambda_j(t-2T)}\,. \tag{2.121}$$

The inequality found proves that the absolute values of the terms of the series

$$\sum_{j\in\mathbb{N}}{}' e^{-\lambda_j t} D^\alpha_x\phi_j(x) D^\beta_y\phi^*_j(y) \tag{2.122}$$

are x, y-uniformly bounded, for $(x,y,t) \in I_u \times I_u \times (2T, +\infty)$, by terms of the convergent series

$$\begin{aligned} &\sum_{j\in\mathbb{N}}{}' e^{-\lambda_j(t-2T)} P^{(\alpha,\beta)}_T \\ &\qquad = P^{(\alpha,\beta)}_T \int_{\mathcal{M}} d\mu_g(z)\ \{K(t-2T,z,z|\mathcal{A}) - P(z,z|\mathcal{A})\} \\ &\qquad\quad = P^{(\alpha,\beta)}_T\, \mathrm{Tr}\left\{K_{(t-2T)} - P\right\}\,. \end{aligned} \tag{2.123}$$

This holds for any choice of the multiindices α, β and this entails the regular convergence of the series in the right-hand side of (2.114), as well as the possibility of passing the derivative operators under the symbol of summation and thus the validity of (2.114). Then, (2.115) and (2.116) are simple consequences of (2.118) and the possibility of passing the derivative under the symbol of summation. □

The above theorem and lemma are useful for dealing with the following definitions, which we shall employ later when discussing the zeta-function regularization of the stress-energy tensor.

Theorem 2.11 *Under the general hypotheses on $\mathcal{M}$ and $A \geq 0$, take $\ell > 0$ and refer the derivatives below to coordinates in a sufficiently small neighborhood I_u of any $u \in \mathcal{M}$. For any pair of multiindices α, β, it holds*

$$D_x^\alpha D_y^\beta \zeta(s, x, y|\ell^2 A) =$$
$$\int_0^{+\infty} d\left(\frac{t}{\ell^2}\right)\left(\frac{t}{\ell^2}\right)^{s-1} D_x^\alpha D_y^\beta \frac{\{K(t,x,y|A) - P_0(x,y|A)\}}{\Gamma(s)} . \tag{2.124}$$

Above $\mathrm{Re}\, s > D/2 + |\alpha| + |\beta|$ *and $x, y \in I_u$ and the right-hand side defines an s-analytic function jointly s, x, y-continuous with all its s-derivatives.*

Proof. The proof of this theorem and of the following Theorems 2.12-2.14 is a straightforward generalization of the proof of the corresponding parts of Theorem 2.7, and thus we will just sketch it. As in the proof of Theorem 2.7, the main idea is to use the decomposition (2.62). Then, one studies the possibility of computing the derivative of the off-diagonal zeta function by passing D_x^α and D_y^β under the symbol of integration in both integrals in the right-hand side of (2.62). This is possible, provided the absolute values of the differentiated integrand are (x, y)-uniformly bounded by integrable functions (dependent on α, β in general). This implies also the continuity of the derivatives, because the derivatives of the integrands are continuous functions. The analyticity in s can be proven by checking the Cauchy-Riemann conditions passing the derivative under the symbol of integration once again. The s-derivatives of the integrand at any order can be proven to be bounded by means of a similar procedure. Then the proof proceeds similarly to the proof of Theorem 2.7. One uses Lemma 2.2 and (2.116) (choosing $2T < \ell_0^2$) in place of the corresponding formula (2.65), to prove that the last integral in the right-hand side of (2.62) can be differentiated under the symbol of integration, producing an s-analytic function which is continuous, with all of its s derivatives, for $s \in \mathbb{C}$ and $x, y \in I_u$. The former integral can be studied employing the expansion in Theorem 2.4. The requirement in the point **(a)** $\text{Re}\, s > D/2 + |\alpha| + |\beta|$ arises by the term with $j = 0$ in the heat-kernel expansion when all the derivatives, either in x and in y, act on the exponential producing a factor $t^{-|\alpha|-|\beta|}$ and posing $x = y$ in the end. Equation (2.66) and the subsequent ones have to be changed, by employing $O_\eta^{(\alpha,\beta)}$ in place of O_η and $t^{s-1+N-D/2-|\alpha|-|\beta|}$ in place of $t^{s-1+N-D/2}$. □

Definition 2.7 Under the general hypotheses on $\mathcal{M}$ and $A \geq 0$, take $\ell > 0$ and refer the derivatives to coordinates in a sufficiently small neighborhood I_u of any $u \in \mathcal{M}$. If $x, y \in I_u$, $\text{Re}\, s > D/2 + |\alpha| + |\beta|$, the off-diagonal differentiated local zeta function of $\mathcal{A}$ is

$$\zeta^{(\alpha,\beta)}(s, x, y|\ell^2 \mathcal{A}) \stackrel{\text{def}}{=} D_x^\alpha D_y^\beta \zeta(s, x, y|\ell^2 \mathcal{A}) \,, \tag{2.125}$$

and the differentiated local zeta function of $\mathcal{A}$ is

$$\zeta^{(\alpha,\beta)}(s, x|\ell^2 \mathcal{A}) \stackrel{\text{def}}{=} \left\{D_x^\alpha D_y^\beta \zeta(s, x, y|\ell^2 \mathcal{A})\right\}_{x=y} \,. \tag{2.126}$$

Let us consider a generalization of Theorem 2.7 to the case of differentiated zeta functions. We leave the proofs to the reader, since they are very similar to that of the relevant part of Theorem 2.7, with the suggestion given in the proof of Theorem 2.10. Notice only, that the requirement $D_x^\alpha D_y^\beta P(x, y|\mathcal{A}) = 0$ is simply due to the divergence of the integral $\int_0^{\ell_0^2} dt t^{s-1}$ for $s \leq 0$ and (2.128) is essentially due to the presence of the factor $1/\Gamma(s)$ in all the considered integrals, which vanishes as a simple zero when $s \to 0$.

The first theorem concerns the off-diagonal differentiated zeta function. From now on, we define $\mathcal{D}_{I_u} \stackrel{\text{def}}{=} \{(x, y) \in I_u \times I_u | x = y\}$.

Theorem 2.12 *Under the general hypotheses on $\mathcal{M}$ and $A \geq 0$, take $\ell > 0$, any pair of multiindices α, β and refer the derivatives to coordinates in a sufficiently small neighborhood I_u of any $u \in \mathcal{M}$. If $x \neq y$ in I_u,*

(a) *$\zeta^{(\alpha,\beta)}(s, x, y|\ell^2 \mathcal{A})$ can be analytically continued in the variable s into an entire function on $\mathbb{C}$. It also defines a $C^\infty(\mathbb{C} \times \{(I_u \times I_u) \setminus \mathcal{D}_{I_u}\})$, when varying x, y.*

(b) *for* $\text{Re}\, s > 0$, *or everywhere provided $D_x^\alpha D_y^\beta P(x, y|\mathcal{A}) = 0$, (2.124) holds also when the left-hand side is replaced by the s-continued function $\zeta^{(\alpha,\beta)}$.*

A second theorem concerns the analytic continuation of the on diagonal differentiated zeta function.

Theorem 2.13 *With the same hypotheses as above, if $x = y \in I_u$, then*

(a) *$\zeta^{(\alpha,\beta)}(s, x|\ell^2 \mathcal{A})$ can be analytically continued in the variable s in the complex plane defining a meromorphic function with possible poles at the points, for $j = 0, 1, 2, \ldots$*

$$\begin{aligned} s_j^{(\alpha,\beta)} &= D/2 + |\alpha| + |\beta| - j, \text{ if } D \text{ is odd, or} \\ s_j^{(\alpha,\beta)} &= D/2 + |\alpha| + |\beta| - j, \text{ with } j \leq D/2 - 1 + |\alpha| + |\beta|, \text{ if } D \text{ is even.} \end{aligned}$$

The residues

$$R_N(s,x)_{\ell_0^2} \stackrel{def}{=}$$
$$\frac{1}{(4\pi)^{D/2}\ell^{2s}\Gamma(s)} \sum_{j=0}^{N} \int_0^{\ell_0^2} dt\, t^{s-1-D/2+j} \left\{ D_x^\alpha D_y^\beta e^{-\sigma(x,y)/2t} a_j(x,y|A) \right\}_{x=y},$$

defined for $x \in I_u$ and Re $s > D/2 + |\alpha| + |\beta|$ and continued in the s-complex plane ($\ell_0 > 0$ disappears in the residues).

(b) *varying $x \in I_u$, the s-continued function with all its s-derivatives is jointly continuous in $(\mathbb{C} \setminus \mathcal{P}^{(\alpha,\beta)}) \times \mathcal{M}$, $\mathcal{P}^{(\alpha,\beta)}$ being the set of the actual poles (each for some x);*

(c) *if $k = 1, \ldots D$, $(s, x) \in \{\mathbb{C} \setminus (\mathcal{P}^{(\alpha,\beta)} \cup \mathcal{P}^{(\alpha+1_k,\beta)} \cup \mathcal{P}^{(\alpha,\beta+1_k)})\} \times I_u$, $(\partial/\partial x^k)\zeta^{(\alpha,\beta)}(s, x|\ell^2 \mathcal{A})$ is jointly continuous with all of its s derivatives, analytic in the variable s and*

$$\begin{aligned} &\frac{\partial}{\partial x^k}\zeta^{(\alpha,\beta)}(s, x|\ell^2 \mathcal{A}) \\ &\quad = \zeta^{(\alpha+1_k,\beta)}(s, x|\ell^2 \mathcal{A}) + \zeta^{(\alpha,\beta+1_k)}(s, x|\ell^2 \mathcal{A}), \end{aligned} \tag{2.127}$$

$\zeta^{(\alpha,\beta)}$ being the analytic continuation of (2.126).

The last theorem generalizes the statement **(c)** of Theorem 2.7.

Theorem 2.14 *With the same hypotheses as above, for $x, y \in I_u$, the analytic continuations of the differentiated zeta functions are well-defined in a neighborhood of $s = 0$,*

$$\left[D_x^\alpha D_y^\beta \zeta(s, x, y|\ell^2 \mathcal{A})\right]_{s=0} + D_x^\alpha D_y^\beta P(x, y)$$

$$= \delta_D \delta_{x,y} \lim_{s \to 0} R_N(s,x)_{\ell_0^2} , \tag{2.128}$$

where subscript equality $s = 0$ means the analytic continuation from $\text{Re}\, s > D/2+|\alpha|+|\beta|$ *to $s = 0$ of the considered function, and $N > D/2+2|\alpha|+2|\beta|$ being integer. The result does not depend on the values of $\ell_0 > 0$ and $\ell > 0$.*

Comments.

(1) The right-hand side of (2.128), for $x = y$ and when D is even, has the form

$$(4\pi)^{-D/2} D_x^\alpha D_y^\beta a_{D/2}(x,y|A) + \cdots ,$$

where the dots indicate a finite number of further terms consisting of derivatives of product of heat-kernel coefficients and powers of $\sigma(x,y)$, computed in the coincidence limit of the arguments.

(2) The right-hand side of (2.128) proves that the procedures of s-continuing $D_x^\alpha D_y^\beta \zeta(s,x,y|\ell^2 \mathcal{A})$ and that of taking the coincidence limit of arguments x, y *generally do not commute.* This means that, understanding both sides in the sense of analytic continuation, in general

$$\zeta^{(\alpha,\beta)}(s,x,y|\ell^2 \mathcal{A})|_{x=y} \neq \zeta^{(\alpha,\beta)}(s,x|\ell^2 \mathcal{A}) . \tag{2.129}$$

Here the coincidence limit is taken *after* the analytic continuation. Obviously, whenever $\text{Re}\, s > D/2 + |\alpha| + |\beta|$,

$$\zeta^{(\alpha,\beta)}(s,x,y|\ell^2 \mathcal{A})|_{x=y} = \zeta^{(\alpha,\beta)}(s,x|\ell^2 \mathcal{A}) . \tag{2.130}$$

Our last theorem concerns the possibility to compute the differentiated local zeta functions by employing a series instead of an integration.

Theorem 2.15 *Under the general hypotheses on $\mathcal{M}$ and $A \geq 0$, take $\ell > 0$, any pair of multiindices α, β and refer the derivatives to coordinates in a sufficiently small neighborhood I_u of any $u \in \mathcal{M}$. In the sense of the pointwise convergence, the following equations hold*

$$\begin{aligned} \zeta^{(\alpha,\beta)}(s,x,y|\ell^2 \mathcal{A}) &= \sum_{j\in\mathbb{N}}{}' \left(\ell^2 \lambda\right)^{-s} D_x^\alpha \phi_j(x) D_y^\beta \phi_j^*(y) , \\ \zeta^{(\alpha,\beta)}(s,x|\ell^2 \mathcal{A}) &= \sum_{j\in\mathbb{N}}{}' \left(\ell^2 \lambda\right)^{-s} D_x^\alpha \phi_j(x) D_x^\beta \phi_j^*(x) , \end{aligned} \tag{2.131}$$

provided $\text{Re}\, s > 3D/2 + |\alpha| + |\beta|$ *and $(x,y) \in I_u \times I_u$.*

Proof. For $\mathrm{Re}\, s > D/2 + |\alpha| + |\beta|$ both the differentiated zeta functions are continuous in all variables and

$$\zeta^{(\alpha,\beta)}(s,x,y|\ell^2\mathcal{A})|_{x=y} = \zeta^{(\alpha,\beta)}(s,x|\ell^2\mathcal{A})\,, \tag{2.132}$$

therefore, we may perform the proof in the general case $x \neq y$ and finally consider the coincidence limit of the arguments. By Theorem 2.11, for $\mathrm{Re}\, s > D/2 + |\alpha| + |\beta|$, one has

$$\begin{aligned}
&\zeta^{(\alpha,\beta)}(s,x,y|\ell^2\mathcal{A}) \\
&= \frac{1}{\ell^{2s}\Gamma(s)}\int_0^{\ell_0^2} dt\, t^{s-1} D_x^\alpha D_y^\beta \left\{K(t,x,y|\mathcal{A}) - P(x,y|\mathcal{A})\right\} \\
&+ \frac{1}{\ell^{2s}\Gamma(s)}\int_{\ell_0^2}^{+\infty} dt\, t^{s-1} D_x^\alpha D_y^\beta \left\{K(t,x,y|\mathcal{A}) - P(x,y|\mathcal{A})\right\}. \quad (2.133)
\end{aligned}$$

The second integral can also be written as

$$\frac{1}{\ell^{2s}\Gamma(s)}\int_{\ell_0^2}^{+\infty} dt\, t^{s-1} {\sum_{j\in\mathbb{N}}}' D_x^\alpha \phi_j(x) D_y^\beta \phi_j^*(y) e^{-\lambda_j t}\,, \tag{2.134}$$

by Lemma 2.2. We want to show that it is possible to interchange the symbol of series with that of integration. Afterwords we prove the same for the former integral in (2.133), then the well-known formula (2.53) will complete the proof.

To prove the possibility of interchanging the integration with the summation in the integral (2.134), it is sufficient to show that the argument of summation is integrable in the measure $\int dt \times \sum_j$, then Fubini's theorem allows one to interchange the integrations. Lemma 2.2 and Theorem 2.11 entail that, for $t > 2T > 0$

$$\begin{aligned}
{\sum_{j\in\mathbb{N}}}' |t^{s-1} D_x^\alpha \phi_j(x) D_y^\beta \phi_j^*(y) e^{-\lambda_j t}| &\leq P_T^{(\alpha,\beta)} {\sum_{j\in\mathbb{N}}}' t^{\mathrm{Re}\,(s-1)} e^{-\lambda_j(2t-2T)} \\
&\leq Q_T^{(\alpha,\beta)} t^{\mathrm{Re}\, s-1} e^{-\lambda(t-2T)}\,, \quad (2.135)
\end{aligned}$$

where λ is the first strictly positive eigenvalue of A. Fix $T < \ell_0^2/2$. The t-integration in $[\ell_0^2, +\infty)$ of the last line above is finite for any $s \in \mathbb{C}$. Thus, a part of Fubini's theorem prove that the function after the summation symbol in (2.134) is integrable in the product measure.

Let us perform a similar proof for the first integral in the right-hand side of (2.133). The integral can be written as

$$\frac{1}{\ell^{2s}\Gamma(s)}\int_0^{\ell_0^2} dt\, t^{s-1}{\sum_{j\in\mathbb{N}}}' D_x^\alpha\phi_j(x)D_y^\beta\phi_j^*(y)e^{-\lambda_j t}\,. \tag{2.136}$$

We now show that it is possible to interchange the symbol of series with that of integration. By Lemma 2.2, posing $T = t/4$ we have, for $t \in (0, \ell_0^2]$

$$\begin{aligned}&{\sum_{j\in\mathbb{N}}}' |t^{s-1}D_x^\alpha\phi_j(x)D_y^\beta\phi_j^*(y)e^{-\lambda_j t}| \\ &\leq P_{t/4}^{(\alpha,\beta)}t^{\mathrm{Re}\,s-1}\mathrm{Tr}\left\{K_{t/2}-P\right\}\,,\end{aligned} \tag{2.137}$$

where P_T has been defined in (2.118). Employing the expansion in Theorem 2.4 and taking into account of the finite volume of the manifold one finds that there is a positive constant C such that, for $t \in (0, \ell_0^2]$,

$$P_T^{(\alpha,\beta)} \leq CT^{-D-|\alpha|-|\beta|}\,. \tag{2.138}$$

This is due to the leading order for $t \to 0^+$ of the heat-kernel expansion in Theorem 2.4. The same expansion gives $\mathrm{Tr}\left\{K_{t/2}-P\right\} \leq Gt^{-D/2}$ for some constant $G > 0$. This upper bound with that in (2.138) with $T = t/4$, inserted in (2.137) entail

$${\sum_{j\in\mathbb{N}}}' |t^{s-1}D_x^\alpha\phi_j(x)D_y^\beta\phi_j^*(y)e^{-\lambda_j t}| \leq Bt^{\mathrm{Re}\,s-1-3D/2-|\alpha|-|\beta|} \tag{2.139}$$

for some positive constant B. The last function of t is absolutely integrable on $[0, \ell_0^2]$ for $\mathrm{Re}\,s > 3D/2 + |\alpha| + |\beta|$. Finally, dealing with as in the previously considered case, by Lebesgue's theorem, for $\mathrm{Re}\,s > 3D/2+|\alpha|+|\beta|$, we can interchange the symbols of the integral and the series, also in the second integral of (2.133), and then (2.53) entails the thesis. □

2.5 Noncompact Manifolds and Manifolds with a Boundary

What remains of all these results when the hypotheses of a compact manifold is dropped? An introductory but general discussion on the heat kernel in the case of noncompact manifolds (including vectorial and tensorial fields and more general connections than the metric one) can be found in [S.A. Fulling (1991)].

The main problem is related to the fact that $A = -\Delta^{(X)} + V$ may fail to be essentially self-adjoint and several self-adjoint extension and heat-kernel candidates may appear. On the physical ground, this should imply the existence of different quantum states related to the Green functions of different self-adjoint extensions after the Wick rotation. However, if A is positive or bounded from below, for instance on $C^\infty(\mathcal{M})$, one may take advantage of the self-adjoint Friedrichs' extension. Given any self-adjoint extension $\mathcal{A}$ of A, one may consider the semigroup $e^{-t\mathcal{A}}$, $t > 0$. If the semigroup admits a smooth heat kernel $K(t, x, y|\mathcal{A})$ it must be unique and the main part of the zeta function theory developed above can be implemented, at least in the case of $\mathcal{A} > 0$ (to overcome the large t-integration problems as said below) provided a small t-expansion as in Theorem 2.4. The uniqueness property is easily proven. For a kernel K,

$$(e^{-t\mathcal{A}}\phi)(x) = \int_{\mathcal{M}} d\mu_g(y)\, K(t, x, y|\mathcal{A})\phi(y)\,,$$

$$(e^{-t\mathcal{A}}\phi)(x) = \int_{\mathcal{M}} d\mu_g(y)\, K(t, x, y|\mathcal{A})\phi(y)\,,$$

and $\phi \in L^2(\mathcal{M}, \mu_g)$ on the smooth Euclidean noncompact manifold $(\mathcal{M}, g)$, $t \in (0, +\infty)$, define $H \stackrel{\text{def}}{=} K - K'$. It must hold

$$\left| \int_{\mathcal{M}} \int_{\mathcal{M}} d\mu_{\mathbf{g}}(x) \otimes d\mu_{\mathbf{g}}(y)\, H(t, x, y|\mathcal{A}) \frac{\chi_E(x)\chi_{E'}(y)}{\mu(E)\mu(E')} \right| = \left| \frac{(\chi_E, \chi_{E'})}{\mu(E)\mu(E')} \right| = 0$$

for all pairs of measurable sets E, E' with a finite measure. Shrinking E, E' about x, y, respectively, and taking the smoothness of H into account, one gets $H(t, x, y|\mathcal{A}) = 0$.

Several results of the heat-kernel theory for the pure Laplacian, with trivial extensions to the case $\mathcal{A} = -\Delta + m^2$ can be found in the literature (see [R.M. Wald (1979); I. Chavel (1984); E.B. Davies (1989)]). In particular it is well-known that $-\Delta$ (but it straightforwardly generalizes to $-\Delta + m^2$ with $m^2 \geq 0$ constant) turns out to be essentially self-adjoint on $C^\infty(\mathcal{M})$ if the Euclidean manifold $\mathcal{M}$ is geodesically complete [P. Chernoff (1973); E.B. Davies (1989)]. In that case, also in the presence of $m^2 > 0$, it is known that the generated semigroup admits a smooth heat kernel with the usual small t expansion [R.M. Wald (1979)]. Quite a complete theory has been developed in [I. Chavel (1984)] by using another approach, for the case $V \equiv 0$, considering noncompact manifolds and without employing self-adjoint extensions, but realizing a noncompact manifold as a suitable limit of compact manifolds (generally with boundaries) and looking for heat kernels directly.

Another crucial remark is that, in general, the lack of compactness of the manifold produces the failure of expansions such as those in (2.16), because the spectrum of $\mathcal{A}$ becomes continuous in general. Anyway, in several cases one finds out that it is possible to replace the sum in (2.16) with integrals, dealing with opportune spectral measures,

$$K(t,x,y|\mathcal{A}) = \int_{\text{Spec}(\mathcal{A})} \sum_j e^{-\lambda t}\phi_{j\lambda}(x)\phi^*_{j\lambda}(y)d\mu_{\mathcal{A}}(\lambda)\ , \tag{2.140}$$

the functions $\phi_{j\lambda}$ being eigenfunctions (in some weak sense) of $\mathcal{A}$ with common eigenvalue λ. We expect that this is a general result. Similar expansions can be produced also for the local zeta function, which can still be defined by (2.50) provided the corresponding integral converges. As we said, a sufficient condition is that $\mathcal{A} > 0$. In fact, assume that $e^{-t\mathcal{A}}$ be represented by a smooth heat kernel $K(t,x,y|\mathcal{A})$ if $t>0$ on the smooth Euclidean noncompact manifold $(\mathcal{M},g)$. If $E, E' \subset \mathcal{M}$ are measurable with finite measure, then

$$\left|\int_{\mathcal{M}}\int_{\mathcal{M}} d\mu_g(x)\otimes d\mu_g(y)\, K(t,x,y|\mathcal{A})\frac{\chi_E(x)\chi_{E'}(y)}{\mu(E)\mu(E')}\right| = \left|\frac{(\chi_E, e^{-t\mathcal{A}}\chi_{E'})}{\mu(E)\mu(E')}\right| .$$

Cauchy-Schwarz's inequality bounds the right-hand side by $e^{-\Lambda t}$, where $0 < \Lambda \stackrel{\text{def}}{=} \inf \text{Spec}(\mathcal{A})$. Shrinking E and E' about x and y respectively, from the smoothness of $K(t,x,y|\mathcal{A})$ we get

$$|K(t,x,y|\mathcal{A})| \leq e^{-\Lambda t}$$

for all $(t,x,y) \in (0,+\infty)\times\mathcal{M}\times\mathcal{M}$. This bound assures the convergence of the integral in (2.50) for large t (for $P \equiv 0$ in our hypotheses). The convergence as $t \to 0^+$ and the analytic properties of the local zeta function depend on the validity of some small t expansion similar to (2.19) at least for $\alpha = \beta = 0$, which, as said, holds at least in the case A is the unique self-adjoint extension of $-\Delta + m^2$ on $C^\infty(\mathcal{M})$ on geodesically complete manifolds.

In recent years, the theory of heat kernels and zeta functions in symmetric manifolds has been developed on mathematical and physical grounds, also proving the validity of (2.140) in noncompact but symmetric manifolds [A.A. Bytsenko et al. (1996)].

There exists a wide theoretical-physics literature on applications of these topics [A.A. Bytsenko et al. (1996)] in quantum field theory in noncompact/singular curved backgrounds, concerning particular cases and also higher spin of the field (see [D. Iellici and V. Moretti (1996); V. Moretti and D. Iellici (1997)] for the case of photons and gravitons spaces containing conical singularities), in particular, in the presence of noncompact manifolds containing conical singularities, very important within quantum field theory in the presence of a black hole [G. Cognola et al. (1996)]. (Problems related to the heat kernel and the zeta function in the case of a compact manifold containing conical singularities is not trivial on a mathematical point of view, it was treated by J. Cheeger in [J. Cheeger (1983)].) The existence of the integrated zeta function is much more difficult to study in the general case of a noncompact manifold, and outside very particular situations (e.g. the Euclidean section of anti de Sitter spacetime), the integrated zeta function does not exist and one needs some volume cutoffs. Recently, other ways to overcome these shortcomings have been pointed out [W. Müller (1998)]. The problem of the existence of the integrated zeta function is dramatically important in the issue of the computation of thermodynamical quantities of fields propagating in the spacetime around a black hole (and thus in the general issue of the black hole entropy). In that case, horizon divergences also appear and different integrated renormalization procedures disagree among themselves [G. Cognola et al. (1996); D. Iellici and V. Moretti (1996); V. Moretti and D. Iellici (1997); V. Moretti (1997a); V.P. Frolov and D.V. Fursaev (1998); D. Iellici (1998)]. Finally, the issue whether or not $-(2\beta)^{-1}\zeta'(0,x|\ell^2\mathcal{A})$ defines the true local density of free energy still remains an open question (see the discussion in [V. Moretti (1997b)]).

The presence of boundaries, maintaining the compactness, changes the results obtained in the case above ("closed" manifold) only for the presence of further terms in the heat-kernel expansion (2.19) depending on the boundaries [I. Chavel (1984)] and with noninteger powers of t. These terms can be interpreted as distributions concentrated on the boundary of the manifold (see [E. Elizalde et al. (1994); E. Elizalde (1995)] for the corresponding bibliography and physical applications). Obviously, in this case

A is not essentially self-adjoint and the considered self-adjoint extension depends on the boundary conditions imposed (Dirichlet/Neumann/Mixed problems) and the choice is related to the particular quantum state one is investigating.

2.6 The Stress-Energy Tensor and Field-Fluctuation Regularization

To conclude the chapter, we consider a few recent developments of the local zeta function theory. Besides the effective Lagrangian and the effective action, further important one-loop quantities are the (quantum) field fluctuation and the averaged (quantum) stress-energy tensor. In the following, we extend the technique based on the local spectral zeta function to encompass the regularization of those physical objects [V. Moretti (1997b); V. Moretti (1999a); V. Moretti (1999b)].

Remark. In this Section, we shall deal with real scalar Klein-Gordon fields ϕ, and the potential V in (2.6) is assumed to be of the form

$$V = m^2 + \xi R(x) + V'(x) \,, \tag{2.141}$$

where V' is any smooth function. All the results presented here can be straightforwardly generalized to the case of complex fields and Klein-Gordon operators (2.4), with obvious changes.

2.6.1 *The stress-energy tensor*

We remind the reader that the *conformal coupling* $\xi = \xi_D$ of the field is defined by putting [N.D. Birrell and P.C.W. Davies (1982); S.A. Fulling (1991); R.M. Wald (1984); R.M. Wald (1994)]

$$\xi_D \stackrel{\text{def}}{=} (D-2)/[4(D-1)] \,. \tag{2.142}$$

The (Euclidean) *classical* stress-energy tensor $T_{ab}(x)$, is defined (e.g. see [R.M. Wald (1984)]) as the locally quadratic form of the field obtained by the usual functional derivative once the field ϕ is fixed

$$T_{ab}[\phi, g](x) \stackrel{\text{def}}{=} \frac{2}{\sqrt{g}} \frac{\delta}{\delta g_{ab}(x)} \left(\frac{1}{2} \int_{\mathcal{I}} d^D x \sqrt{g(x)} \phi A[g] \phi \right) \,. \tag{2.143}$$

This functional derivative can be rigorously understood in terms of a Gâteaux derivative for functionals on real $C^\infty(\mathcal{I})$ *symmetric* tensor fields g_{ab} and the integration above is performed in the open set $\mathcal{I}$ containing x where the coordinate system considered is defined (the choice of $\mathcal{I}$ does not affect the final results). (2.143) means that, and this is the rigorous definition of the *symmetric* tensor field $T_{ab}[\phi, g](x)$, for any C^∞ *symmetric* tensor field h_{ab} with compact support contained in $\mathcal{I}$

$$2\frac{d}{d\alpha}S_{\mathcal{I}}[g+\alpha h]|_{\alpha=0} = \frac{1}{2}\int_{\mathcal{I}} d^Dx\,\sqrt{g(x)}T_{ab}[\phi,g](x)h^{ab}(x)\,, \tag{2.144}$$

where the functional derivative is computed on the actual value of the metric and the classical Euclidean action associated with $\mathcal{I}$ is

$$S_{\mathcal{I}}[\phi,g] \stackrel{\text{def}}{=} \frac{1}{2}\int_{\mathcal{I}} d^Dx\,\sqrt{g(x)}\phi A[g]\phi\,. \tag{2.145}$$

A direct computation of $T_{ab}[\phi, g](x)$ yields:

$$\begin{gathered}\nabla_a\phi(x)\nabla_b\phi(x) - \frac{1}{2}g_{ab}(x)\left[\nabla_c\phi(x)\nabla^c\phi(x) + \left(m^2+V'(x)\right)\phi^2(x)\right] \\ +\xi\left[\left(R_{ab}(x)-\frac{1}{2}g_{ab}(x)R(x)\right)\phi^2(x) + g_{ab}(x)\nabla_c\nabla^c\phi^2 - \nabla_a\nabla_b\phi^2(x)\right]\,.\end{gathered} \tag{2.146}$$

As is well known, T_{ab} given in (2.146) and evaluated for a particular ϕ is *conserved* ($\nabla_a T^{ab} \equiv 0$) provided ϕ is a sufficiently smooth (customary C^∞) solution of the Euclidean motion, namely, $\mathcal{A}\phi \equiv 0$, and $V' \equiv 0$. More generally for solution of Euclidean motion, in local coordinates and for any point $x \in \mathcal{M}$ one finds

$$\nabla_a T^{ab}[\phi,g](x) = -\frac{1}{2}\phi^2(x)\nabla^b V'(x)\,. \tag{2.147}$$

Another important classical property is the following. Whenever the field ϕ is massless and conformally coupled (i.e. $V'(x) \equiv m^2 = 0$), the Euclidean action $S_{\mathcal{M}}$ is invariant under local conformal transformations and it holds also

$$g_{ab}T^{ab}[\phi,g](x) = 0 \tag{2.148}$$

everywhere, for smooth fields ϕ which are solution of the (Euclidean) motion equations. Equations (2.147) and (2.148) can be checked for the tensor in (2.146) directly and our general hypotheses hold.

Passing to quantum averaged quantities, and adopting Schwinger's point of view [J. Schwinger (1951b)] the averaged *one-loop stress-energy tensor* for the quantum state determined by the Feynman propagator obtained by Green function of some self-adjoint extension of A_0, A, can be *formally* defined by [S.W. Hawking (1977); R.M. Wald (1979); N.D. Birrell and P.C.W. Davies (1982)]

$$\begin{aligned} \langle T_{ab}(x|\mathcal{A})\rangle &\stackrel{\text{def}}{=} \frac{2}{\sqrt{g}}\frac{\delta}{\delta g_{ab}(x)}W \\ &= \frac{1}{Z[\mathcal{A},g]}\int \mathcal{D}\phi\, e^{-\frac{1}{2}\int_{\mathcal{M}} d\mu_g\, \phi\mathcal{A}[g]\phi}\, T_{ab}[\phi,g](x)\,, \end{aligned} \tag{2.149}$$

where W is the quantum effective action. It is well-known that the right-hand side of (2.149) and the corresponding quantity in the Lorentzian section are affected by divergences, whenever one tries to compute them by direct procedures [N.D. Birrell and P.C.W. Davies (1982); S.A. Fulling (1991); R.M. Wald (1994)]. For instance, proceeding as usual (e.g. see [R.M. Wald (1994)]), interpreting the functional integral of $\phi(x)\phi(y)$ as a Green function of $\mathcal{A}$ (the analytic continuation of the Feynman propagator) $G(x,y)$ and then defining an off-diagonal quantum averaged stress tensor

$$\begin{aligned} \langle T_{ab}(x,y|\mathcal{A})\rangle &= Z[\mathcal{A},\mathbf{g}]^{-1}\int \mathcal{D}\phi\, e^{-\frac{1}{2}\int_{\mathcal{M}} d\mu_g\, \phi\mathcal{A}[g]\phi} O_{ab}(x,y)\phi(x)\phi(y) \\ &= O_{ab}(x,y)G(x,y)\,, \end{aligned} \tag{2.150}$$

where $O_{ab}(x,y)$ is a bi-vectorial differential operator (see [R.M. Wald (1994)]), the limit of coincidence of arguments x and y, necessary to get $\langle T_{ab}(x)\rangle$, trivially diverges. One is therefore forced to remove these divergences *by hand*, and this is nothing but the main idea of the *point-splitting procedure.* Within the point-splitting procedure (2.147) is requested also for the quantum averaged stress tensor at least in the case $V' \equiv 0$. Conversely, the property (2.148) generally does not hold in the case of a conformally coupled massless field: a *conformal anomaly* appears [R.M. Wald (1978); N.D. Birrell and P.C.W. Davies (1982); R.M. Wald (1994); S.A. Fulling (1991)].

Another approach to interpret the left-hand side of (2.149) in terms of the local ζ function was introduced in [V. Moretti (1997b)] and rigorously developed in [V. Moretti (1999a); V. Moretti (1999b)]. Afterwards, the technique has been used by other authors in different contexts: Gödel-like universes [E. Radu and D. Astefanesei (2002); E. Radu (1998b); E. Radu (1998a); E. Radu (1999)] anti-De Sitter spacetime [M.M. Caldarelli (1999)], the metric ground state of the (2+1)-dimensional black hole of Bañados, Teitelboim and Zanelli which contains a naked singularity [D. Binosi et al. (1999a)].

This approach has produced correct results, in agreement with point-splitting procedures, in several concrete cases and it has pointed out to a strong self-consistence and a general agreement with the general axiomatic theory of the stress-tensor renormalization built up by Wald [N.D. Birrell and P.C.W. Davies (1982); R.M. Wald (1994)]. (It is anyway worth stressing that Wald's axiomatic approach concerns the Lorentzian theory and, thus, any comparison involves an analytical continuation of the Euclidean theory. In such a way all the issues related to the locality of the theory cannot be compared directly with the general zeta function approach.) Similarly to the case of the effective Lagrangian, the *definition* of the formal quantity on the left-hand side of (2.149), given in terms of the zeta function and the heat kernel, contains an implicit *infinite renormalization* procedure in the sense that the result is finally free from divergences.

2.6.2 Zeta-function regularization of the stress-energy tensor and the field fluctuation

The key idea of zeta-function regularization of the stress-energy tensor [V. Moretti (1997b)] concerns the extension of the use of the zeta function from the effective action to the stress tensor, by employing some manipulations of the series involved in this technique. Formally, (2.71) can be understood as

$$W[\mathcal{A}]_\ell = \frac{1}{2}\frac{d}{ds}\left\{-\sum_{j\in\mathbb{N}}{}' \left(\ell^2\lambda_j\right)^{-s}\right\}\Bigg|_{s=0} . \tag{2.151}$$

Actually, the identity above holds true in the sense of the analytic continuation. Then, one can try to give some meaning to the following formal passages

$$\langle T_{ab}(x|\mathcal{A})\rangle_\ell = \frac{2}{\sqrt{g(x)}}\frac{\delta}{\delta g_{ab}(x)}\frac{1}{2}\frac{d}{ds}\left\{-\sum_{j\in\mathbb{N}}{}' \left(\ell^2\lambda_j\right)^{-s}\right\}\Bigg|_{s=0}$$

$$= \frac{1}{2}\frac{d}{ds}\left\{-\sum_{j\in\mathbb{N}}{}' \frac{2}{\sqrt{g(x)}}\frac{\delta}{\delta g_{ab}(x)}\left(\ell^2\lambda_j\right)^{-s}\right\}\Bigg|_{s=0}$$

$$= \frac{1}{2}\frac{d}{ds}\left\{s\ell^2\sum_{j\in\mathbb{N}}{}' \left(\ell^2\lambda_j\right)^{-(s+1)}\frac{2}{\sqrt{g(x)}}\frac{\delta\lambda_j}{\delta g_{ab}(x)}\right\}\Bigg|_{s=0} . \tag{2.152}$$

If the series in the last line of (2.152) converges and it can be analytically continued into a regular function $Z_{ab}(s,x|\ell^2\mathcal{A})$ in a neighborhood of $s=0$, one can *define* the renormalized, averaged one-loop stress tensor as

$$\langle T_{ab}(x|\mathcal{A})\rangle_\ell \stackrel{\text{def}}{=} \frac{1}{2}\frac{d}{ds}\bigg|_{s=0} Z_{ab}(s,x|\ell^2\mathcal{A}) . \tag{2.153}$$

A closed form for a candidate function $Z_{ab}(s,x|\ell^2\mathcal{A})$ is the following [V. Moretti (1997b)]

$$Z_{ab}(s,x|\ell^2\mathcal{A}) = 2s\ell^2\zeta_{ab}(s+1,x|\ell^2\mathcal{A}) + s g_{ab}(x)\zeta(s,x|\ell^2\mathcal{A}) , \tag{2.154}$$

where $\zeta_{ab}(s,x|\ell^2\mathcal{A})$ is the analytic continuation of the series

$$\sum_{j\in\mathbb{N}}{}' \left(\ell^2\lambda_j\right)^{-s} T_{ab}[\phi_j,\phi_j^*,g](x) , \tag{2.155}$$

and

$$T_{ab}[\phi,\phi^*,g](x) \stackrel{\text{def}}{=} \frac{2}{\sqrt{g(x)}}\frac{\delta_g}{\delta g_{ab}(x)}\frac{1}{2}\int_{\mathcal{I}} d\mu_g\, \phi^*\mathcal{A}[g]\phi . \tag{2.156}$$

We show how to achieve (2.154). The crucial point is to obtain an explicit form for the functional derivative of λ_j with respect to $g^{ab}(x)$, to insert in the right-hand side of (2.152). At least formally, it is possible to give a closed expression for it [V. Moretti (1997b)]. Suppose that, under our general hypotheses, ϕ_j is the *normalized* eigenvector of $\mathcal{A}$ associated with the eigenvalues λ_j. Changing g (preserving the topology of the manifold), λ_j and ϕ_j change. Assuming $\mathcal{I}=\mathcal{M}$, it holds

$$\lambda_j = 2S_{\mathcal{M}}[\phi_j,\phi_j^*] := \int_{\mathcal{M}} d\mu_g\, \phi_j^*\mathcal{A}[g]\phi_j ,$$

and so

$$\frac{1}{2}\frac{\delta\lambda_j}{\delta g^{ab}(x)} = \frac{\delta_g S_{\mathcal{M}}}{\delta g^{ab}(x)} + \int_{\mathcal{M}} d^D y \, \frac{\delta S_{\mathcal{M}}}{\delta\phi_j^*(y)}\frac{\delta\phi_j^*(y)}{\delta g^{ab}(x)} + \int_{\mathcal{M}} d^D y \, \frac{\delta S_{\mathcal{M}}}{\delta\phi_j(y)}\frac{\delta\phi_j(y)}{\delta g^{ab}(x)} \, .$$

Using

$$\frac{\delta S_{\mathcal{M}}[\phi,\phi^*]}{\delta\phi(x)} = \sqrt{g(x)}\,(\mathcal{A}\phi)(x)$$

and the analogue for ϕ^*, the identity above can be written as

$$\frac{\delta\lambda_j}{\delta g^{ab}(x)} = 2\frac{\delta_g S_{\mathcal{M}}}{\delta g^{ab}(x)} + \lambda_j \int_{\mathcal{M}} d^D y \, \sqrt{g(y)}\frac{\delta\phi_j^*(y)\phi_j(y)}{\delta g^{ab}(x)} \, .$$

The integral in the right-hand side can be decomposed as

$$\frac{\delta}{\delta g^{ab}(x)}\int_{\mathcal{M}} d^D y \, \sqrt{g(y)}\phi_j^*(y)\phi_j(y) - \int_{\mathcal{M}} d^D y \, \frac{\delta\sqrt{g(y)}}{\delta g^{ab}(x)}\phi_j^*(y)\phi_j(y)$$

and the former term vanishes due to the normalization of eigenvectors which is supposed to be preserved by the variational procedure, while a few elementary calculations give rise to the well-known result

$$\frac{\delta\sqrt{g(y)}}{\delta g^{ab}(x)} = -\frac{1}{2}g_{ab(x)}\sqrt{g(x)}\delta(x,y) \, .$$

Putting all together, we get finally

$$\frac{\delta\lambda_j}{\delta g^{ab}(x)} = \frac{\lambda_j}{2}g_{ab(x)}\,\sqrt{g(x)}\,\phi_j^*(x)\phi_j(x) + 2\frac{\delta_g S_{\mathcal{M}}[\phi_j,\phi_j^*]}{\delta g^{ab}(x)} \, .$$

This identity gives rise to (2.154) straightforwardly. Notice that no rigor was employed in the passages above and thus there is no guarantee of obtaining meaningful results in such a way. In the next sections, we show that, actually, the method works very well. Concerning the field fluctuations, starting from the formal identity,

$$\langle\phi^2(x|\mathcal{A})\rangle = -\frac{2}{Z[\mathcal{A},g]\sqrt{g(x)}}\frac{\delta}{\delta J(x)}\int \mathcal{D}\phi \, e^{-\frac{1}{2}\int_{\mathcal{M}} d\mu_g \, \phi\mathcal{A}[J]\phi}\Big|_{J\equiv 0} \, , \quad (2.157)$$

where $\mathcal{A}[J] = \mathcal{A} + J$ and $\mathcal{A} = \mathcal{A}[0]$, and following a way very similar to that above [D. Iellici and V. Moretti (1998); V. Moretti (1999a)], one formally finds

$$\left.\frac{\delta\lambda_j[J]}{\delta J(x)}\right|_{J=0} = \phi_j(x)\phi_j^*(x)\sqrt{g(x)}\,.$$

Similarly to the case of the stress-energy tensor, this leads to the tentative, formal identity

$$\langle\phi^2(x|\mathcal{A})\rangle_\ell \stackrel{\text{def}}{=} \left.\frac{d}{ds}\Phi(s,x|\ell^2\mathcal{A})\right|_{s=0}, \tag{2.158}$$

where

$$\Phi(s,x|\ell^2\mathcal{A}) \stackrel{\text{def}}{=} \ell^2 s\zeta(s+1,x|\ell^2\mathcal{A})\,. \tag{2.159}$$

Under our general hypotheses, differently from the case of Z_{ab}, by Theorem 1.7, we know that $s \mapsto \zeta(s+1,x|\ell^2\mathcal{A})$ has a simple pole in $s=0$. That pole is cancelled out by the factor s in (2.159) so that the final left-hand side of (2.159) is well-defined as a function of x.

2.6.3 *The regularized stress tensor and its properties*

The precise form of $T_{ab}[\phi,\phi^*,g](x)$ reads in our case

$$\begin{aligned} T_{ab}[\phi,\phi^*,g](x) &= \frac{1}{2}(\nabla_a\phi(x)\nabla_b\phi^*(x)+\nabla_a\phi^*(x)\nabla_b\phi(x)) \\ &\quad -\frac{1}{2}g_{ab}(x)\left[\nabla_c\phi(x)\nabla^c\phi^*(x)+\left(m^2+V'(x)\right)|\phi|^2(x)\right] \\ &\quad +\xi\left[\left(R_{ab}(x)-\frac{1}{2}g_{ab}(x)R(x)\right)|\phi|^2(x)\right. \\ &\quad \left.+\,g_{ab}(x)\nabla_c\nabla^c|\phi|^2(x)-\nabla_a\nabla_b|\phi|^2(x)\right]. \end{aligned} \tag{2.160}$$

A few trivial manipulations which make use of $\mathcal{A}\phi_j=\lambda_j\phi_j$ lead us to a more simple expression

$$\begin{aligned} T_{ab}[\phi_j,\phi_j^*,g](x) &= \frac{1}{2}(\nabla_a\phi_j(x)\nabla_b\phi_j^*(x)+\nabla_a\phi_j^*(x)\nabla_b\phi_j(x)) \\ &\quad -\xi\nabla_a\nabla_b|\phi_j|^2+\left(\xi-\frac{1}{4}\right)g_{ab}(x)\Delta|\phi_j|^2(x) \\ &\quad +\xi R_{ab}(x)|\phi_j|^2-\frac{g_{ab}(x)}{2}\lambda_j|\phi_j|^2(x)\,. \end{aligned} \tag{2.161}$$

Following the insights in the section above, we can give the definition:

Definition 2.8 *Within the general hypotheses on $\mathcal{M}$ and $A \geq 0$ given by (2.6), the local zeta function of the stress-energy tensor is the symmetric tensor field defined in local coordinates as*

$$Z_{ab}(s, x|\ell^2 \mathcal{A}) \stackrel{def}{=} 2\ell^2 s \zeta_{ab}(s+1, x|\ell^2 \mathcal{A}) + s g_{ab}(x) \zeta(s, x|\ell^2 \mathcal{A}) , \qquad (2.162)$$

where $\zeta_{ab}(s, x|\ell^2 \mathcal{A})$ is defined as the sum of the series, in a neighborhood I_u of any $u \in \mathcal{M}$ and for Re $s > 3D/2 + 2$:

$$\sum_{j\in\mathbb{N}}' \left(\ell^2 \lambda_j\right)^{-s} T_{ab}[\phi_j, \phi_j^*, g](x) , \qquad (2.163)$$

and $T_{ab}[\phi_j, \phi_j^*, g](x)$ is defined in (2.156) and (2.161). The local zeta function of the field fluctuation

$$\mathbb{C} \times \mathcal{M}(s, x) \mapsto \Phi(s, x|\ell^2 \mathcal{A}) ,$$

is defined in (2.159).

Comments

(1) Concerning the stress-energy tensor, the above definition makes sense since the relevant series converges for *Re* $s > 3D/2 + 2$ because of Theorem 2.15. Moreover the definition does not depend on the base of smooth orthogonal normalized eigenvectors of $\mathcal{A}$ (take the finite dimension of each eigenspace into account as stated in Theorem 2.2).

(2) The fact that the coefficients $Z_{ab}(s, x|\ell^2 \mathcal{A})$ do define a tensor is a direct consequence of (2.161) and (2.163). This can be easily proven for *Re* $s > 3D/2 + 2$ where one can make use of the series (2.163) which clearly defines a tensor, since all terms of the sum are separately the components of a tensor. Then, the proven property remains unchanged after analytic continuation to values of s where the series does not converge.

(3) The final expression of T_{ab}, and thus Z_{ab} itself, does not contain either m^2 or V' explicitly.

(4) Due to the uniqueness theorem for analytic functions and Theorem 2.11, each component of Z_{ab} can be built up, in convenient regions, employing the heat kernel in the fashion of Theorem 2.11. Following this route, by straightforward application of Theorems 2.11-2.14, one proves that the symmetric tensorial field $Z_{ab}(s, x|\ell^2 \mathcal{A})$ is continuous, together with all of its s derivatives, as a function of (s, x). In particular, each component defines a meromorphic function of the variable s, whenever x is fixed. Therefore, let us consider the *simple* poles which may appear in the s-continued components of $Z_{ab}(s, x|\ell^2 \mathcal{A})$. One has to rewrite each component of this tensorial field in terms of simple zeta functions and differentiated zeta functions, via Theorems 2.11-2.14 and Theorem 2.15. Once this accomplished, we see that the only functions which really appear in Z_{ab} are zeta functions, and each function is evaluated at $s+1$ (see (2.164)). The local zeta function evaluated at $s+1$ admits possible simple poles at the points s_j with $s_j = D/2 - j - 1$, where $j = 0, 1, \cdots$ whenever D is odd, otherwise $j = 0, 1, \cdots D/2 - 1$, whenever D is even. Anyhow, the factor s in (2.162) cancels out the possible pole at $s = 0$, which may appear in $\zeta(s+1)$ when D is even. The functions $\zeta^{(1_c, 1_d)}(s+1, x|\ell^2 \mathcal{A})$ have been considered in Theorems 1.10-1.13 (Theorem 1.12 in particular) and their possible simple poles may arise in s_j with $s_j = D/2 + |1_a| + |1_b| - 1 - j = D/2 - j + 1$ and $j = 0, 1, \cdots$ whenever D is odd, otherwise $j = 0, 1, \cdots D/2 + 1$ whenever D is even. Once again, because of the factor s in the right-hand side of (2.162), any possible simple pole at $s = 0$ is cancelled out.

The last comment above can be stated as a theorem.

Theorem 2.16 *In the general hypotheses on $\mathcal{M}$ and $A \geq 0$,*

(a) *each component of $Z_{ab}(s, x|\ell^2 \mathcal{A})$ can be analytically continued into a meromorphic function of s whenever x is fixed. In particular, in the sense of analytic continuation, if x belongs to a sufficiently small neighborhood of any point $u \in \mathcal{M}$, it holds that*

$$Z_{ab}(s, x|\ell^2 \mathcal{A}) = \ell^2 s \left[\zeta^{(1_a, 1_b)}(s+1, x|\ell^2 \mathcal{A}) + \zeta^{(1_b, 1_a)}(s+1, x|\ell^2 \mathcal{A}) \right]$$
$$+ 2\ell^2 s \left[\left(\xi - \frac{1}{4} \right) g_{ab}(x) \Delta + \xi R_{ab}(x) - \xi \nabla_a \nabla_b \right] \zeta(s+1, x|\ell^2 \mathcal{A}) . \tag{2.164}$$

(b) *The possible poles of each component of $Z_{ab}(s, x|\ell^2 \mathcal{A})$, which are simple poles only, are situated at the points*

$$\begin{aligned} s_j &= D/2 - j + 1, \quad j = 0, 1, 2, \cdots \quad \textit{if } D \textit{ is odd;} \\ s_j &= D/2 - j + 1, \quad j = 0, 1, 2, \cdots D/2 \quad \textit{if } D \textit{ even.} \end{aligned}$$

(c) *Varying $x \in I_u$ and $s \in \mathbb{C}$, the s-analytically continued symmetric tensor field $(s, x) \mapsto Z_{ab}(s, x|\ell^2 \mathcal{A})$ defines an s-analytic tensorial field of $C^0((\mathbb{C} \setminus \mathcal{P}) \times I_u)$, together with all its s-derivatives, where $\mathcal{P}$ is the set of the actual poles (each for some x) among the points listed above.*

Proof. The proof has already been sketched. □

Remark

Equation (2.164) could be used as an independent definition of the ζ function of the stress tensor. The important point is that it does not refer to any series of eigenvectors. It could be considered as the starting point for the generalization of this theory in the case when the spectrum of the operator $\mathcal{A}$ is continuous, provided that the functions in the right-hand side of (2.164) are defined in terms of t integration of derivatives of the heat kernel.

Definition 2.9 *Under the general hypotheses on $\mathcal{M}$ and $A \geq 0$ and for $x \in I_u$, where I_u is a sufficiently small neighborhood of $u \in \mathcal{M}$, the one-loop renormalized stress tensor is defined, in a local coordinate system in I_u, by the set of functions $(a, b = 1, \cdots, D)$,*

$$\langle T_{ab}(x|\mathcal{A})\rangle_\ell \stackrel{def}{=} \frac{1}{2}\frac{d}{ds}Z_{ab}(s, x|\ell^2\mathcal{A})\Big|_{s=0}, \tag{2.165}$$

where the tensorial field Z_{ab} is the s-analytic continuation given in Definition 2.8, and $\ell > 0$ is any fixed constant with the dimensions of a length. Similarly, the one-loop renormalized field fluctuations are defined as

$$\langle \phi^2(x|\mathcal{A})\rangle_\ell \stackrel{def}{=} \frac{d}{ds}\Phi(s, x|\ell^2\mathcal{A})\Big|_{s=0}. \tag{2.166}$$

We state and prove the most important properties of $\langle T_{ab}(x|\mathcal{A})\rangle_\ell$ and $\langle \phi^2(x|\mathcal{A})\rangle_\ell$ in the following theorem.

Theorem 2.17 *Under the general hypotheses on $\mathcal{M}$ and $A \geq 0$, the functions $x \mapsto \langle T_{ab}(x|\mathcal{A})\rangle_\ell$ and $x \mapsto \langle \phi^2(x|\mathcal{A})\rangle_\ell$ satisfy the following properties.*

(a) *The functions $x \mapsto \langle T_{ab}(x|\mathcal{A})\rangle_\ell$ $(a, b = 1, 2, \cdots, D)$ define a C^∞ symmetric tensorial field on $\mathcal{M}$ and $x \mapsto \langle \phi^2(x|\mathcal{A})\rangle_\ell$ defines a C^∞ scalar field.*

(b) *$\langle T_{ab}(x|\mathcal{A})\rangle_\ell$ is conserved for $V' \equiv 0$, and more generally*

$$\nabla^a\langle T_{ab}(x|\mathcal{A})\rangle_\ell = -\frac{1}{2}\langle \phi^2(x|\mathcal{A})\rangle_\ell \, \nabla_b V'(x) \tag{2.167}$$

everywhere in $\mathcal{M}$.

(c) *Under the rescaling $\ell^2 \to \alpha\ell^2$, where real $\alpha > 0$,*

$$\langle T_{ab}(x|\mathcal{A})\rangle_\ell \to \langle T_{ab}(x|\mathcal{A})\rangle_{\sqrt{\alpha}\ell} = \langle T_{ab}(x|\mathcal{A})\rangle_\ell - (\log\alpha) t_{ab}(x|\mathcal{A}) \,, \quad (2.168)$$

where $t_{ab}(x|\mathcal{A}) = Z_{ab}(0, x|\mathcal{A})/2$, which coincides also with the residue of the pole of $\zeta_{ab}(s+1, x|\mathcal{A})$ at $s = 0$. $t_{ab}(x|\mathcal{A})$ is a symmetric tensor (conserved for $V' \equiv 0$), which does not depend on ℓ, built up as a linear combination of product of the metric, curvature tensors, $V'(x)$ and their covariant derivatives, evaluated at the point x. In general, it satisfies

$$\nabla^a t_{ab}(x|\mathcal{A}) = -\delta_D \frac{a_{D/2-1}(x, x|A)}{2(4\pi)^{D/2}} \nabla_b V'(x) \,, \quad (2.169)$$

where $\delta_D = 0$ when D is odd and $\delta_D = 1$ otherwise. In terms of heat-kernel coefficients one also has

$$t_{ab}(x|\mathcal{A}) = \frac{\delta_D}{(4\pi)^{D/2}} \left\{ a_{D/2-1,(ab)}(x, x|A) + \frac{1}{2} g_{ab}(x) a_{D/2}(x, x|A) \right.$$
$$+ \left. \left[\left(\xi - \frac{1}{4} \right) g_{ab}(x)\Delta + \xi R_{ab}(x) - \xi \nabla_a \nabla_b \right] a_{D/2-1}(x, x|A) \right\} \,, \quad (2.170)$$

where we have employed the notations (using the same coordinate system for x and y)

$$a_{j,(ab)}(x, x|A) \stackrel{def}{=} \frac{1}{2} \left[\left(\nabla_{(x)a} \nabla_{(y)b} + \nabla_{(y)a} \nabla_{(x)b} \right) a_j(x, y|A) \right] |_{x=y} \,. \quad (2.171)$$

(d) *One has,*

$$g^{ab}(x)\langle T_{ab}(x|\mathcal{A})\rangle_\ell = \left(\frac{\xi_D - \xi}{4\xi_D - 1} \Delta - m^2 - V'(x) \right) \langle \phi^2(x|\mathcal{A})\rangle_\ell$$
$$+ \delta_D \frac{a_{D/2}(x, x|A)}{(4\pi)^{D/2}} - P(x, x|\mathcal{A}) \,.$$

Sketch of Proof. But for the issue concerning smoothness, the property (**a**) is a trivial consequence of the corresponding one for $Z_{ab}(s, x|\ell^2\mathcal{A})$ discussed in Comment (2) after Definition 2.8. The tensorial field belongs to C^∞, because of the C^∞ smoothness of the functions $(s, x) \mapsto Z_{ab}(s, x|\ell^2\mathcal{A})$ for $(s, x) \in J_0 \times I_u$, where J_0 and I_u are, respectively, a neighborhood of $s = 0$ in $\mathbb{C}$ and of $u \in \mathcal{M}$. Indeed, first of all, no pole at $s = 0$ arises in the functions $(s, x) \mapsto Z_{ab}(s, x|\ell^2\mathcal{A})$ and in their x derivatives. This is because, considering (2.162) and (**b**)-(**c**) of Theorem 2.13, we notice that if any pole appears in the various $\zeta^{(\alpha,\beta)}$ functions used building up Z_{ab} it has to be a simple pole. Anyhow, the factor s makes the global functions Z_{ab} regular at $s = 0$. Using recursively (**b**)-(**c**) of Theorem 2.13 one has that each function $x \mapsto Z_{ab}(s, x|\ell^2 \mathcal{A})$ is C^∞ in a neighborhood of u for any fixed $u \in \mathcal{M}$ and $s = 0$. More generally, these results hold for s fixed in neighborhood of 0 because, by (**a**) of Theorem 2.13, one has that no pole can arise in an open disk centered at $s = 0$, with radius $\rho = 1/2$. The functions Z_{ab} and all their x derivatives are also s-analytic, for x fixed in a neighborhood of 0. Then, we can conclude that any function $(s, x) \mapsto Z_{ab}(s, x|\ell^2 \mathcal{A})$ is C^∞ in a neighborhood of $(0, u)$ for any fixed $u \in \mathcal{M}$. The C^∞ smoothness of the stress tensor then follows trivially from (2.165) directly. The smoothness of $\mapsto \langle \phi^2(x|\mathcal{A}) \rangle_\ell$ can be proven similarly.

Property (**b**) can be proven as follows. From (**b**)-(**c**) of Theorem 2.13 and taking into account (**a**) of Theorem 2.16 and the definition (2.165), we have that (**b**) holds true if $\nabla^a Z_{ab}(s, x|\ell^2 \mathcal{A}) = -\Phi(s, x|\ell^2 \mathcal{A}) \nabla_b V'(x)$ for the considered point x and $s \in \mathbb{C}$ away from the poles, the function Z in the rhs being that of the field fluctuations. By resorting to the theorem on the uniqueness of the analytic continuation, if one is able to prove such an identity for Re s sufficiently large, this implies also the validity of $\nabla^a Z_{ab}(s, x|\ell^2 \mathcal{A}) = -\Phi(s, x|\ell^2 \mathcal{A}) \nabla_b V'(x)$ everywhere in the variable s. Therefore, let us prove that there is an $M > 0$ such that $\nabla^a Z_{ab}(s, x|\ell^2 \mathcal{A}) = -\Phi(s, x|\ell^2 \mathcal{A}) \nabla_b V'(x)$ for Re $s > M$ and this will be enough to prove point (**b**). To reach this goal, we represent $\nabla^a Z_{ab}(s, x|\ell^2 \mathcal{A})$ employing (2.164), for each function Z_{ab}. Then, we make recursive use of (2.127) and obtain $\nabla^a Z_{ab}(s, x|\ell^2 \mathcal{A})$ written as a linear combination of the functions $\zeta^{(\alpha,\beta)}(s + 1, x|\ell^2 \mathcal{A})$. Finally we can expand all these functions in series of the form (2.131) of Theorem 2.15, provided Re $s > M$ for an opportune $M > 0$. Taking into account the comment (4) after Definition 2.8, the explicit expression of the final series of $\nabla^a Z_{ab}(s, x|\ell^2 \mathcal{A})$ reads, for Re $s > M$,

$$
\begin{aligned}
\nabla^a Z_{ab}(s, x|\ell^2 \mathcal{A}) \quad = \quad & s {\sum_{j \in \mathbb{N}}}' \frac{2}{\lambda_j} (\ell^2 \lambda_j)^{-s} \nabla^a \Big\{ T_{ab}[\phi_j, \phi_j^*, g](x) \\
& + \frac{1}{2} \lambda_j g_{ab}(x) \phi_j(x) \phi_j^*(x) \Big\} . \qquad (2.172)
\end{aligned}
$$

Finally, using the zeta form of function Φ, one has to prove that for any

$x \in \mathcal{M}$,

$$\nabla^a \left\{ T_{ab}[\phi_j, \phi_j^*, g](x) + \frac{1}{2}\lambda_j g_{ab}(x)\phi_j(x)\phi_j^*(x) \right\}$$
$$= -\frac{1}{2}\phi_j(x)\phi_j^*(x)\nabla_b V'(x) \,. \tag{2.173}$$

This is nothing but the generalized "conservation law" of the stress tensor for the action

$$S_j[\phi, \phi^*] = \frac{1}{2}\int_{\mathcal{M}} d\mu_g \left[\phi \mathcal{A}[g]\phi^* - \lambda_j \phi\phi^*\right] . \tag{2.174}$$

Indeed, (2.173) holds when the field ϕ satisfies $\mathcal{A}\phi = \lambda_j \phi$ i.e., for eigenfunctions of $\mathcal{A}$ ϕ_j with eigenvalue λ_j. Therefore, (2.173) holds true and (**b**) is proven.

Concerning the point (**c**), (2.168) with $t_{ab}(x|\mathcal{A}) = Z_{ab}(0, x|\mathcal{A})/2$ arises as a direct consequence of the Definition (2.165), noticing that $\Phi(s, x|\ell^2\mathcal{A})$ is analytic at $s = 0$ (this is a straightforward consequence of the pole structure of $\zeta(s, x|\ell^2\mathcal{A})$ Theorem 2.7) and that (2.165) can be written down also

$$\langle T_{ab}(x|\mathcal{A})\rangle_\ell = \frac{1}{2}\left.\frac{d}{ds}\right|_{s=0} Z_{ab}(s, x|\mathcal{A}) - \frac{1}{2}Z_{ab}(0, x|\mathcal{A})\log\ell^2 \,, \tag{2.175}$$

from (2.162). Similarly, from Definition 2.8, one also sees that $Z(s, x|\ell^2\mathcal{A})/2$ evaluated at $s = 0$ takes contribution only from the possible pole at $s = 1$ of the function $s \mapsto \zeta_{ab}(s, x|\ell^2\mathcal{A})$ (the remaining simple zeta function is regular for $s = 0$) and coincides with the value of residue of the pole of this function at $s = 0$. When D is odd, no pole of $s \mapsto \zeta_{ab}(s, x|\ell^2\mathcal{A})$ arises at $s = 1$ because of (**c**) of Theorem 2.13, this is the reason for the δ_D in the right-hand side of (2.170). The form (2.170) of t_{ab} assures that it is built up as a linear combination of products of the metric, curvature tensors, $V'(x)$ and their covariant derivatives, everything evaluated at the same point x. The property (2.169) is a consequence of $\nabla^a Z_{ab}(s, x|\ell^2\mathcal{A}) = -\Phi(s, x|\ell^2\mathcal{A})\nabla_b V'(x)$ proven during the proof of (**b**), employing the known pole structure of the function Φ. Notice also that t_{ab} is symmetric by construction. Therefore, we have to prove (2.170) and this conclude the proof of (**c**). It is sufficient to show that

$$\lim_{s\to 0} s\zeta_{ab}(s + 1, x|\mathcal{A}) =$$

$$\begin{aligned}
& \frac{\delta_D}{(4\pi)^{D/2}} \left\{ a_{D/2-1,(ab)}(x,x|A) + \frac{1}{2} g_{ab}(x) a_{D/2}(x,x|A) \right. \\
+ & \left. \left[\left(\xi - \frac{1}{4} \right) g_{ab}(x) \Delta + \xi R_{ab}(x) - \xi \nabla_a \nabla_b \right] a_{D/2-1}(x,x|A) \right\} .
\end{aligned}$$

From (**a**) of Theorem 2.16 this is equivalent to

$$\begin{aligned}
& \lim_{s\to 0} \frac{s}{2} \left[\zeta^{(1_a,1_b)}(s+1,x|\mathcal{A}) + \zeta^{(1_b,1_a)}(s+1,x|\mathcal{A}) \right] \\
+ & \lim_{s\to 0} s \left[\left(\xi - \frac{1}{4} \right) g_{ab}(x) \Delta + \xi R_{ab}(x) - \xi \nabla_a \nabla_b \right] \zeta(s+1,x|\ell^2 \mathcal{A}) \\
= & \frac{\delta_D}{(4\pi)^{D/2}} \left\{ a_{D/2-1,(ab)}(x,x|A) + \frac{1}{2} g_{ab}(x) a_{D/2}(x,x|A) \right. \\
+ & \left. \left[\left(\xi - \frac{1}{4} \right) g_{ab}(x) \Delta + \xi R_{ab}(x) - \xi \nabla_a \nabla_b \right] a_{D/2-1}(x,x|A) \right\} .
\end{aligned} \tag{2.176}$$

The proof of this identity is quite straightforward, thus we shall sketch it only. Using Theorems 2.11-2.14, it is sufficient to consider the decomposition for large Re s (and a similar decomposition interchanging a with b)

$$\begin{aligned}
& s\zeta^{(1_a,1_b)}(s+1,x,y|\ell^2\mathcal{A})|_{x=y} \\
= & \frac{s}{\Gamma(s+1)} \int_0^{+\infty} dt\, t^s D_x^{1_a} D_y^{1_b} \left[K(t,x,y|\mathcal{A}) - P(x,y|\mathcal{A}) \right] |_{x=y} \\
= & \frac{s}{\Gamma(s+1)} \left(\int_0^{\ell_0^2} + \int_{\ell_0^2}^{+\infty} \right) dt\, D_x^{1_a} D_y^{1_b} \{\ldots\}|_{x=y} .
\end{aligned} \tag{2.177}$$

Then, we can expand the integrand of the first integral in the second line of (2.177), using (2.19), and we can continue both integrals in the second line of (2.177) as far as $s = 0$. A direct computation proves that, because of the presence of the factor s, only the first integral in (2.177), expanded as said above, gives a contribution. This contribution arises from the terms of the heat-kernel expansion which, once integrated in t (taking into account the factor t^s in the integrand), have a pole for $s = 0$. This pole is cancelled out by the factor s giving a finite result. The terms which have no pole at $s = 0$ vanish, due to the factor s, in the limit $s \to 0$. A similar procedure can be employed concerning the second limit in (2.176). In performing the calculations, it is worth to remind that $\nabla_{(x)a}\sigma(x,y)$ and $\nabla_{(y)b}\sigma(x,y)$ *vanish* in the limit $x \to y$, and furthermore $\nabla_{(x)a}\nabla_{(y)b}\sigma(x,y)|_{x=y} = -g_{ab}(y)$. Summing all contributions one obtains (2.170).

Concerning the point (**d**), the proof can be dealt with as follows. Starting from (2.161), one finds

$$g^{ab}T_{ab}[\phi_j,\phi_j^*,g] = \nabla_c\phi_j\nabla^c\phi_j^*$$

$$+\left\{\xi R + \left[\xi\,(D-1) - \frac{D}{4}\right]\Delta\right\}|\phi_j|^2 - \frac{D}{2}\lambda_j|\phi_j|\,. \tag{2.178}$$

Then, employing the identities $2\nabla_c\phi^*\nabla^c\phi = \Delta|\phi|^2 - \phi\Delta\phi^* - \phi^*\Delta\phi$ and $(-\Delta + \xi R + m^2 + V')\phi_j = \lambda_j\phi_j$, we have

$$g^{ab}T_{ab}[\phi_j,\phi_j^*,g] = \left[\xi\,(D-1) - \frac{D-2}{4}\right]\Delta|\phi_j|^2$$

$$-(m^2+V')|\phi|^2 + \frac{2-D}{2}\lambda_j|\phi_j|\,.$$

Since $\xi_D = (D-2)/[4(D-1)]$, we have finally

$$g^{ab}T_{ab}[\phi_j,\phi_j^*,g] = \left[\frac{\xi_D-\xi}{4\xi_D-1}\Delta - m^2 - V'\right]|\phi_j|^2 + \frac{2-D}{2}\lambda_j|\phi_j|\,.$$

Definition 2.8 entails that, for Re s sufficiently large

$$\begin{aligned} g^{ab}Z_{ab}(s,x|\ell^2\mathcal{A}) &= 2\ell^2\left[\frac{\xi_D-\xi}{4\xi_D-1}\Delta - m^2 - V'\right]s\zeta(s+1,x|\ell^2\mathcal{A}) \\ &+ 2s\zeta(s,x|\ell^2\mathcal{A})\,. \end{aligned}$$

From Definition 2.9 and (2.179) below the formula in (**d**) follows. □

Comments

(**1**) Concerning the point (**b**) which generalizes the classical law (2.147), we stress that this result is very non-trivial. We have not put this result somewhere "by hand" in the definitions and hypotheses we have employed. Notice also that, in the case $V' \equiv 0$, the tensor $T_{ab}[\phi_j,\phi_j^*,g]$, we have used in the definitions is not conserved. Nevertheless, the final stress tensor is conserved. This should mean that the local zeta function approach is indeed quite a deep procedure.

(2) Concerning the point (**c**), we notice that this result is in agreement with Wald's axioms [R.M. Wald (1994)] and, on purely mathematical grounds, it reduces the ambiguity allowed by Wald's theorem. Indeed, Wald's theorem involves at least two arbitrary terms, dependent on two free parameters. The point (**d**) proves that the corresponding ambiguity related to the field fluctuations is consistent with that which arises from the stress tensor. Assuming the zeta function procedure, the only ambiguity remaining is just that related to the initial arbitrary length scale ℓ. On the other hand, there is no physical evidence that the zeta function procedure is the physically correct one, and thus one cannot conclude that this method gets rid of the ambiguity pointed out by Wald et al.

(3) Concerning the point (**d**), we notice that, in the case $\xi = \xi_D$ and $V' \equiv m^2 = 0$ the usual conformal anomaly [N.D. Birrell and P.C.W. Davies (1982); R.M. Wald (1978)] arises provided D is even and the kernel of $\mathcal{A}$ is trivial. Anyhow, in the case Ker $\mathcal{A}$ is non-trivial, the trace anomaly takes a contribution from the null modes also when D is odd. In any cases, for the anomalous term, it holds from (**c**) in Theorem 2.7 that

$$\delta_D \frac{a_{D/2}(x,x|A)}{(4\pi)^{D/2}} - P(x,x|\mathcal{A}) = \zeta(0,x|\ell^2 A) \tag{2.179}$$

also for $\xi \neq \xi_D$.

To conclude, a few comments on the relation with point-splitting procedure are in order. In [V. Moretti (1999b)], it is proven that, similarly to Theorem 2.8, the field fluctuation $\langle\phi^2(x|\mathcal{A})\rangle_\ell$ and stress-energy tensor $\langle T_{ab}(x|\mathcal{A})\rangle_\ell$ produced by the local zeta function technique can also be obtained by means of point-splitting procedures. To illustrate this result, we start by noticing that (2.146) can be re-written in the more convenient form:

$$\begin{aligned}
T_{ab}[\phi,g](x) &= (1-2\xi)\left[\nabla_a\phi(x)\nabla_b\phi(x) + \phi(x)\nabla_a\nabla_b\phi(x)\right] \\
&+ \left(2\xi - \frac{1}{2}\right) g_{ab}(x)\left[\nabla_c\phi(x)\nabla^c\phi(x) + \phi(x)\Delta\phi(x)\right] \\
&+ \left[\frac{g_{ab}(x)}{D}\phi(x)\Delta\phi(x) - \phi(x)\nabla_a\nabla_b\phi(x)\right] \\
&- \xi\left[R_{ab}(x) - \frac{g_{ab}(x)}{D}R(x)\right]\phi^2(x)
\end{aligned}$$

$$- \frac{V'(x)+m^2}{D} g_{ab}(x)\phi^2(x) \, . \tag{2.180}$$

The right-hand side can be viewed as the action on $\phi(x)\phi(y)$ of a non-local operator $\mathcal{D}_{ab}(x,y)$ which "splits" the point y followed by the coincidence limit $x \to y$. To this end, define

$$\begin{aligned}
\mathcal{D}_{ab}(x,y) &= \frac{1-2\xi}{2}\Big[I_a^{a'}\nabla_{(x)a'}\nabla_{(y)b} + I_b^{b'}\nabla_{(x)b'}\nabla_{(y)a} + \nabla_{(y)a}\nabla_{(y)b} \\
&+ I_a^{a'}I_b^{b'}\nabla_{(x)a'}\nabla_{(x)b'}\Big] \\
&+ \left(2\xi - \frac{1}{2}\right)\frac{g_{ab}(y)}{2}\Big[2I_c^{c'}\nabla_{(x)c'}\nabla^{(y)c} + \Delta_x + \Delta_y\Big] \\
&+ \frac{1}{2}\left[\frac{g_{ab}(y)}{D}(\Delta_x + \Delta_y) - \nabla_{(y)a}\nabla_{(y)b} - I_a^{a'}I_b^{b'}\nabla_{(x)a'}\nabla_{(x)b'}\right] \\
&+ \xi\left[R_{ab}(y) - \frac{g_{ab}(y)}{D}R(y)\right] \\
&- \frac{V'(y)+m^2}{D}g_{ab}(y) \, .
\end{aligned} \tag{2.181}$$

Above $I_a^{b'} = I_{(y)a}^{(x)b'}(y,x)$ is a generic component of the bitensor of parallel displacement from y to x, so the (co)tangent space at the point x is identified with the fixed (co)tangent space at the point y. As we said, it is a trivial task to prove that *the action of $\mathcal{D}_{ab}(x,y)$ on $\phi(x)\phi(y)$ produces $T_{ab}(y)$ in the limit $x \to y$.*

Exactly as in the case of the Lagrangian regularization, to get $\langle T_{ab}(y|\mathcal{A})\rangle_\ell$ by employing $\mathcal{D}_{ab}(x,y)$ within a similar point-splitting procedure, one expects that:

(1) $\phi(x)\phi(y)$ has to be replaced by the Green function

$$G(x,y|\mathcal{A}) \stackrel{\text{def}}{=} \ell^2\zeta(1,x,y|\ell^2\mathcal{A}),$$

(remember that G does not depend on ℓ) and

(2) the divergence which arises as $x \to y$ should be rectified by a suitable subtraction of divergent terms.

Concerning the field fluctuation, the idea is similar but simpler, because one expect that $\langle\phi^2(x|\mathcal{A})\rangle_\ell$ can be equivalently obtained in the coincidence limit of $G(x,y|\mathcal{A})$, after subtraction of the singular part.

Theorem 2.18 *Under our general hypotheses on $\mathcal{M}$ and $A \geq 0$,*

$$\langle \phi^2(y|\mathcal{A})\rangle_\ell = \lim_{x\to y} \{G(x,y|\mathcal{A}) - H_\ell(x,y|A)\} \,, \qquad (2.182)$$

where $\delta_D = 0$ if D is odd and $\delta_D = 1$ if D is even, and more remarkably

$$\begin{aligned}\langle T_{ab}(y|\mathcal{A})\rangle_\ell &= \lim_{x\to y} \mathcal{D}_{ab}(x,y)\,\{\, G(x,y|\mathcal{A}) - H_\ell(x,y|A)\,\} \\ &\quad + \frac{g_{ab}(y)}{D}\left(\delta_D \frac{a_{D/2}(y,y|A)}{(4\pi)^{D/2}} - P(y,y|\mathcal{A})\right), \qquad (2.183)\end{aligned}$$

where $P(y,y|\mathcal{A})$ is the C^∞ integral kernel of the projector on the kernel of $\mathcal{A}$ and $H_\ell(x,y|A)$ is defined as (the summation appears for $D \geq 4$ only)

$$\begin{aligned}H_\ell(x,y|A) &= \sum_{j=0}^{D/2-2} (D/2-j-2)! \left(\frac{2}{\sigma}\right)^{D/2-j-1} \frac{a_j(x,y|A)}{(4\pi)^{D/2}} \\ &- \frac{a_{D/2-1}(x,y|A)}{(4\pi)^{D/2}} (2\gamma - \log \ell^2) \\ &- \frac{2a_{D/2-1}(x,y|A) - a_{D/2}(x,y|A)\sigma}{2(4\pi)^{D/2}} \log\left(\frac{\sigma}{2}\right), \qquad (2.184)\end{aligned}$$

if D is even, and (the summation appears for $D \geq 5$ only)

$$\begin{aligned}H_\ell(x,y|A) &= \sum_{j=0}^{(D-5)/2} \frac{(D-2j-4)!!\sqrt{\pi}}{2^{(D-3)/2-j}} \left(\frac{2}{\sigma}\right)^{D/2-j-1} \frac{a_j(x,y|A)}{(4\pi)^{D/2}} \\ &+ \frac{a_{(D-3)/2}(x,y|A)}{(4\pi)^{D/2}} \sqrt{\frac{2\pi}{\sigma}} - \frac{a_{(D-1)/2}(x,y|A)}{(4\pi)^{D/2}} \sqrt{2\pi\sigma}\,, \end{aligned} \qquad (2.185)$$

if D is odd.

Proof. See proofs of Theorem 2.6 in [V. Moretti (1999a)] and Theorem 4.1 in [V. Moretti (1999b)]. □

$H_\ell(x,y|A)$ is called the *Hadamard parametrix.* The result obtained shows that the local zeta function approach is equivalent to a point-splitting procedure. It is possible to show that the procedure is in agreement with Wald's axioms on the renormalized, averaged stress-energy tensor (see discussion in [V. Moretti (1999b)]). It is known [R.M. Wald (1994)] that the point-splitting procedure is not completely determined, due to the presence of arbitrary smooth terms in the divergence structure, which cannot be fixed after one has assumed Wald's axioms. In a sense, the local zeta function procedure fixes these terms, giving a precise form of the Hadamard parametrix (see discussion in [V. Moretti (1999b)]). Since terms of the form "smooth function times $\sigma(x,y)$" added to H_ℓ do not affect the final result (see Comment (1) after Theorem 4.1 in [V. Moretti (1999b)]), the Hadamard parametrix, in the case of D even, can be replaced by

$$\begin{aligned} H_\ell(x,y|A) &= \sum_{j=0}^{D/2-2} (D/2-j-2)! \left(\frac{2}{\sigma}\right)^{D/2-j-1} \frac{a_j(x,y|A)}{(4\pi)^{D/2}} \\ &\quad - \frac{2a_{D/2-1}(x,y|A) - a_{D/2}(x,y|A)\sigma}{2(4\pi)^{D/2}} \log\left(\frac{\gamma\sigma}{\ell^2}\right) . \end{aligned} \tag{2.186}$$

This fact entails that the ambiguous term $w(x,y)$ used in the Hadamard parametrix and selected by the zeta function approach (see discussions in [R.M. Wald (1978); V. Moretti (1999b)]) is completely determined by the choice of the scale ℓ.

2.6.4 *On the physical interpretation*

Let us consider some issues related with the physical interpretations of the above theory. Suppose that $\mathbb{S}^1$ acts as a globally one-parameter isometry group on the Riemannian manifold $\mathcal{M}$, giving rise to closed orbits with period $\beta > 0$. Suppose also that there exists a $D-1$ embedded submanifold Σ which intersects each orbit once and orthogonally. In this case, the Riemannian metric is termed *static*, the parameter of the group τ is called the *Euclidean time* of the manifold with period β, and the submanifold Σ is said to be the *Euclidean space* of the manifold. As we known, β is interpreted as the "statistical mechanics" inverse temperature of the quantum state associated with the unique self-adjoint extension of the operator A, $\mathcal{A}$ (see the first section in the present chapter). Anyway, β has no direct physical meaning, because it is not preserved by rescaling the normalization of the Killing vector K associated with the isometry group. The physical temperature which, in principle, may be measured by a thermometer is the local-rescaling invariant temperature $T_T \stackrel{\text{def}}{=} 1/[\sqrt{(K,K)}\beta]$, called the *Tolman's temperature*.

If $(x^1, \cdots, x^{D-1}) \equiv \mathbf{x}$ are local coordinates in Σ, they define *local static coordinates* in $\mathcal{M}$, $(\tau, \mathbf{x})$, $\tau \in (0,\beta)$ being the parameter associated with the integral curves of K. It is possible to define a similar coordinate patch in a neighborhood of every point $p \in \mathcal{M}$ (assuming that $\tau \in (-\epsilon, \beta - \epsilon)$ if $p \in \mathcal{M}$) except for the point of a set $\mathcal{F} \subset \mathcal{M}$ which contains the fixed points of the group. In the coordinates $(\tau, \mathbf{x})$, $\partial_\tau g_{ab} = 0$ and $g_{\tau\alpha} = 0$ for $\alpha = 1, \cdots, D-1$.

The important result is that, under our general hypotheses, the stress-energy tensor *depends on* $\mathbf{x}$ *only* of local static coordinates if $\mathcal{M}$ is static and V' does not depend on τ. Under these same hypotheses, it also holds, in local static coordinates $(\tau = x^0, x^1, \ldots, x^{D-1})$,

$$\langle T_{\tau\alpha}(\mathbf{x}|\mathcal{A})\rangle_\ell = \langle T_{\alpha\tau}(\mathbf{x}|\mathcal{A})\rangle_\ell = 0 \tag{2.187}$$

for $\alpha = 1, \cdots, D-1$. As a consequence, one is allowed to look for analytic continuations towards Lorentzian metrics, performing the analytic continuation $\tau \to it$ without encountering imaginary components of the continued stress tensor. This fact is crucial from a physical point of view. Notice that also $\langle \phi^2(x|\mathcal{A})\rangle_\ell$ and the effective Lagrangian $\mathcal{L}_{\text{eff}}(x|\mathcal{A})_\ell$ (Definition 2.6) do not depend on the temporal coordinate and that, moreover, all results contained in Theorem 2.17 hold true in the Lorentzian section of the manifold, considering the trivial analytical continuations of all the terms which appear in the thesis. More precisely, one has:

Theorem 2.19 *Within the general hypotheses on $\mathcal{M}$ and $A \geq 0$, suppose that $\mathcal{M}$ is static with Euclidean time $\tau \in (0, \beta)$, V' is invariant under Euclidean time displacements, and the set of fixed points under Euclidean time displacements has negligible measure.*

(a) *For every $\ell > 0$, $\beta > 0$ and every $x \in \mathcal{M}$,*

$$\langle T_{ab}(x|\mathcal{A})\rangle_\ell K^a(x)\sigma^b(x) = 0\,, \tag{2.188}$$

where K is the Killing vector field associated to the time τ and $\sigma(x)$ is any vector orthogonal to K in x.

(b) *Denoting the Lie derivative along K by $\pounds_K$,*

$$\pounds_{Kc}\langle T^{ab}(x|\mathcal{A})\rangle_\ell = 0\,, \tag{2.189}$$

$$\pounds_K\langle \phi^2(x|\mathcal{A})\rangle_\ell = 0\,, \tag{2.190}$$

$$\pounds_K\mathcal{L}_{\mathit{eff}}(x|\mathcal{A})_\ell = 0\,, \tag{2.191}$$

for every $x \in \mathcal{M}$.

(c) *Finally, all the results of Theorem 2.17 hold in the Lorentzian section obtained by local analytic continuation $\tau \mapsto i\tau$.*

Proof. Proof of (**a**) and (**b**). First of all, we notice that the set $\mathcal{F}$ of fixed points of the Euclidean-time isometry group is closed. Under the given hypotheses, take $x \in \mathcal{M} \setminus \mathcal{F}$. Choose now a local static coordinate system $(x^0 = \tau, (x^1, \ldots, x^{D-1}) = \mathbf{x})$ about x. We want to prove (**a**) and (**b**) in these coordinates. The proof for the case $x \in \mathcal{F}$ is a straightforward consequence of the continuity of the involved functions because, if $x \in \mathcal{F}$, there is a sequence $x_n \to x$ as $n \to \infty$ with $x_n \in \mathcal{M} \setminus \mathcal{F}$ since Int $\mathcal{F} = \emptyset$ in our hypotheses.

(2.188) is equivalent to (2.187) in the above coordinates, since $K^a = \delta^a_0$ and $((\partial/\partial x^\alpha), K) = 0$ if $\alpha = 1, \ldots, D-1$. Let us prove (2.187). To this end, consider the zeta function of the stress-energy tensor $Z_{ab}(s, x|\ell^2 \mathcal{A})$. Since $g_{0\alpha}(x) = 0$ and $\partial_\tau g_{ab}(x) = 0$, only the first line of (2.160) and the last term in the last line may produce the components $_{0\alpha}$ of the stress-energy tensor. Actually, the dependence from τ of the eigenfunctions ϕ_j of the operator A has the form $e^{i\omega\tau}$ with $\omega \in \mathbb{R}$ as proven below. As a consequence, the components $_{0\alpha}$ of the last term in the last line of (2.160) vanish, because the argument of the covariant derivatives (which commute on scalar fields) does not depend on τ. Moreover, since $\mathcal{A}$ is real and thus ϕ_j and ϕ_j^* are associated with the same eigenvalue, the contribution coming from the first line of (2.160) computed for $b \neq a = 0$ and $a \neq b = 0$ vanishes if one takes the summation over j to get the stress-tensor zeta function. (2.189), (2.190), (2.191) can be proven similarly, working in local static coordinates where the Lie derivative reduces to the ordinary τ derivative. The crucial point is that the various zeta functions which have to be differentiated with respect to τ turn out to be constant with respect to τ itself. This is because the τ dependence arises from products of (derivatives of) ϕ_j and ϕ_j^* and $\phi_j = \varphi_j(\mathbf{x})e^{i\omega\tau}$.

It remains to be shown that $\phi_j = \varphi_j(\mathbf{x})e^{i\omega_j\tau}$, for every j and corresponding $\omega_j \in \mathbb{R}$. Each eigenvector of $\mathcal{A}$ is a $C^\infty(\mathcal{M})$ eigenfunction of A and the various functions used to define the differential operator A do not depend on τ. As a consequence,

$$\mathcal{A}\partial_\tau \phi_{jk_j} = \partial_\tau A\phi_{jk_j} = \lambda_j \partial_\tau \phi_{jk_j} \ , \tag{2.192}$$

where ϕ_{jk_j} is an eigenvector of A with eigenvalue λ_j. (2.192) holds in the considered locally static coordinates about $x \in \mathcal{M} \setminus \mathcal{F}$ and thus (2.192) holds *almost* everywhere in $\mathcal{M}$ provided $\partial_\tau \phi_{jk_j}$ is interpreted as the $C^\infty(\mathcal{M})$ scalar field $(K, \nabla\phi_{jk_j})$. Remembering that the dimension of each eigenspace d_j is finite (Theorem 2.2), it must be

$$\partial_\tau \phi_{jk_j}(x) = \sum_{l_j=1}^{d_j} c_{k_j l_j} \phi_{jl_j}(x) \,, \tag{2.193}$$

almost everywhere. As locally $\partial_\tau g_{ab} = 0$ and, since $g_{0\alpha} = 0$, $g = (K, K)h$ where h is the determinant of the metric induced in Σ, from (2.193), one finds that

$$\begin{aligned} c^*_{k_j h_j} + c_{h_j k_j} &= \int_{\mathcal{M}} d\mu_g(x)\, \partial_\tau \left\{ \phi^*_{jk_j}(x)\phi_{jh_j}(x) \right\} \\ &= \int_0^\beta \frac{d}{d\tau} \int_\Sigma d\tau\, d\nu(p)\, \phi^*_{jk_j}(\tau, p)\phi_{jh_j}(\tau, p)(K(p), K(p))^{1/2} \,, \end{aligned}$$

where p is any point of the submanifold Σ and ν is the (finite) Riemannian measure induced by the metric of $\mathcal{M}$. We have passed the derivative through the symbol of integration, employing Fubini's theorem and Lebesgue's dominate convergence theorem. The right-hand side of the identity above must vanish because, for any fixed p,

$$\lim_{\tau \to 0^+} \phi(\tau, p)_{jk_j} = \lim_{\tau \to \beta^-} \phi(\tau, p)_{jk_j} \,, \tag{2.194}$$

the orbits of the coordinate τ being closed and the functions being continuous in the whole manifold. As a consequence the matrix of the coefficients c_{pq} must be anti-Hermitian. To conclude, we can choose an orthogonal basis of smooth normalized eigenfunctions in the considered eigenspace, where the matrix of coefficients c_{pq} is diagonal with eigenvalues $i\omega_{jl_j}$, $\omega_{jl_j} \in \mathbb{R}$ and $l_j = 1, 2, \cdots, d_j$. In that basis and in local coordinates, (2.193) reads

$$\partial_\tau \phi_{jk_j}(x) = i\omega_{jk_j} \phi_{jk_j}(x) \,, \tag{2.195}$$

and thus, by (2.194),

$$\phi_{jk_j}(x) = e^{i\omega_{jk_j}\tau} \varphi_{jk_j}(x^1, \cdots, x^{D-1}) \,, \tag{2.196}$$

with $\omega_{jk_j} = 2\pi n_{jk_j}/\beta$, $n_{jk_j} = 0, \pm 1, \pm 2 \ldots$.

The simple proof of the last statement in the thesis is left to the reader. The proof can be performed in local coordinates. □

Closing remark

Changes in the period β of the manifold which increase the Riemannian length of the orbits (not only the normalization of the Killing vector) in general produce *conical singularities* in the fixed points of the isometry group, if any. In this case, the obtained manifold fails to be smooth in those points and, in general, the theorems proven in this chapter may not hold true. Recently, the equivalence of the point-splitting regularization procedure and the zeta-function procedure for averaged values of the stress energy operator has been proven in [V. Moretti (2003)] within the general algebraic framework introduced by S. Hollands and R.M. Wald [S. Hollands and R.M. Wald (2001)] in discussing the locally and generally-covariant renormalization procedure in curved spacetime.

Chapter 3

Generalized Spectra and Spectral Functions on Non-commutative Spaces

Quantum field theories on non-commutative spaces generalize the familiar structure of conventional field theories. A motivation for considering these models is their appearance in M-theory and in the theory of strings [A. Connes et al. (1998); M.R. Douglas and C. Hall (1998); N. Seiberg and E. Witten (1999); Y. Cheung and M. Krogh (1998); C.S. Chu and P.M. Ho (1999); V. Schomerus (1999); F. Ardalan et al. (1999)], and also the fact that they are perfectly consistent formulations by themselves. It has been shown that non-commutative gauge theories describe the low energy excitations of open strings on D-branes in a background Neveu-Schwarz two-form field B [A. Connes et al. (1998); M.R. Douglas and C. Hall (1998); N. Seiberg and E. Witten (1999)].

The non-commutative perturbative dynamics on D-dimensional manifolds have been investigated [S. Minwalla et al. (2000)] and the Kaluza-Klein spectrum for interacting scalars and vector fields has been calculated in [J. Gomis et al. (2000b)]; it is consistent with formulae found in [S. Minwalla et al. (2000); Y. Kiem and S. Lee (2000); W. Fischler et al. (2000b)], where a connection between calculations performed in string theory and some others, done in field theory on a non-commutative torus, has been made. The one-loop Casimir energy of scalar and vector fields on a non-commutative space like $\mathbb{R}^{1,d} \times T_\theta^2$ (where $\mathbb{R}^{1,d}$ is a flat $(d+1)$-dimensional Minkowski space and T_θ^2 a two-dimensional non-commutative torus) has been calculated in [S. Nam (2000)]. The higher dimensional non-commutative torus case has been investigated in [W.H. Huang (2001); A.A. Bytsenko et al. (2001b)]; thermal effects in perturbative non-commutative gauge theories have been actively studied in Refs. [R.G. Cai and N. Ohta

(2000); G. Arcioni and M.A. Vazquez-Mozo (2000); W. Fischler et al. (2000a); G. Arcioni et al. (2000); J. Gomis et al. (2000a); K. Landsteiner et al. (2000)].

We will here calculate —using dimensional regularization implemented with zeta-function techniques— the partition function at inverse temperature, β, for massless scalar fields defined on D-dimensional manifolds with compact and non-compact dimensions. This can be considered as *the first* quantum correction to the total partition function. Since we will compute a functional determinant of a one-loop operator, our computation is, as well known, equivalent to summing an infinite number of one-loop diagrams [S. Minwalla et al. (2000)]. The relevance of this contribution will be discussed later.

In the first section of this chapter, we will give explicit formulae for the zeta functions corresponding to bosonic and to fermionic quantum fields living on a non-commutative, partially toroidal spacetime. Formulas for the most general case of the zeta function associated to a quadratic, linear, constant form (in $\mathbb{Z}$) will also be obtained. They provide the analytical continuation of the zeta functions in question to the whole complex s-plane, in terms of series of Bessel functions (of exponentially fast convergence), thus being extended Chowla-Selberg (CS) formulae. As well known, this is the most convenient expression that can be found for the analytical continuation of a zeta function; in particular, the residua of the poles and their finite parts are explicitly given there. An important novelty is the fact that simple poles show up at $s = 0$, as well as in other places (simple or double, depending on the number of compactified, non-compactified and non-commutative dimensions of the spacetime), where they had never appeared before. This poses a challenge to the zeta-function regularization procedure.

Indeed, for its application *in practice*, the zeta-function regularization method relies on the existence of simple formulae that give the analytical continuation of the zeta function, from the region of the complex plane extending to the right of the abscissa of convergence, Re $s > s_0$, to the rest of the complex plane [E. Elizalde et al. (1994); E. Elizalde (1995); A.A. Bytsenko et al. (1996); M. Bordag et al. (1996a); M. Bordag et al. (1997); E. Elizalde et al. (1998e)]. These are not only the reflection formula of the corresponding zeta function in each case, but also some other, even more fundamental expressions, as the Jacobi theta-function identity,

Poisson's and Plana's resummation formulae, and the Chowla-Selberg formula. However, some of these powerful expressions are often restricted to specific zeta functions, and their explicit derivation is usually quite involved. For instance, until very recently, the Chowla-Selberg formula was only known to exist for the homogeneous, two-dimensional Epstein zeta function. Ultimate extensions of this to more general zeta functions in any number of dimensions were given in the seminal papers [E. Elizalde (1998); E. Elizalde (2000); E. Elizalde (2001a)] and will be reviewed in Section 3.1 of the present chapter.

3.1 Extended Chowla-Selberg Formulae and Arbitrary Spectral Forms

A fundamental property shared by all zeta functions is the existence of a reflection formula. For the Riemann zeta function it is: $\Gamma(s/2)\zeta(s) = \pi^{s-1/2}\Gamma(1-s/2)\zeta(1-s)$. For a generic zeta function, $\mathfrak{Z}(s)$, we may write it as: $\mathfrak{Z}(\omega - s) = F(\omega, s)\mathfrak{Z}(s)$. It allows for its analytic continuation in a very easy way —what is, in simple cases, the whole story of the zeta-function regularization procedure. But the analytically continued expression thus obtained is just another series, which has again a slow convergence behavior, of power series type [C.M. Bender and S.A. Orszag (1978)] (actually the same that the original series had, on its convergence domain). Some years ago, S. Chowla and A. Selberg found a formula, for the Epstein zeta-function in the two-dimensional case [S. Chowla and A. Selberg (1949)], that yields *exponentially quick convergence everywhere*, not just in the reflected domain. They were very proud of it. In [E. Elizalde (1994c)], a first attempt was done to try to extend this expression to inhomogeneous zeta functions (very important for physical applications, see [E. Elizalde (1994d)]), but remaining always in *two* dimensions, for this was commonly believed to be an insurmountable restriction of the original formula (see, for instance, Ref. [S. Iyanaga and Y. Kawada (1977)]). More recently, extensions to an *arbitrary* number of dimensions [E. Elizalde (1998); E. Elizalde (1999)], both for the homogeneous (quadratic form) and nonhomogeneous (quadratic plus affine form) cases were constructed. However, some of the new formulae (remarkably the ones corresponding to the zero-mass case, e.g., the original Chowla–Selberg framework!) are *not explicit*, since they involve solving a rather non-trivial recurrence. (By the way, these explains why the Chowla–Selberg formula had not been extended to higher-dimensional Epstein zeta functions before.)

We will here provide explicit, Chowla-Selberg-like extended formulae for *all* possible cases involving forms of the very general type: quadratic plus linear plus constant form [E. Elizalde (1998)]. Then, we will move to specific applications of these formulae in non-commutative field theory. In particular, we will obtain the explicit analytic continuation of the zeta functions corresponding to scalar and vector fields defined on a quite general, partially non-commutative toroidal manifold. Their pole structure will be discussed in detail. The existence of simple poles at $s = 0$ comes as a novelty in the zeta-function regularization method in this case, confirming a result obtained in [A.A. Bytsenko et al. (2001b)]. In other places, up to double poles will be shown to appear. The corresponding residua and finite parts at the poles are immediately obtained from these expressions.

Let A is a positive-definite elliptic pseudo-differential operator of positive order $m \in \mathbb{R}$, acting on the space of smooth sections of the vector bundle over a closed D-dimensional manifold $\mathcal{M}$. Recall that the *zeta function* $\zeta(s|A)$ is defined for Re $s > D/m \equiv s_0$, where $s_0 = \dim \mathcal{M}/\mathrm{ord}\, A$ is called the ***abscissa of convergence*** of $\zeta(s|A)$. Under these conditions, it can be proven that $\zeta(s|A)$ has a meromorphic continuation to the whole complex plane $\mathbb{C}$ (regular at $s = 0$), provided that the principal symbol of A (that is $a_m(x, \xi)$) admits a *spectral cut*: $L_\theta = \{\lambda \in \mathbb{C}; \mathrm{Arg}\, \lambda = \theta,\ \theta_1 < \theta < \theta_2\}$, $\mathrm{Spec}\, A \cap L_\theta = \emptyset$ (Agmon-Nirenberg condition). The definition of $\zeta(s|A)$ depends on the position of the cut L_θ. The only possible singularities of $\zeta(s|A)$ are *simple poles* at $s_k = (D-k)/m,\ k = 0, 1, 2, \ldots, n-1, n+1, \ldots$. M.Kontsevich and S. Vishik have managed to extend this definition to the case when $m \in \mathbb{C}$ (no spectral cut exists) [M. Kontsevich and S. Vishik (1993)].

Consider now the following zeta function (Re $s > p/2$):

$$\begin{aligned} \zeta_{\mathbf{A},\mathbf{c},q}(s) &= \sum_{\mathbf{n}\in\mathbb{Z}^p}{}' \left[\frac{1}{2}(\mathbf{n}+\mathbf{c})^T \mathbf{A}\, (\mathbf{n}+\mathbf{c}) + q\right]^{-s} \\ &\equiv \sum_{\mathbf{n}\in\mathbb{Z}^p}{}' \left[Q\,(\mathbf{n}+\mathbf{c}) + q\right]^{-s}. \end{aligned} \tag{3.1}$$

The prime on a summation sign means that the point $\mathbf{n} = \mathbf{0}$ is to be excluded from the sum. As we shall see, this is irrelevant when q or some component of $\mathbf{c}$ is non-zero but, on the contrary, it becomes an inescapable condition in the case when $c_1 = \cdots = c_p = q = 0$. Note that, alternatively, we can view the expression inside the square brackets of the zeta function as a sum of a quadratic, a linear, and a constant form, namely, $Q(\mathbf{n}+\mathbf{c})+q = Q(\mathbf{n}) + L(\mathbf{n}) + \bar{q}$.

Our aim is to obtain a formula that gives (the analytic continuation of) this multidimensional zeta function in terms of an exponentially convergent series, and which is valid in the whole complex plane, exhibiting the singularities (poles) of the meromorphic continuation —with the corresponding residua— explicitly. The only condition on the matrix A is that it correspond to a (non-negative) quadratic form, which we call Q. The vector $\mathbf{c}$ is arbitrary, while q will be (for the moment) a positive constant. As we shall see, the solution to this problem will depend very much (its explicit form) on the fact that q and/or $\mathbf{c}$ are zero or not. According to this, we will have to distinguish different cases, leading to unrelated final formulae, all to be viewed as different non-trivial extensions of the Chowla–Selberg formula.

Use of the Poisson resummation formula in Eq. (3.1) yields [E. Elizalde (1998); E. Elizalde (1999)]

$$\zeta_{\mathbf{A},\mathbf{c},q}(s) = \frac{(2\pi)^{p/2} q^{p/2-s}}{\sqrt{\det \mathbf{A}}} \frac{\Gamma(s-p/2)}{\Gamma(s)} + \frac{2^{s/2+p/4+2}\pi^s q^{-s/2+p/4}}{\sqrt{\det \mathbf{A}}\,\Gamma(s)} \times \sum_{\mathbf{m}\in\mathbb{Z}^p_{1/2}}{}' \cos(2\pi \mathbf{m}\cdot\mathbf{c}) \left(\mathbf{m}^T\mathbf{A}^{-1}\mathbf{m}\right)^{s/2-p/4} K_{p/2-s}\left(2\pi\sqrt{2q\,\mathbf{m}^T\mathbf{A}^{-1}\mathbf{m}}\right), \tag{3.2}$$

where K_ν are the modified Bessel functions and the subindex 1/2 in $\mathbb{Z}^p_{1/2}$ means that in this sum, only half of the vectors $\mathbf{m} \in \mathbb{Z}^p$ enter. That is, if we take an $\mathbf{m} \in \mathbb{Z}^p$ we must then exclude $-\mathbf{m}$ (as a simple criterion one can, for instance, select those vectors in $\mathbb{Z}^p \backslash \{\mathbf{0}\}$ whose first non-zero component is positive). Equation (3.2) fulfills *all* the requirements of a Chowla–Selberg formula. But it is very different from the original one, constituting a non-trivial extension to the case of a quadratic+linear+constant form, in any number of dimensions, with the constant term being non-zero.

It is notorious how the only pole of this inhomogeneous Epstein zeta function appears, explicitly, at $s = p/2$, where it belongs. Its residue is given by:

$$\text{Res}_{s=p/2}\zeta_{\mathbf{A},\mathbf{c},q}(s) = \frac{(2\pi)^{p/2}}{\Gamma(p/2)} (\det \mathbf{A})^{-1/2}. \tag{3.3}$$

Limit $q \to 0$.

After some work, one can obtain the limit of expression (3.2) as $q \to 0$ (for simplicity we also set $\mathbf{c} = \mathbf{0}$)

$$\zeta_{\mathbf{A},\mathbf{0},0}(s) = 2^{1+s} a^{-s} \zeta(2s) + \sqrt{\frac{\pi}{a}} \frac{\Gamma(s-1/2)}{\Gamma(s)} \zeta_{\Xi_{p-1},\mathbf{0},0}(s-1/2)$$
$$+ \frac{4\pi^s}{a^{s/2+1/4}\,\Gamma(s)} \sum_{\mathbf{n}_2 \in \mathbb{Z}^{p-1}}{}' \sum_{n_1=1}^{\infty} \cos\left(\frac{\pi n_1}{a} \mathbf{b}^T \mathbf{n}_2\right) n_1^{s-1/2}$$
$$\times \left(\mathbf{n}_2^T \Xi_{p-1} \mathbf{n}_2\right)^{1/4-s/2} K_{s-1/2}\left(\frac{2\pi n_1}{\sqrt{a}} \sqrt{\mathbf{n}_2^T \Xi_{p-1} \mathbf{n}_2}\right). \qquad (3.4)$$

In Eqs. (3.2) and (3.4), $\mathbf{A}$ is a $p \times p$ symmetric matrix $\mathbf{A} = (a_{ij})_{i,j=1,2,\ldots,p} = \mathbf{A}^T$, $\mathbf{A}_{p-1}$ the $(p-1) \times (p-1)$ reduced submatrix $\mathbf{A}_{p-1} = (a_{ij})_{i,j=2,\ldots,p}$, a the first component, $a = a_{11}$, $\mathbf{b}$ the $p-1$ vector $\mathbf{b} = (a_{21}, \ldots, a_{p1})^T = (a_{12}, \ldots, a_{1p})^T$, and Ξ_{p-1} is the following $(p-1) \times (p-1)$ matrix: $\Xi_{p-1} = \mathbf{A}_{p-1} - (1/4a)\mathbf{b} \times \mathbf{b}^T$. More precisely, what one actually obtains by taking the limit is *the reflected formula*, as one would get after using the Epstein zeta function [P. Epstein (1903)] (see Eq. (3.48) below). Finally, it can be written as (3.4). (It is a rather non-trivial exercise to perform this calculation.) Note that Eq. (3.4) has all the properties demanded from a Chowla–Selberg formula, but it is actually *not explicit*. It is in fact a recurrence, rather lengthy to solve as it stands. In fact, it can be viewed as the straightforward extension of the original Chowla–Selberg formula to higher dimensions. It was the top result of work on this subject, for the case $q = c_1 = \cdots = c_p = 0$ [E. Elizalde (1998); E. Elizalde (1999)], till very recently [E. Elizalde (2000); E. Elizalde (1998); E. Elizalde (2001a)].

Using a different approach, this recurrence can be solved explicitly, in a much more simple way [E. Elizalde (1998)]. Indeed, let us proceed in a complementary way, namely, by doing the inversion provided by the Poisson resummation formula (or the Jacobi identity), with respect to $p-1$ of the indices (say, $j = 2, 3, \ldots, p$). This leaves us with three sums, corresponding to positive, zero, and negative values of the remaining index (n_1, in this case). The zero value of n_1 (in correspondence with the rest of the n_i's not being all zero) classifies the number of different situations (according to the values of the c_i's an q being all zero or not) into just two cases. It is immediate that, from start, all c_i's can be taken to be between 0 and 1:

$0 \le c_i < 1$, $i = 1, 2, \ldots, p$.

(i) The first case is, thus, when at least one of the c_i's or $q \ge 0$ is not zero. Since the case $q \neq 0$ has already been solved, we will mean by this case now that, say $c_1 \neq 0$.

(ii) The second case is when all $q = c_1 = \cdots = c_p = 0$.

The case $q = 0, \quad c_1 \neq 0$

The non-diagonal subcase. By doing the inversion provided by the Poisson resummation formula (or the Jacobi identity), with respect to $p-1$ of the indices (here, $j = 2, 3, \ldots, p$), we readily obtain:

$$\begin{aligned}
\zeta_{\mathbf{A}_p,\mathbf{c},0}(s) = {} & \frac{2^s}{\Gamma(s)} (\det \mathbf{A}_{p-1})^{-1/2} \Big\{ \pi^{(p-1)/2} (a_{11} - \mathbf{a}_{p-1}^T \mathbf{A}_{p-1}^{-1} \mathbf{a}_{p-1})^{(p-1)/2-s} \\
& \times \Gamma\left(s - (p-1)/2\right) \left[\zeta_H(2s-p+1, c_1) + \zeta_H(2s-p+1, 1-c_1)\right] \\
& + 4\pi^s \left(a_{11} - \mathbf{a}_{p-1}^T \mathbf{A}_{p-1}^{-1} \mathbf{a}_{p-1}\right)^{(p-1)/4-s/2} \sum_{n_1 \in \mathbb{Z}} \\
& \sum_{\mathbf{m} \in \mathbb{Z}_{1/2}^{p-1}}{}' \cos\left[2\pi \mathbf{m}^T \left(\mathbf{c}_{p-1} + \mathbf{A}_{p-1}^{-1} \mathbf{a}_{p-1} (n_1 + c_1)\right)\right] \\
& \times |n_1 + c_1|^{(p-1)/2-s} \left(\mathbf{m}^T \mathbf{A}_{p-1}^{-1} \mathbf{m}\right)^{s/2-(p-1)/4} \\
& \times K_{(p-1)/2-s} \left(2\pi |n_1 + c_1| \sqrt{(a_{11} - \mathbf{a}_{p-1}^T \mathbf{A}_{p-1}^{-1} \mathbf{a}_{p-1}) \mathbf{m}^T A_{p-1}^{-1} \mathbf{m}} \right) \Big\} \\
& - \left(\frac{1}{2} \mathbf{c}^T \mathbf{A} \mathbf{c} \right)^{-s}. \qquad (3.5)
\end{aligned}$$

Here, and in what follows, $\mathbf{A}_{p-1}$ is (as before) the submatrix of $\mathbf{A}_p$ composed of the last $p-1$ rows and columns. Moreover, a_{11} is the first diagonal component of $\mathbf{A}_p$, while $\mathbf{a}_{p-1} = (a_{12}, \ldots, a_{1p})^T = (a_{21}, \ldots, a_{p1})^T$, and $\mathbf{m} = (n_2, \ldots, n_p)^T$. Note that this is an *explicit formula*, that the only pole at $s = p/2$ appears also explicitly, and that the second term of the right-hand side is an exponentially-fast convergent series. It has, therefore (as Eq. (3.2)), all the properties required to qualify as a Chowla–Selberg formula.

The diagonal subcase. In this very common, particular case the preceding expression reduces to the more simple form:

$$\begin{aligned}\zeta_{\mathbf{A}_p,\mathbf{c},0}(s) &= \frac{2^s}{\Gamma(s)}(\det \mathbf{A}_{p-1})^{-1/2}\Big\{\pi^{(p-1)/2}a_1^{(p-1)/2-s}\Gamma(s-(p-1)/2) \\ &\qquad\times \left[\zeta_H(2s-p+1,c_1)+\zeta_H(2s-p+1,1-c_1)\right] \\ &+4\pi^s a_1^{(p-1)/4-s/2}\sum_{n_1\in\mathbb{Z}}\sum_{\mathbf{m}\in\mathbb{Z}_{1/2}^{p-1}}{}' \cos\left(2\pi\mathbf{m}^T\mathbf{c}_{p-1}\right)\,|n_1+c_1|^{(p-1)/2-s} \\ &\times\left(\mathbf{m}^T\mathbf{A}_{p-1}^{-1}\mathbf{m}\right)^{s/2-(p-1)/4}K_{(p-1)/2-s}\left(2\pi|n_1+c_1|\sqrt{a_1\mathbf{m}^T\mathbf{A}_{p-1}^{-1}\mathbf{m}}\right)\Big\} \\ &\qquad\qquad -\left(\frac{1}{2}\mathbf{c}^T\mathbf{A}\mathbf{c}\right)^{-s}. \end{aligned} \tag{3.6}$$

The case $c_1 = \cdots = c_p = q = 0$

The non-diagonal subcase. As remarked in [E. Elizalde (1998); E. Elizalde (1999)], it was quite difficult to obtain a closed formula, and for some time only a (rather non-trivial) recurrence existed, Eq. (3.4), relating the p-dimensional case with the $(p-1)$-dimensional one. Recently, it was realized [E. Elizalde (2001a)] that one can actually still proceed as if we had in fact $c_1 = 1 \neq 0$, both for positive and for negative values of n_1. A sum, though, remains with $n_1 = 0$ —and the rest of the n_i's not all being zero— what yields, once more, the same zeta function of the beginning, but corresponding to $p-1$ indices. That is,

$$\begin{aligned}\zeta_{\mathbf{A}_p,\mathbf{0},0}(s) &= \zeta_{\mathbf{A}_{p-1},\mathbf{0},0}(s)+\frac{2^{1+s}}{\Gamma(s)}(\det \mathbf{A}_{p-1})^{-1/2}\Big\{\pi^{(p-1)/2} \\ &\times\left(a_{11}-\mathbf{a}_{p-1}^T\mathbf{A}_{p-1}^{-1}\mathbf{a}_{p-1}\right)^{(p-1)/2-s}\Gamma\left(s-(p-1)/2\right)\zeta_R(2s-p+1) \\ &\qquad +4\pi^s\left(a_{11}-\mathbf{a}_{p-1}^T\mathbf{A}_{p-1}^{-1}\mathbf{a}_{p-1}\right)^{(p-1)/4-s/2}\sum_{n=1}^{\infty} \\ &\sum_{\mathbf{m}\in\mathbb{Z}_{1/2}^{p-1}}{}'\cos\left(2\pi\mathbf{m}^T\mathbf{A}_{p-1}^{-1}\mathbf{a}_{p-1}n\right)n^{(p-1)/2-s}\left(\mathbf{m}^T\mathbf{A}_{p-1}^{-1}\mathbf{m}\right)^{s/2-(p-1)/4} \\ &\qquad\times K_{(p-1)/2-s}\left[2\pi n\sqrt{(a_{11}-\mathbf{a}_{p-1}^T\mathbf{A}_{p-1}^{-1}\mathbf{a}_{p-1})\mathbf{m}^T\mathbf{A}_{p-1}^{-1}\mathbf{m}}\right]\Big\}. \end{aligned} \tag{3.7}$$

This is also a recurrent expression, but of a different kind. It is an alternative to (3.4), obtained with the help of a different strategy.

Remarkably enough, it is easy to resolve this recurrence explicitly, and indeed to obtain a *closed formula* for this case (we shall write the dimensions of the submatrices of $\mathbf{A}$ as subindices). The result is [E. Elizalde (2001a)]

$$\begin{aligned}
\zeta_{\mathbf{A}_p}(s) \equiv \zeta_{\mathbf{A}_p,\mathbf{0},0}(s) = & \frac{2^{1+s}}{\Gamma(s)} \sum_{j=1}^{p} (\det \mathbf{A}_{p-j})^{-1/2} \Big\{ \pi^{(p-j)/2} \\
& \times \left(a_{jj} - \mathbf{a}_{p-j}^T \mathbf{A}_{p-j}^{-1} \mathbf{a}_{p-j}\right)^{(p-j)/2-s} \Gamma(s-(p-j)/2)\, \zeta_R(2s-p+j) \\
& + 4\pi^s \left(a_{jj} - \mathbf{a}_{p-j}^T \mathbf{A}_{p-j}^{-1} \mathbf{a}_{p-j}\right)^{(p-j)/4-s/2} \sum_{n=1}^{\infty} \\
\sum_{\mathbf{m}_{p-j} \in \mathbb{Z}_{1/2}^{p-j}}{}' & \cos\left(2\pi \mathbf{m}_{p-j}^T \mathbf{A}_{p-j}^{-1} \mathbf{a}_{p-j} n\right) n^{(p-j)/2-s} \left(\mathbf{m}_{p-j}^T \mathbf{A}_{p-j}^{-1} \mathbf{m}_{p-j}\right)^{(2s-p+j)/4} \\
& \times K_{(p-j)/2-s}\left[2\pi n \sqrt{(a_{jj} - \mathbf{a}_{p-j}^T \mathbf{A}_{p-j}^{-1} \mathbf{a}_{p-j}) \mathbf{m}_{p-j}^T \mathbf{A}_{p-j}^{-1} \mathbf{m}_{p-j}}\right] \Big\}.
\end{aligned} \tag{3.8}$$

With a similar notation as above, here $\mathbf{A}_{p-j}$ is the submatrix of $\mathbf{A}_p$ composed of the last $p-j$ rows and columns. Moreover, a_{jj} is the j-th diagonal component of $\mathbf{A}_p$, while $\mathbf{a}_{p-j} = (a_{jj+1}, \dots, a_{jp})^T = (a_{j+1j}, \dots, a_{pj})^T$, and $\mathbf{m}_{p-j} = (n_{j+1}, \dots, n_p)^T$. Again, this is an extension of the Chowla-Selberg formula to the case in question. It exhibits all the same good properties. Physically, it corresponds to the homogeneous, massless case. It is to be viewed, in fact, as the genuine multidimensional extension of the Chowla-Selberg formula.

Diagonal subcase. Let us consider once again the diagonal case, with $\mathbf{c} = \mathbf{0}$, which is quite important in practice and gives rise to more simple expressions. For the recurrence, we have

$$\begin{aligned}
\zeta_{\mathbf{A}_p}(s) = & \zeta_{\mathbf{A}_{p-1}}(s) + \frac{2^{1+s}}{\Gamma(s)} (\det \mathbf{A}_{p-1})^{-1/2} \Big[\pi^{(p-1)/2} a_1^{(p-1)/2-s} \\
& \Gamma(s-(p-1)/2)\, \zeta_R(2s-p+1) + 4\pi^s a_1^{(p-1)/4-s/2} \\
& \sum_{n=1}^{\infty} \sum_{\mathbf{m} \in \mathbb{Z}^{p-1}}{}' n^{(p-1)/2-s} \left(\mathbf{m}^T \mathbf{A}_{p-1}^{-1} \mathbf{m}\right)^{s/2-(p-1)/4}
\end{aligned}$$

$$\times K_{(p-1)/2-s}\left(2\pi n\sqrt{a_1 \mathbf{m}^T \mathbf{A}_{p-1}^{-1}\mathbf{m}}\right)\Bigg] . \tag{3.9}$$

As above, we can solve this finite recurrence and obtain the following simple and explicit formula for this case:

$$\zeta_{\mathbf{A}_p}(s) = \frac{2^{1+s}}{\Gamma(s)} \sum_{j=0}^{p-1} (\det \mathbf{A}_j)^{-1/2} \Bigg[\pi^{j/2} a_{p-j}^{j/2-s} \Gamma(s-j/2) \zeta_R(2s-j)$$
$$+ 4\pi^s a_{p-j}^{j/4-s/2} \sum_{n=1}^{\infty} \sum_{\mathbf{m}_j \in \mathbb{Z}^j}{}' n^{j/2-s} \left(\mathbf{m}_j^t \mathbf{A}_j^{-1} \mathbf{m}_j\right)^{s/2-j/4}$$
$$\times K_{j/2-s}\left(2\pi n\sqrt{a_{p-j} \mathbf{m}_j^t \mathbf{A}_j^{-1}\mathbf{m}_j}\right)\Bigg] , \tag{3.10}$$

where $\mathbf{A}_p = \text{diag}\,(a_1, \ldots, a_p)$, $\mathbf{A}_j = \text{diag}\,(a_{p-j+1}, \ldots, a_p)$, $\mathbf{m}_j = (n_{p-j+1}, \ldots, n_p)^T$, and ζ_R is the Riemann zeta function. Note again the fact that this and Eq. (3.8) are *explicit* expression for the multidimensional, generalized Chowla-Selberg formula and, in this way, they go beyond any result obtained previously.

It is immediate to see that the term for $j = 0$ in the sum yields the last term, $\zeta_{\mathbf{A}_1}(s)$, of the recurrence, that is:

$$\zeta_{\mathbf{A}_1}(s) = \sum_{n_1=-\infty}^{+\infty}{}' \left(\frac{a_1}{2} n_1^2\right)^{-s} = 2^{1+s} a_1^{-s} \zeta_R(2s). \tag{3.11}$$

It exhibits a pole, at $s = 1/2$ which is spurious —it is actually *not* a pole of the whole function. It cancels, in fact, with another one with opposite sign coming from the next term, with further cancellations of this kind going on, each term with the next. Concerning the pole structure of the resulting zeta function, as given by Eq. (3.10), it is not difficult to see that *only* the pole at $s = p/2$ is actually there (as it should). It is in the last term, $j = p - 1$, of the sum, and it has the correct residue (3.3). The rest of the poles at $s = (p - j)/2$ are not such: they compensate among themselves, one term of the sum with the next, adding pairwise to zero.

Summing up, this formula, Eq. (3.10), provides a convenient analytic continuation of the zeta function to the whole complex plane, with its only simple pole showing up explicitly [E. Elizalde (1998)]. Aside from this, the finite part of the first sum in the expression is quite easy to obtain, and the remainder —an awfully looking multiple series— is in fact an extremely-fast convergent sum, yielding a small contribution, as happens in the Chowla–Selberg formula. In fact, since it corresponds to the case $q = 0$, this expression should be viewed as *the* extension of the original Chowla-Selberg formula —for the zeta function associated with a homogeneous quadratic form in two dimensions— to an arbitrary number, p, of dimensions. The rest of the formulae above provide also extensions of the original Chowla–Selberg expression.

The general case of a quadratic+linear+constant form has thus been completed [E. Elizalde (1998); E. Elizalde (2001a)]. As we clearly see, the main formulae corresponding to the three different subcases, namely Eq. (3.2), Eq. (3.5), and Eq. (3.8), are in fact quite distinct and one cannot directly go from one to another by adjusting some parameters [E. Elizalde (1998)].

For the sake of completeness, we must mention the following. Notice that all cases considered here correspond to having a non-identically-zero quadratic form Q. For Q identically zero, that is, the linear+constant (or affine) case, the formulae for the analytic continuation are again quite different from the ones above. The corresponding zeta function is called Barnes' zeta function. This case has been thoroughly studied in [E. Elizalde (1999)].

3.2 Barnes and Related Zeta Functions

For completeness of the mathematical scheme, we will here consider the case when the quadratic part in the quadratic+linear+constant form considered before in general, is identically equal to zero. One is left with the linear part (plus a possible constant term), and this yields an absolutely different case from the point of view of the analytical continuation to perform. Both the strategy and the final formulae have nothing to do with those in the previous cases. There is a wide mathematical literature dealing with this subject [E. Elizalde (1999)].

3.2.1 The two-dimensional case

Consider the Barnes zeta function in two dimensions [E.W. Barnes (1903)]

$$\zeta_B(s;a|\mathbf{r}) = \sum_{n_1,n_2=0}^{\infty} (a + r_1 n_1 + r_2 n_2)^{-s}, \quad \mathrm{Re}\, s > 2, \quad r_1, r_2 > 0, \tag{3.12}$$

and the related zeta function

$$\zeta(s;a|\mathbf{r}) = \sum_{n_1,n_2\in Z}{}' (a + r_1 n_1 + r_2 n_2)^{-s}, \tag{3.13}$$

where the prime means that the term with $n_1 = n_2 = 0$ is absent from the sum (actually, we could have defined the Barnes zeta function in this way too, in order to allow for the particular case $a = 0$, but that would not be the usual definition). The two functions are related, in fact:

$$\begin{aligned}\zeta(s;a|\mathbf{r}) &= \zeta_B(s;a|\mathbf{r}) + \zeta_B(s;a|-\mathbf{r}) + \zeta_B(s;a|(r_1,-r_2)) \\ &\quad + \zeta_B(s;a|(-r_1,r_2)) - \sum_{i=1}^{2}\left[r_i^{-s}\zeta_H(s,a/r_i)\right. \\ &\quad \left. + (-r_i)^{-s}\zeta_H(s,-a/r_i)\right],\end{aligned} \tag{3.14}$$

where ζ_H is the Hurwitz zeta function. Recall Hermite's formula for the Hurwitz zeta function:

$$\zeta_H(s,a) = \frac{a^{1-s}}{s-1} + \frac{a^{-s}}{2} + 2\int_0^\infty dy\, \frac{\sin[s\arctan(y/a)]}{(a^2+y^2)^{s/2}(e^{2\pi y}-1)}. \tag{3.15}$$

This expression exhibits the singularity structure of $\zeta_H(s,a)$ explicitly, the integral being an analytic function of s, $\forall s \in \mathbb{C}$ (it is uniformly convergent for $|s| \leq R$, for any $R > 0$). If we expand $\zeta_H(s,a)$ around $s = 0$, we obtain, at $s = 0$, the formula:

$$\zeta_B(0;a|\mathbf{r}) = \frac{r_1}{2r_2} B_2(a/r_1) + \frac{1}{2}\left(\frac{1}{2} - \frac{a}{r_1}\right) + \frac{r_2}{12 r_1}, \tag{3.16}$$

where $B_2(x)$ is the Barnes polynomial of order two. Taking the derivative with respect to s, at $s = 0$, we obtain

$$\begin{aligned}\zeta_B'(0;a|\mathbf{r}) &= -\zeta_B(0;a|\mathbf{r})\log r_2 + \left(1 - \log\frac{r_1}{r_2}\right)\frac{r_1}{2r_2}B_2(a/r_1) \\ &\quad - \frac{r_1}{r_2}\zeta_H'(-1,a/r_1) + \left(\frac{a}{2r_1} - \frac{1}{4}\right)\log\frac{r_1}{r_2} + \frac{1}{2}\log\Gamma(a/r_1)\end{aligned}$$

$$-\frac{1}{4}\log(2\pi) - \frac{r_2}{12r_1}\left[\log(r_1/r_2) + \psi(a/r_1)\right] + R(a|\mathbf{r}), \qquad (3.17)$$

where

$$R(a|\mathbf{r}) \begin{cases} = \dfrac{2r_1}{r_2}\displaystyle\sum_{n=0}^{\infty}(n + a/r_1)\left[\int_0^{\infty} \frac{\arctan x \, dx}{\exp\left[2\pi(r_1/r_2)x\,(n + a/r_1)\right] - 1} \right. \\ \qquad\qquad \left. -\dfrac{1}{24}\left(\dfrac{a}{r_2} + \dfrac{r_1}{r_2}n\right)^{-2}\right], \\ \sim \dfrac{1}{\pi}\displaystyle\sum_{k=1}^{\infty}(-1)^k(2k)!\,\zeta(2k+2)\left(\frac{r_2}{2\pi r_1}\right)^{2k+1} \zeta_H(2k+1, a/r_1). \end{cases} \qquad (3.18)$$

These are two possible expressions for the remainder term $R(a|\mathbf{r})$. The first is valid for $r_2 \leq r_1 \leq a$. It is very quickly convergent in this region and, therefore, appropriate for numerical calculations. The second is an explicit asymptotic series for $r_2 \ll 2\pi a$.

For the other two-dimensional zeta function considered at the beginning:

$$\begin{aligned} \zeta'(0; a|\mathbf{r}) &= \sum_{\alpha=1}^{4} \zeta_B'(0; a|\mathbf{r}_\alpha) + \sum_{i=1}^{2}\left[\left(\frac{1}{2} - \frac{a}{r_i}\right)\log r_i + \left(\frac{1}{2} + \frac{a}{r_i}\right)\log(-r_i) \right. \\ &\quad \left. - \log\Gamma\left(\frac{a}{r_i}\right) - \log\Gamma\left(\frac{a}{-r_i}\right) + \log(2\pi)\right], \end{aligned} \qquad (3.19)$$

where

$$\mathbf{r}_1 = (r_1, r_2), \quad \mathbf{r}_2 = (r_1, -r_2), \quad \mathbf{r}_3 = (-r_1, r_2), \quad \mathbf{r}_4 = (-r_1, -r_2). \quad (3.20)$$

Note that this evaluation involves an analytic continuation on the values of r_1 and r_2 —which are strictly positive in the case of the Barnes zeta function— to negative and, in general, complex values $r_1, r_2 \in \mathbb{C}$. In this way we elude the apparently unsurmountable problems that a direct interpretation of the sum in Eq. (3.13) poses, even for positive $r_1, r_2 \in \mathbb{R}$. In particular, for $r_1 = r_2$ it develops an infinite number of zero and constant modes, for an infinite number of constants, what renders a direct interpretation of the series extremely problematic. Naive analytical continuation followed by a natural transportation of the negative signs of the indices n_i

to the parameters r_i, yields

$$\zeta'(0;a|\mathbf{r}) = -\log\left[\Gamma\left(\frac{a}{r_2}\right)\Gamma\left(\frac{a}{-r_2}\right)\right] + \frac{a\pi i}{r_2} - \log\frac{a}{r_2} \pm i\frac{\pi}{2} + \log 2\pi. \tag{3.21}$$

Note that the breaking of the initial symmetry under interchange of r_1 and r_2 has finally resulted in the disappearance of one of the two parameters from the end result. Remarkable is the fact, however, that the final formula actually preserves the modular invariance of the initial one, under the parameter change on the other variable, here r_2

$$a \longrightarrow a + kr_2, \qquad k \in \mathbb{Z},$$

since it transforms as

$$\zeta'(0;a|\mathbf{r}) \longrightarrow \zeta'(0;a|\mathbf{r}) + 2\pi k,$$

so that the modular symmetry of a with respect to r_2 is preserved. The r_1 dependence has disappeared completely, the reason for this is the following. The formula for the Barnes zeta function for positive r_1 and r_2 obtained in the preceding section cannot be analytically continued to both r_1 and r_2 negative in this naive way, because, for any value of s (and a, r_1, r_2 fixed), once we break the symmetry and do the analytical continuation say in r_2, then the analytical continuation in r_1 is restricted to the only possibility for n_1 of being zero. In fact, put it in other words, when we try to pull out of the r_1 real axis, for any value of $n_1 \neq 0$ we encounter a pole of the initial zeta function, taking n_2 big enough. Now, it is immediate to see that one can interpret the final formula (3.21) as the true analytical continuation for the restriction $n_1 = 0$. In principle there is a way to perform a true analytic continuation in the two variables taking into account the contribution of the poles when crossing the axis (a whole series of them appear). This issue is still under investigation.

3.2.2 *The D-dimensional case*

Using once more Hermite's formula, we obtain the following recurrence for the Barnes zeta function in D dimensions

$$\zeta_B^{(D)}(s;a|\mathbf{r}) = \sum_{\mathbf{n}_D=\mathbf{0}}^{\infty} (a + \mathbf{r}_D \cdot \mathbf{n}_D)^{-s}, \qquad \operatorname{Re} s > D, \tag{3.22}$$

in terms of corresponding ones in $D-1$ dimensions

$$\begin{aligned}
\zeta_B^{(D)}(s;a|\mathbf{r}) &= \frac{r_D^{-1}}{s-1}\zeta_B^{(D-1)}(s-1;a|\mathbf{r}_{D-1}) + \frac{1}{2}\zeta_B^{(D-1)}(s;a|\mathbf{r}_{D-1}) \\
+r_D^{-1}\sum_{\mathbf{n}_{D-1}=\mathbf{0}}^{\infty} & (a+\mathbf{r}_{D-1}\cdot\mathbf{n}_{D-1})^{1-s}\int_0^\infty dx\,\frac{(1+x^2)^{-s/2}\sin(s\arctan x)}{e^{2\pi r_D^{-1}(a+\mathbf{r}_{D-1}\cdot\mathbf{n}_{D-1})x}-1}.
\end{aligned} \tag{3.23}$$

Proceeding as before, we take the derivative at $s=0$, with the result:

$$\begin{aligned}
\zeta_B^{(D)'}(0;a|\mathbf{r}) &= -r_D^{-1}\zeta_B^{(D-1)}(-1;a|\mathbf{r}_{D-1}) - r_D^{-1}\zeta_B^{(D-1)'}(-1;a|\mathbf{r}_{D-1}) \\
+\frac{1}{2}\zeta_B^{(D-1)'}(0;a|\mathbf{r}_{D-1}) &+ \frac{1}{2\pi}\sum_{k=0}^{[D/2-1]}(-1)^k\left(\frac{r_D}{2\pi}\right)^{2k+1}(2k)!\,\zeta(2k+2) \\
&\times \mathrm{FP}\left[\zeta_B^{(D-1)}(2k+1;a|\mathbf{r}_{D-1})\right] + R_D(0;a|\mathbf{r}),
\end{aligned} \tag{3.24}$$

where FP[ ] means the finite part of, and for the remainder term we get the two different expressions:

$$R_D(0;a|\mathbf{r})\begin{cases} = r_D^{-1}\displaystyle\sum_{\mathbf{n}_{D-1}=\mathbf{0}}^{\infty}(a+\mathbf{r}_{D-1}\cdot\mathbf{n}_{D-1})\left[\int_0^\infty dx\,\frac{\arctan x}{e^{2\pi r_D^{-1}(a+\mathbf{r}_{D-1}\cdot\mathbf{n}_{D-1})x}-1}\right. \\ \qquad \left. -\displaystyle\sum_{k=0}^{[D/2-1]}(-1)^k\left(\frac{r_D}{2\pi}\right)^{2k+2}(2k)!\,(a+\mathbf{r}_{D-1}\cdot\mathbf{n}_{D-1})^{-2(k+1)}\right], \\ \sim \dfrac{1}{2\pi}\displaystyle\sum_{k=[D/2]}^{\infty}(-1)^k\left(\frac{r_D}{2\pi}\right)^{2k+1}(2k)!\,\zeta(2k+2)\zeta_B^{(D-1)}(2k+1;a|\mathbf{r}_{D-1}). \end{cases} \tag{3.25}$$

In order to obtain $\zeta_B^{(D)'}(0;a|\mathbf{r})$ we need two derivatives of the $D-1$ zeta function, namely $\zeta_B^{(D-1)'}(0;a|\mathbf{r}_{D-1})$ and $\zeta_B^{(D-1)'}(-1;a|\mathbf{r}_{D-1})$. The derivative at $s=-1$ can be calculated in the same way, just by performing first the expansion of the Barnes zeta function around $s=-1$. For the case $D=2$ (needed in order to initiate the recurrence), the result is the following:

$$\zeta_B^{(2)'}(-1;a|\mathbf{r}) = -r_1\log r_1\;\zeta_B^{(2)}(-1;a|\mathbf{r}) + r_1\left[-\frac{r_1}{4r_2}\zeta_H(-2,a/r_1)\right.$$

$$-\frac{r_1}{2r_2}\zeta_H'(-2,a/r_1)+\frac{1}{2}\zeta_H'(-1,a/r_1)+\frac{r_2}{12r_1}\zeta_H(0,a/r_1)$$
$$-\frac{r_2}{12r_1}\zeta_H'(0,a/r_1)\Bigg]+R_1(-1;a|\mathbf{r}), \qquad (3.26)$$

the remainder $R_1(-1;a|\mathbf{r})$ being given by the asymptotic expansion:

$$R_1(-1;a|\mathbf{r})=\frac{r_1}{\pi}\sum_{k=1}^{\infty}(-1)^k(2k-1)!\zeta(2k+2)\left(\frac{r_2}{2\pi r_1}\right)^{2k+1}\zeta_H(2k,a/r_1), \qquad (3.27)$$

or by the corresponding expression as the difference of an integral and the term $k=0$ of the series (as above). This allows for the calculation of any derivative of the Barnes zeta function in d dimensions recursively (with an accurate numerical determination of the remainder).

3.3 Spectral Zeta Functions for Scalar and Vector Fields on a Spacetime with a Non-commutative Toroidal Part

We shall now consider the physical example of a quantum system consisting of scalars and vector fields on a D-dimensional non-commutative manifold $\mathcal{M}$ of the form $\mathbb{R}^{1,d}\times\mathbb{T}^p_\theta$ (thus $D=d+p+1$). $\mathbb{T}^p_\theta$ is a p-dimensional non-commutative torus. Here θ is the non-commutative parameter.

This physical system has attracted much interest recently, in connection with M-theory and with string theory [A. Connes et al. (1998); M.R. Douglas and C. Hall (1998); N. Seiberg and E. Witten (1999); Y. Cheung and M. Krogh (1998); C.S. Chu and P.M. Ho (1999); V. Schomerus (1999); F. Ardalan et al. (1999)], and also because of the fact that those are perfectly consistent theories by themselves, which could lead to brand new physical situations. It has been shown, in particular, that non-commutative gauge theories describe the low energy excitations of open strings on D-branes in a background Neveu-Schwarz two-form field [A. Connes et al. (1998); M.R. Douglas and C. Hall (1998); N. Seiberg and E. Witten (1999)].

This interesting system provides us with a quite non-trivial case where the formulae derived above are indeed useful. For one, the zeta functions corresponding to bosonic and fermionic operators in this system are of a different kind, never considered before. And, moreover, they can be most conveniently written in terms of the zeta functions in the previous section.

What is also nice is the fact that a unified treatment (with just *one* zeta function) can be given for both cases, the nature of the field appearing there as a simple parameter, together with those corresponding to the numbers of compactified, non-compactified, and non-commutative dimensions of the spacetime.

3.3.1 *Poles of the zeta function*

The spectral zeta function for the corresponding (pseudo)differential operator can be written in the form [A.A. Bytsenko et al. (2001b)]

$$\begin{aligned}\zeta_\alpha(s) &= \frac{V}{(4\pi)^{(d+1)/2}} \frac{\Gamma(s-(d+1)/2)}{\Gamma(s)} \sum_{\mathbf{n}\in\mathbb{Z}^p}{}' Q(\mathbf{n})^{(d+1)/2-s} \\ &\quad \times \left[1+\Lambda\theta^{2-2\alpha}Q(\mathbf{n})^{-\alpha}\right]^{(d+1)/2-s}, \qquad (3.28)\end{aligned}$$

where $V = \mathrm{Vol}\,(\mathbb{R}^{d+1})$, the volume of the non-compact part, $Q(\mathbf{n}) = \sum_{j=1}^{p} a_j n_j^2$, a diagonal quadratic form, and the compactification radii $R_j = a_j^{-1/2}$. Moreover, the value of the parameter $\alpha = 2$ for scalar fields and $\alpha = 3$ for vectors, distinguishes between the two different fields. In the particular case when we set all the compactification radii equal to R, we obtain:

$$\begin{aligned}\zeta_\alpha(s) &= \frac{V}{(4\pi)^{(d+1)/2}} \frac{\Gamma(s-(d+1)/2)}{\Gamma(s)R^{d+1-2s}} \sum_{\mathbf{n}\in\mathbb{Z}^p}{}' I(\mathbf{n})^{(d+1)/2-s} \\ &\quad \times \left[1+\Lambda\theta^{2-2\alpha}R^{2\alpha}I(\mathbf{n})^{-\alpha}\right]^{(d+1)/2-s}, \qquad (3.29)\end{aligned}$$

being now the quadratic form: $I(\mathbf{n}) = \sum_{j=1}^{p} n_j^2$.

After some calculations, this zeta function can be written in terms of the Epstein zeta function of the previous section, with the result:

$$\begin{aligned}\zeta_\alpha(s) &= \frac{V}{(4\pi)^{(d+1)/2}} \sum_{l=0}^{\infty} \frac{\Gamma(s+l-(d+1)/2)}{l!\,\Gamma(s)} (-\Lambda\theta^{2-2\alpha})^l \\ &\quad \times \zeta_{Q,0,0}(s+\alpha l-(d+1)/2), \qquad (3.30)\end{aligned}$$

which reduces, in the particular case of equal radii, to

$$\begin{aligned}\zeta_\alpha(s) &= \frac{V}{(4\pi)^{(d+1)/2}R^{d+1-2s}} \sum_{l=0}^{\infty} \frac{\Gamma(s+l-(d+1)/2)}{l!\,\Gamma(s)} (-\Lambda\theta^{2-2\alpha})^l \\ &\quad \times \zeta_E(s+\alpha l-(d+1)/2), \qquad (3.31)\end{aligned}$$

where we use here the notation $\zeta_E(s) \equiv \zeta_{I,\mathbf{0},0}(s)$, e.g., the Epstein zeta function for the standard quadratic form.

The pole structure of the resulting zeta function deserves a careful analysis. It differs, in fact, very much from all cases that were known in the literature till now. This is not difficult to understand, from the fact that the pole of the Epstein zeta function at $s = p/2 - \alpha k + (d+1)/2 = D/2 - \alpha k$, when combined with the poles of the gamma functions, yields a very rich pattern of singularities for $\zeta_\alpha(s)$, on taking into account the different possible values of the parameters involved. The pole structure is straightforwardly found from the explicit expressions for the zeta functions given above.

Having already given the formula (3.30) above —that contains everything needed to perform such calculation of pole position, residua and finite part— for its importance for the calculation of the determinant and the one-loop effective action from the zeta function, we will here start by specifying what happens at $s = 0$. Remarkably enough, a pole appears in many cases (depending on the values of the different parameters). This will also serve as an illustration of what one has to expect for other values of s. The general case will be left for the following subsection.

It is convenient to classify the different possible subcases according to the values of d and $D = d + p + 1$. We obtain, at $s = 0$, the pole structure given in Table 3.3.1.

$$\text{For } d = 2k \quad \left\{ \begin{array}{lll} \text{if } D \neq \dot{\overline{2\alpha}} & \Longrightarrow & \zeta_\alpha(0) = 0, \\ \text{if } D = \dot{\overline{2\alpha}} & \Longrightarrow & \zeta_\alpha(0) = \text{finite.} \end{array} \right.$$

$$\text{For } d = 2k-1 \left\{ \begin{array}{lll} \text{if } D \neq \dot{\overline{2\alpha}} \left\{ \begin{array}{ll} \text{finite,} & \text{for } l \leq k \\ 0, & \text{for } l > k \end{array} \right\} & \Longrightarrow & \zeta_\alpha(0) = \text{finite,} \\ & & \\ \text{if } D = 2\alpha l \left\{ \begin{array}{ll} \text{pole,} & \text{for } l \leq k \\ \text{finite,} & \text{for } l > k \end{array} \right\} & \Longrightarrow & \zeta_\alpha(0) = \text{pole.} \end{array} \right.$$

Table 3.1 Pole structure of the zeta function $\zeta_\alpha(s)$, at $s = 0$, according to the different possible values of d and D ($\dot{\overline{2\alpha}}$ means *multiple* of 2α).

Here l is the summation index in Eq. (3.30). The appearance of a pole of the zeta function $\zeta_\alpha(s)$, for both values of α, at $s = 0$ is, let us repeat, an absolute novelty, bound to have important physical consequences for the regularization process. It is necessary to observe, that this fact is *not* in contradiction with the well known theorems on the pole structure of a (elliptic) differential operator [P.B. Gilkey (1995)]. The situation that appears in the non-commutative case is completely different. (i) To begin with, we do not have a standard differential operator, but a strictly *pseudodifferential* one, from the beginning. (ii) Moreover, the new spectrum is *not* perturbatively connected (for $\theta \to 0$) with the corresponding one for the commutative case. (iii) And also to be remarked is the fact that one is now working with non-local operators.

3.3.2 *Explicit analytic continuation of* $\zeta_\alpha(s)$

Substituting the corresponding formula, from the preceding section, for the Epstein zeta functions in Eq. (3.30), we obtain the following explicit analytic continuation of $\zeta_\alpha(s)$ ($\alpha = 2, 3$), for bosonic and fermionic fields, to the *whole* complex s-plane:

$$\begin{aligned}
\zeta_\alpha(s) \;&=\; \frac{2^{s-d}V}{(2\pi)^{(d+1)/2}\Gamma(s)} \sum_{l=0}^{\infty} \frac{\Gamma(s+l-(d+1)/2)}{l!\,\Gamma(s+\alpha l-(d+1)/2)} \left(-2^\alpha \Lambda \theta^{2-2\alpha}\right)^l \\
&\times \sum_{j=0}^{p-1} (\det \mathbf{A}_j)^{-1/2} \left[\pi^{j/2} a_{p-j}^{-s-\alpha l+(d+j+1)/2} \right. \\
&\quad \times \Gamma(s+\alpha l-(d+j+1)/2)\, \zeta_R(2s+2\alpha l-d-j-1) \\
&\quad + 4\pi^{s+\alpha l-(d+1)/2} a_{p-j}^{-(s+\alpha l)/2-(d+j+1)/4} n^{(d+j+1)/2-s-\alpha l} \\
&\qquad \sum_{n=1}^{\infty} \sum_{\mathbf{m}_j \in \mathbb{Z}^j}{}' \left(\mathbf{m}_j^t \mathbf{A}_j^{-1} \mathbf{m}_j\right)^{(s+\alpha l)/2-(d+j+1)/4} \\
&\quad \left. \times K_{(d+j+1)/2-s-\alpha l}\left(2\pi n \sqrt{a_{p-j} \mathbf{m}_j^t \mathbf{A}_j^{-1} \mathbf{m}_j}\right)\right]. \qquad (3.32)
\end{aligned}$$

As discussed in the previous subsection in detail, the non-spurious poles of this zeta function are to be found in the terms corresponding to $j = p-1$. With the knowledge we have gained from the analytical continuation of the Epstein zeta functions, the final analysis can be here completed at once. Note that the situation here corresponds to the diagonal case, with $c_1 = \cdots = c_p = q = 0$, of Section 3.1.

To be remarked again is that, what we have in the end, by using our method, is an exponentially fast convergent series of Bessel functions together with a first, finite part, where a pole (simple or double, as we shall see) may show up, for specific values of the dimensions of the different parts of the manifold, depending also on the nature (scalar vs. vectorial) of the fields (the value of α, see Table 1 and Eq. (3.32)).

To summarize the discussion at the end of Section 3.1, the pole structure of Eq. (3.32) is in fact best seen from Eq. (3.30) (for $s = 0$ it has been analyzed in Table 1 already). For a fixed value of the summation index l, the contribution to the only pole of the zeta function $\zeta_E(s+\alpha l-(d+1)/2)$, at $s = D/2 - \alpha l$, comes from the last term of the j-sum only, namely from $j = p-1$. It is easy to check that it yields the corresponding residuum, associated to the second sum in Eq. (3.32). Combined now with the poles of the gamma functions, and taking into account the first series in l, this yields the following expression for the residua of the zeta function $\zeta_\alpha(s)$ at the poles $s = D/2 - \alpha l$, $l = 0, 1, 2, \ldots$

$$\begin{aligned} \text{Res}\ \zeta_\alpha(s)|_{s=D/2-\alpha l} &= \frac{2^{p/2-d}\pi^{(p-d-1)/2}V}{\Gamma(p/2)} \left(\det \mathbf{A}_p\right)^{-1/2} \\ &\times \frac{(-\Lambda\theta^{2-2\alpha})^l}{l!} \frac{\Gamma(p/2+(1-\alpha)l)}{\Gamma(D/2-\alpha l)}, \qquad l = 0, 1, 2, \ldots \end{aligned} \tag{3.33}$$

Actually, depending on D and p being even or odd, completely different situations arise, for different values of l: from the disappearance of the pole, giving rise to a finite contribution, to the appearance of a simple or a double pole. We shall distinguish four different situations and, to simplify the notation, we will denote by U the whole factor in the expression (3.33) for the residuum, that multiplies the last fraction of two gamma functions (in short, Res $\zeta_\alpha = U\,\Gamma_1/\Gamma_2$).

(1) For $D - 2\alpha l = -2h$, $h = 0, -1, -2, \ldots$

  (a) for $p/2 + (1-\alpha)l \neq 0, -1, -2, \ldots \quad \Longrightarrow \quad$ finite,
  Res $\zeta_\alpha = -h!\ U\ \Gamma(p/2 + (1-\alpha)l)$;
  (b) for $p = 2(\alpha - 1)l - 2k$, $k = 0, -1, -2, \ldots \quad \Longrightarrow \quad$ pole,
  Res $\zeta_\alpha = (h!/k!)\ U$.

(2) For $D - 2\alpha l \neq -2h$, $h = 0, -1, -2, \ldots$

  (a) for $p/2 + (1-\alpha)l \neq 0, -1, -2, \ldots \quad \Longrightarrow \quad$ pole,
  Res $\zeta_\alpha = U\ \Gamma(p/2 + (1-\alpha)l)/\Gamma(D/2 + \alpha l)$;

(b) for $p = 2(\alpha - 1)l - 2k$, $k = 0, -1, -2, \ldots \implies$ double pole,
Res $\zeta_\alpha = (-1/k!)\, U\, /\Gamma(D/2 + \alpha l)$.

Note that we here just quote the *generic* situation that occurs for l *large* enough in each case. For instance, if $p = 2$ a double pole appears for $l = 1, 2, \ldots$. For $p = 4$, a double pole appears for $l = 1, 2, \ldots$, if $\alpha = 3$, but only for $l = 2, 3, \ldots$, if $\alpha = 2$. For $p = 6$, a double pole appears for $l = 2, 3, \ldots$, if $\alpha = 3$, but only for $l = 3, 4, \ldots$, if $\alpha = 2$, and so on. The case with both D and p even (what implies d odd) is the most involved one. For $p = 2$ and $D = 4$, for instance, there is a transition from a pole for $l = 0$ corresponding to the zeta function factor, to a pole for $l = 1$ and higher, corresponding to the gamma function in the numerator (the compensation of the pole of the zeta function factor with the one coming from the gamma function in the denominator prevents the formation of a double pole). In any case, the explicit analytic continuation of $\zeta_\alpha(s)$ given by Eq. (3.32) contains *all* the information one needs for calculating the poles and corresponding residua in a straightforward way.

The pole structure can be summarized as in Table 3.3.2.

p \ D	even	odd
odd	(1a) *pole* / finite ($l \geq l_1$)	(2a) *pole* / pole
even	(1b) *double pole* / pole ($l \geq l_1, l_2$)	(2b) *pole* / double pole ($l \geq l_2$)

Table 3.2 General pole structure of the zeta function $\zeta_\alpha(s)$, according to the different possible values of D and p being odd or even. In italics, the type of behavior corresponding to lower values of l is quoted, while the behavior shown in roman characters corresponds to larger values of l.

An application of these formulae to the calculation of the one-loop partition function corresponding to quantum fields at finite temperature, on a non-commutative flat spacetime, will be given in the next Section.

3.4 Applications to Quantum Field Theory in Non-commutative Space

In this Section we will describe an application of the formalism we have developed considering quantum fields propagating in non-commutative spaces.

We shall make use of the analytical regularization illustrated in Chapter 1 and finite-temperature effects will be investigated [A.A. Bytsenko et al. (2001a)].

3.4.1 *Finite-temperature partition function*

As discussed in Chapter 1, in ultrastatic spacetimes, finite-temperature effects can be easily described with the help of the imaginary time formalism —in which the fields are assumed to be periodic with period β. For this reason, we will consider manifolds with topology $\mathcal{M} = S^1 \times \mathbb{R}^{d-1} \times X_\theta^p$, where $p = D-d$ and $\dim \mathcal{M} = D$. In the non compact case $X_\theta^p = \mathbb{R}_\theta^p$, while in the compact case $X_\theta^p = T_\theta^p$ is a non-commutative torus. If we denote by x^μ the non-commutative coordinates, we assume that

$$[x^\mu, x^\nu] = i\theta^{\mu\nu}, \tag{3.34}$$

with $\mu, \nu = 1, ..., p$, and $\theta^{\mu\nu} = \theta\sigma^{\mu\nu}$ is a real, non-singular, antisymmetric matrix with entries ± 1, being θ the non-commutative parameter. X_θ^p may be regarded as a non-commutative associative algebra, with elements given by ordinary continuous functions on X_θ^p whose product is given by the Moyal bracket or $(\star)-$product of functions:

$$(F \star G)(x) = \exp\left(\frac{i}{2}\theta^{\mu\nu}\frac{\partial}{\partial\alpha^\mu}\frac{\partial}{\partial\beta^\nu}\right) F(x+\alpha)G(x+\beta)|_{\alpha=\beta=0}. \tag{3.35}$$

The action of the massless interacting scalar fields we are considering has the form

$$S_{(\text{scalar})} = \int d^D x \left(\frac{1}{2}(\partial\phi)^2 + \frac{\lambda}{r!}\phi \star \phi \star ... \star \phi\right). \tag{3.36}$$

Since we have to deal with a renormalizable scalar field theory, the number r of scalar fields in the $(\star)-$product should be chosen as an integer, given by $r = 2D/(D-2)$. Clearly, the only possible choices are $D = 3$, $r = 6$, $D = 4, r = 4$, and $D = 6$, $r = 3$.

We are interested in the one-loop approximation for a theory with the above action. It has been shown that in the massless case, the operator A, related to the one-loop contribution, has the formal aspect [S. Minwalla et al. (2000)]:

$$A = k^2 + \frac{\Lambda}{k \circ k} + M^2, \tag{3.37}$$

where k^2 is the D-dimensional Laplacian on $\mathbb{R}^D$ and $k \circ k = \theta^2 k_\mu k_\mu$. The effective mass term M^2 depends on λ and corresponds to the planar diagram contributions, while the non-local λ term represents the non-planar diagram contribution.

We are also assuming that the time coordinate is commutative (unitarity of the field theory) and we denote by $\tau = -it$ the compactified imaginary time with period β. Thus, the finite-temperature one-loop operator is

$$A = L_d + L_p + \Lambda(\theta^2 L_p)^{-1}, \tag{3.38}$$

where $L_d = -\partial_\tau^2 + L_{d-1} = -\partial_\tau^2 + k_j^2 + M^2$ and $L_p = k_\mu^2$ are the Laplacian-like operators on the spaces $S^1 \times \mathbb{R}^{d-1}$ and X_θ^p, respectively.

We observe that in the *massive* case, the one-loop effective operator is much more complicated. It reads

$$A = k^2 + m^2 + \frac{2\Lambda m}{(k \circ k)^{3/2}} K_1(2m(k \circ k)^{-1/2}) + M^2. \tag{3.39}$$

This appears as an intractable form. Since $K_1(z) \to 1/z$ as $z \to 0$, in the massless limit, one obtains Eq. (3.37). If we take the commutative limit, namely $\theta \to 0$ in Eq. (3.39), one gets $A = k^2 + m^2 + M^2$, which is the correct one-loop operator in the commutative case. In [S. Minwalla et al. (2000)], an approximate expression valid for small m has been considered and a mixing between UV and IR divergences has been noticed. It should be noted that the massless limit and the limit $\theta \to 0$ do not commute, namely it is not possible to consider the $\theta \to 0$ in Eq. (3.37). Here, for the sake of simplicity, we shall consider the massless case only, bearing in mind possible limitations of this choice.

We shall make use of the zeta function regularization discussed in Chapter 1. This means that the one-loop effective partition function is given by means of Eq. (1.16) or (1.27), with $\nu = 1/2$, since we are dealing with neutral scalar fields.

We are going to compute a functional determinant, and we are effectively adding up an infinite number of one-loop diagrams, planar and non-planar ones. Our calculation should be viewed as an attempt at an implementation of the background field method to the non-commutative case.

In order to go on, we need the evaluation of the analytic continuation of the zeta function $\zeta(s|A)$. The results of Chapter 1 give

$$\zeta(s|A) = \frac{\beta}{(4\pi)^{\frac{1}{2}}\Gamma(s)}\Bigg(\Gamma(s-\frac{1}{2})\zeta(s-\frac{1}{2}|L_Y) + 2\sum_{n=1}^{\infty}\int_0^{\infty} dt t^{s-\frac{3}{2}} e^{-\frac{n^2\beta^2}{4t}}\omega(t|L_Y)\Bigg). \tag{3.40}$$

Here $L_Y = L_{d-1}+L_X$. In order to evaluate the high-temperature expansion the second term on the r.h.s. of Eq. (3.40) can be conveniently rewritten as a Mellin-Barnes integral. As a result, we have

$$\zeta(s|A) = \frac{\beta}{(4\pi)^{\frac{1}{2}}\Gamma(s)}\Bigg(\Gamma(s-\frac{1}{2})\zeta(s-\frac{1}{2}|L_Y) + \frac{1}{2\pi i}\int_{\mathrm{Re}\, z=\frac{D}{2}} dz \left(\frac{\beta}{2}\right)^{-z}\Gamma(\frac{z}{2})\zeta_R(z)\Gamma(s+\frac{z-1}{2})\zeta(s+\frac{z-1}{2}|L_Y)\Bigg), \tag{3.41}$$

where $\zeta_R(z)$ is the Riemann zeta function. In the equations above, the first part represents the vacuum contribution, while the second one is the statistical sum contribution. It is also clear that one needs to know the meromorphic structure of the zeta function $\zeta(s|L_Y)$, which can be obtained from $\zeta(s|L_X)$. Making an expansion in terms of the effective mass M^2, a direct computation gives

$$\zeta(z|L_Y) = \frac{\mathrm{Vol}(\mathbb{R}^{d-1})}{(4\pi)^{\frac{d-1}{2}}}\sum_{k=0}^{\infty}(-M^2)^k\frac{\Gamma(z+k-\frac{d-1}{2})}{k!\Gamma(z)}\zeta(z+k-\frac{d-1}{2}|L_X). \tag{3.42}$$

3.4.2 *The spectral zeta function and the regularized vacuum energy*

First of all, let us consider the simple case of a non-compact, non-commutative manifold $X_\theta^p = \mathbb{R}_\theta^p$ (i.e., the non-commutative Euclidean space). The heat-kernel trace reads

$$\mathrm{Tr} e^{-t|L_X} = \frac{2\mathrm{Vol}(\mathbb{R}^p)}{(4\pi)^{p/2}\Gamma(p/2)} a^{p/2}K_{p/2}(2at), \tag{3.43}$$

where $K_\nu(z)$ is the MacDonald function (or modified Bessel function of the third kind), and $a^2 = \Lambda\theta^{-2}$.

It is easy to show that the short-t asymptotics of this heat-kernel trace, for p even contains logarithmic terms in t. Thus, the associated zeta function $\zeta(z|L_Y)$ may not be regular at the origin. In fact, the Mellin transform leads to

$$\zeta(z|L_X) = \frac{\mathrm{Vol}(\mathbb{R}^p)}{2(4\pi)^{p/2}} \frac{\Gamma(\frac{2z+p}{4})\Gamma(\frac{2z-p}{4})}{\Gamma(p/2)\Gamma(z)} a^{p/2-z}. \tag{3.44}$$

Equation (3.44) gives the analytic continuation of the zeta function.

In the case of a compact non-commutative manifold (i.e., the non-commutative tori) $X_\theta^p = T_\theta^p$, for large Re s, the spectral zeta-function associated with the operator L_X reads [J. Gomis et al. (2000a); S. Nam (2000); A.A. Bytsenko et al. (2001b)]:

$$\zeta(s|L_X) = \sum_{\mathbf{n}\in\mathbb{Z}^p/\{\mathbf{0}\}} \varphi(\mathbf{n})^{-s}\left[1+\Lambda\theta^{-2}R^4\varphi(\mathbf{n})^{-2}\right]^{-s}, \tag{3.45}$$

where R is the compactification radius. The analytic continuation can be achieved by using binomial expansion ([E. Elizalde (1994b); E. Elizalde et al. (1994); E. Elizalde (1995); A.A. Bytsenko et al. (1996)]). As a result, for $C = a^2R^4 < 1$ one has

$$\zeta(s|L_X) = R^{2s}\sum_{\ell=0}^{\infty}\frac{(-C)^\ell\Gamma(s+\ell)}{\ell!\Gamma(s)} Z_p\left|\begin{matrix}\mathbf{0}\\ \mathbf{0}\end{matrix}\right|(2s+4\ell), \tag{3.46}$$

where $Z_p\left|{}^{\mathbf{g}}_{\mathbf{h}}\right|(s,\varphi)$ is the p-dimensional Epstein zeta function associated with the quadratic form $\varphi[a(\mathbf{n}+\mathbf{g})] = \sum_j a_j(n_j+g_j)^2$. For Re $s > p$, $Z_p\left|{}^{\mathbf{g}}_{\mathbf{h}}\right|(s,\varphi)$ is given by the formula

$$Z_p\left|\begin{matrix}g_1\dots g_p\\ h_1\dots h_p\end{matrix}\right|(s,\varphi) = {\sum_{\mathbf{n}\in\mathbf{Z}^p}}'\left(\varphi[a(\mathbf{n}+\mathbf{g})]\right)^{-s/2}\exp\left[2\pi i(\mathbf{n},\mathbf{h})\right], \tag{3.47}$$

where g_j and h_j are some real numbers [Bateman (1955)], and the prime means omitting the term with $(n_1, n_2, ..., n_p) = (g_1, g_2, ..., g_p)$ if all of the g_j are integers. The functional equation for $Z_p\left|{}^{\mathbf{g}}_{\mathbf{h}}\right|(s,\varphi)$ reads

$$\begin{aligned} Z_p\left|\begin{matrix}\mathbf{g}\\ \mathbf{h}\end{matrix}\right|(s,\varphi) &= (\det a)^{-1/2}\pi^{\frac{1}{2}(2s-p)}\frac{\Gamma(\frac{p-s}{2})}{\Gamma(\frac{s}{2})}\exp[-2\pi i(\mathbf{g},\mathbf{h})] \\ &\times Z_p\left|\begin{matrix}\mathbf{h}\\ -\mathbf{g}\end{matrix}\right|(p-s,\varphi^*), \end{aligned} \tag{3.48}$$

where $\varphi^*[a(\mathbf{n}+\mathbf{g})] = \sum_j a_j^{-1}(n_j+g_j)^2$. Formulas (3.46) and (3.48) give the analytic continuation of the zeta function.

Alternatively, a powerful expression for the analytic continuation, that extends (to arbitrary number of dimensions and to more general Epstein-like zeta functions) the Chowla-Selberg formula for the homogeneous Epstein zeta function in two dimensions, has been given recently [E. Elizalde (1994c); E. Elizalde (1999); E. Elizalde (1998)]. It provides an expression valid in the whole of the complex plane, exhibiting the poles explicitly, and under the form of a power-fast convergent series. For the compact case, Eq. (3.45), the formula reads:

$$\zeta(s|L_Y) = \frac{2^{s-d+2}\mathrm{Vol}(\mathbb{R}^{d-1})}{(2\pi)^{(d-1)/2}\Gamma(s)} \sum_{\ell=0}^{\infty}\sum_{j=0}^{p-1} \frac{(-4\Lambda\theta^{-2})^\ell \Gamma(s+\ell-\frac{d-1}{2})}{\ell!\,\Gamma(s+2\ell-\frac{d-1}{2})\,(\det \mathbf{A}_j)^{1/2}}$$
$$\times \Bigg[\pi^{j/2} a_{p-j}^{-s-2\ell+(d+j-1)/2} \Gamma(s+2\ell-\frac{d+j-1}{2}) \zeta_R(2s+4\ell-d-j+1)$$
$$+4\pi^{s+2\ell-(d-1)/2} a_{p-j}^{-s/2-\ell-(d+j-1)/4} \sum_{n=1}^{\infty} \sum_{\mathbf{m}_j\in\mathbb{Z}^j}{}' n^{(d+j-1)/2-s-2\ell}$$
$$\times \left(\mathbf{m}_j^t \mathbf{A}_j^{-1}\mathbf{m}_j\right)^{s/2+\ell-(d+j-1)/4} K_{(d+j-1)/2-s-2\ell}\left(2\pi n\sqrt{a_{p-j}\mathbf{m}_j^t\mathbf{A}_j^{-1}\mathbf{m}_j}\right)\Bigg], \tag{3.49}$$

where $a_j = R_j^{-2}$, R_j being the compactification radius corresponding to the jth coordinate (above, and in what follows, only the particular case $R_j = R, \forall j$, is considered). Note that the term $j = 0$ in the second sum reduces to: $a_p^{s-2\ell}\Gamma(s+2\ell)\zeta_R(2s+2\ell)$, which coincides with the result of the 1-dimensional case ($p = 1$), as it should. Here $\mathbf{A}_j$ is the submatrix of $\mathbf{A} = \mathrm{diag}\,(a_1,\ldots,a_p)$ made up of the last j rows and columns: $\mathbf{A}_j = \mathrm{diag}\ (a_{p-j+1},\ldots,a_p)$.

3.4.3 *The regularized vacuum energy*

In order to evaluate the regularized vacuum energy, one needs the meromorphic structure of the zeta functions $\zeta(z|A)$ and $\zeta(z|L_X)$. Since we are interested in the vacuum contribution, only the linear term in β is relevant.

In our situation, i.e. in the case of massless scalar field defined on a D manifold with non-commuting coordinates, the relevant choices are $D = 3$, 4, 6. For the sake of simplicity, let us consider only the non-compact case. Making use of Eqs. (3.42) and (3.44), we get

$$\zeta(s|\ell^2 A) = \frac{\beta \mathrm{Vol}(\mathbb{R}^{D-1})(\ell^2)^s}{(4\pi)^{\frac{D}{2}}\Gamma(s)} \sum_{k=0}^{\infty} \frac{(-M^2)^k}{k!}\Gamma(\frac{2s+2k+p-d}{4}) \times \Gamma(\frac{2s+2k-D}{4}) + \zeta_\beta(s|A) \, , \tag{3.50}$$

where $\zeta_\beta(s|A)$ represents the statistical sum part, which gives a vanishing contribution in the zero-temperature limit. The first part in the above expression gives the analytic continuation of vacuum contribution of the zeta function in an explicit way. We will see that, for particular values of D, p and d, there exists a simple pole at the origin. In these cases, we have to make use of the generalized formula for the evaluation of the vacuum energy. Let us start with the simplest case, using Eq. (1.67) in Chapter 1. For example, if $D = 3,\ p = 2,\ \zeta(s|A)$ is regular at the origin and $\zeta(0|A) = 0$. Thus the vacuum energy is given simply by

$$E_V = \frac{\mathrm{Vol}(\mathbb{R}^2)a^{3/2}}{32\pi^{3/2}} \sum_{k=0}^{\infty} \frac{(-\tilde{M}^2)^k}{k!}\Gamma\left(\frac{2k+1}{4}\right)\Gamma\left(\frac{2k-3}{4}\right) . \tag{3.51}$$

We have introduced $\tilde{M}^2 = M^2/a$ and this factor is dimensionless and independent of the coupling constant.

If $D = 6$, $d = 2$ and $p = 4$, $\zeta(s|A)$ is regular at the origin, but $\zeta(0|A)$ is not vanishing. A direct calculation leads to

$$E_V = -\frac{\mathrm{Vol}(\mathbb{R}^5)a^3}{256\pi^3}\left[\sum_{k=0,\ k\neq 1,3}^{\infty} \frac{(-\tilde{M}^2)^k}{k!}\Gamma\left(\frac{k+1}{2}\right)\Gamma\left(\frac{k-3}{2}\right) -2\tilde{M}^2(\frac{\tilde{M}^4}{6}-1)\log(a\ell^2) + \tilde{M}^2(1-\gamma)\right] . \tag{3.52}$$

The situation changes for $D = 6\ \ (d = 4,\ \ p = 2)$ and $D = 4\ \ (p = d = 2)$. In these cases the zeta function has a pole at the origin and one has to make use of Eq. (1.71), derived in Chapter 1. Let us first consider the case $D = 6,\ d = 4,\ p = 2$. We have

$$E_V = -\frac{\mathrm{Vol}(\mathbb{R}^5)a^3}{256\pi^3}\left[\sum_{k=0,\ k\neq 1,3}^{\infty} \frac{(-\tilde{M}^2)^k}{k!}\Gamma\left(\frac{k-1}{2}\right)\Gamma\left(\frac{k-3}{2}\right)\right.$$

$$+\frac{\tilde{M}^2}{2}\left(C''(0)-2C'(0)\log(a\ell^2)+C(0)(\log a\ell^2)^2\right)\Bigg], \quad (3.53)$$

where

$$C(s)=-\frac{8\Gamma^2(1+s/2)}{\Gamma(1+s)(s-2)}-s\frac{\tilde{M}^4}{6}\Gamma(1+s/2)\,. \quad (3.54)$$

In the other case, $D=4$, $p=d=2$, we get

$$\begin{aligned} E_V &= -\frac{\mathrm{Vol}(\mathbb{R}^3)a^2}{64\pi^2}\Bigg[\sum_{k=1,\ k\neq 2}^{\infty}\frac{(-\tilde{M}^2)^k}{k!}\Gamma\left(\frac{k}{2}\right)\Gamma\left(\frac{k-2}{2}\right) \\ &\quad +\frac{1}{2}(D''(0)-2D'(0)\log(a\ell^2)+D(0)(\log a\ell^2)^2)\Bigg], \end{aligned} \quad (3.55)$$

where now

$$D(s)=\frac{8\Gamma^2(1+s/2)}{\Gamma(1+s)(s-2)}+s\frac{\tilde{M}^4}{2}\Gamma(1+s/2)\,. \quad (3.56)$$

Our evaluation of the functional determinant corresponding to the thermal fluctuation operator has also led to a non-analytic dependence on the coupling constant λ, corresponding to the resummation of all the one-loop diagrams. As is well known, higher-loop diagrams could, in principle, give a contribution of the same order. In any case, our computation is the necessary first step towards the final result.

In the compact case, we only observe that we have qualitatively the same results, but they are much more involved. Here, the sign of vacuum energy is relevant for issues concerning the stabilization mechanism of the compactification radius. However, it is to be noticed that its value depends on the renormalization parameter ℓ and on the number of non-commuting coordinates.

3.4.4 *High-temperature expansion*

To obtain the high-temperature expansion, $\beta\to 0$, one may use Eq. (3.41). We consider only the case $D=4$ since for the other cases, D odd or $D=6$, the high-temperature expansion is standard, in its first leading terms.

For suitable s, we shift the vertical contour in the second term of Eq. (3.41) to the left. There are several simple poles and double poles in z. The simple pole at $z=0$ gives a contribution which cancels with the first term on the r.h.s. of Eq. (3.41). Then there are simple poles at $z = 1$ and $z = 4 - 2r - 2s$. The double poles occur at $z = -4k - 2s$. As a result, the thermal zeta function reads

$$
\begin{aligned}
\zeta(s|A) &= \zeta(s|L_Y) + \frac{\beta}{(4\pi)^{\frac{1}{2}}\Gamma(s)}\left[\left(\frac{\beta}{2}\right)^{-4} A_0(L_Y)\zeta_R(4)\right. \\
&\quad \left. + \left(\frac{\beta}{2}\right)^{-2} A_1(L_Y)\zeta_R(2) + A_2(L_Y)\zeta_R(-1) + \mathcal{O}(s)\right] \\
&\quad - \frac{\beta^{2s+1}B_0(L_Y)}{(2\pi)^{2s+1}\Gamma(s)}\left[-\log\left(\frac{\beta}{2\pi}\right)\Gamma(s+1/2)\zeta_R(1+2s)\right. \\
&\quad \left. -\frac{1}{2}\frac{d}{ds}\Gamma(s+1/2)\zeta_R(1+2s) - \Gamma(s+1/2)\frac{d}{ds}\zeta_R(1+2s)\right] + \frac{\mathcal{O}(\beta^5)}{\Gamma(s)} .
\end{aligned}
\tag{3.57}
$$

Here, $A_j(L_Y)$ are obviously some coefficients, dependent on L_Y. It should be noted that the high-temperature expansion is still singular at $s = 0$, due to the $(d/ds)\zeta_R(1+2s)$ term, and that, strictly speaking, one should make use of the regularized expression in order to compute the first quantum correction to the partition function. However, this term originates a subleading contribution, of order zero in β, the leading contributions being similar to those in the non-commutative case.

As a closing comment, we have to remark that the difficulties posed by the appearance of the non-commutative contributions in the spectral functions (which depend on it non-analytically) should not be underestimated. This calculation should be viewed as an attempt at an implementation of the background field method to the non-commutative case.

On the positive side, this has as a consequence that the quantum theory at finite temperature obtained by 'addition' of the non-commutative dimensions, departs appreciably from the commutative case. This is true concerning the mathematics of the problem: new poles in the zeta function, new asymptotic expansions, as well as the physics of the model. In this chapter we have investigated the changes suffered by the vacuum energy density in the cases considered. After obtaining the general formula for the family of spacetimes at issue, a new prescription for the regularized vacuum energy has been obtained, and this new prescription has been exemplified with the explicit expressions corresponding to dimensions $D = 3$, 4, 6, where changes in the coefficients of the corresponding expressions of the regularized vacuum energy density are notorious.

One should also notice that, in the case of the presence of the pole at the origin for the zeta function, the dependence on the renormalization parameter in the vacuum energy (or in the effective action) is quite complicated, and, as a consequence, the one-loop renormalization group equations will be different from the standard ones. Furthermore, making use of Eq. (A.10), contained in the Appendix of [E. Elizalde et al. (1998d)], it is possible to show that the use of a different regularization, for example the cut-off regularization, leads to the same conclusions. We have also discussed in the above sections how to obtain the high-temperature expansion for the partition function and we have shown that new subleading terms, associated with the non-commutative part of the operator spectrum, are present. An aim for the future would be to try to approach an experimental situation where one could actually check the non-commutative results vs the ones for the ordinary, commutative case.

Chapter 4

Spectral Functions of Laplace Operator on Locally Symmetric Spaces

In the present Chapter we investigate spectral analysis on locally symmetric spaces of rank one. This class of spaces includes several cases of interest with regard to the physical applications. We will treat in detail a specific example, real hyperbolic manifolds, which may be interesting as a cosmological model.

4.1 Locally Symmetric Spaces of Rank One

Let $(\mathcal{M}, g)$ be a locally symmetric Riemannian manifold. Its universal covering $X \to \mathcal{M}$ is a Riemannian symmetric space of rank one. The group of orientation preserving isometries $G^{'}$ of X is a connected semisimple Lie group of real rank one; $X = G^{'}/K^{'}$, where $K^{'}$ is the maximal compact subgroup of $G^{'}$. The fundamental group π of $\mathcal{M}$ acts by covering transformations on X and gives rise to a discrete (co-compact) subgroup $\Gamma \subset G^{'}$. Let G be a linear connected finite covering of $G^{'}$ such that the embedding $\Gamma \hookrightarrow G^{'}$ lifts to an embedding $\Gamma \hookrightarrow G$. Then $X = G/K$, where K is the maximal compact subgroup of G [S. Helgason (1962)]. Up to local isomorphism we can represent X by the following quotients:

$$X = \left[\begin{array}{ll} SO_1(n,1)/SO(n) & \text{(I)} \\ SU(n,1)/U(n) & \text{(II)} \\ SP(n,1)/(SP(n) \otimes SP(1)) & \text{(III)} \\ F_{4(-20)}/Spin(9) & \text{(IV)} \end{array} \right], \tag{4.1}$$

where $D \equiv \dim X = n, 2n, 4n, 16$ respectively. Then (see for detail [F.L. Williams (1997)])

$$SO(p,q) \overset{def}{=} \left\{ g \in GL(p+q,\mathbb{R} \left| \begin{matrix} g^t I_{pq} g = I_{pq} \\ \det g = 1 \end{matrix} \right. \right\}, \tag{4.2}$$

$$SU(p,q) \overset{def}{=} \left\{ g \in GL(p+q,\mathbb{C}) \left| \begin{matrix} g^t I_{pq} \overline{g} = I_{pq} \\ \det g = 1 \end{matrix} \right. \right\}, \tag{4.3}$$

$$SP(p,q) \overset{def}{=} \left\{ g \in GL(2(p+q),\mathbb{C}) \left| \begin{matrix} g^t J_{p+q} g = J_{p+q} \\ g^t K_{pq} \overline{g} = K_{pq} \end{matrix} \right. \right\}, \tag{4.4}$$

where I_n is the identity matrix of order n and

$$I_{pq} = \begin{pmatrix} -I_p & 0 \\ 0 & I_q \end{pmatrix}, \quad J_n = \begin{pmatrix} 0 & I_n \\ -I_n & 0 \end{pmatrix}, \quad K_{pq} = \begin{pmatrix} I_{pq} & 0 \\ 0 & I_{pq} \end{pmatrix}. \tag{4.5}$$

The groups $SU(p,q), SP(p,q)$ are connected; the group $SO_1(p,q)$ is defined as the connected component of the identity in $SO(p,q)$ while $F_{4(-20)}$ is the unique real form of F_4 with Dynkin diagram

$$\bigcirc - \bigcirc = \bigcirc - \bigcirc \tag{4.6}$$

for which the character $(\dim X - \dim K)$ assumes the value (-20) [S. Helgason (1962)]. We assume that if $G = SO(m,1)$ or $SU(q,1)$ then m is even and q is odd.

Let $\mathbf{g} = \mathfrak{k} \oplus \mathfrak{p}$ be a Cartan decomposition of the Lie algebra $\mathbf{g}$ of G. Let $\mathfrak{a} \subset \mathfrak{p}$ be a one-dimensional subspace and let $J = K \bigcap G_{\mathfrak{a}}$ be the centralizer of $\mathfrak{a}$ in K. Fixing a positive root system of $(\mathbf{g}, \mathfrak{a})$ we have an Iwasawa decomposition $\mathbf{g} = \mathfrak{k} \oplus \mathfrak{a} \oplus \mathfrak{n}$. We shall need the real number ρ_0 which corresponds to $1/2$ the sum of the positive real restricted roots of G with respect to a nilpotent factor N of G in an Iwasawa decomposition $G = KAN$. ρ_0 is given by $\rho_0 = (n-1)/2, n, 2n+1, 11$ respectively in the cases (I) to (IV). For details on these matters the reader may consult [S. Helgason (1962)].

The spherical harmonic analysis on X is controlled by the Harish - Chandra's Plancherel density $|C(r)|^{-2}$, a function on the real numbers $\mathbb{R}$, computed by Miatello [R. Miatello (1976); R. Miatello (1979); R. Miatello (1980)], and others, in the rank–one case we are considering. We choose a normalization of Haar measure on G however, which differs from that of [R. Miatello (1976); R. Miatello (1979); R. Miatello (1980)] (see

[F.L. Williams (1998)]). For a suitable constant C_G depending only on G, and for a suitable even polynomial $P(r)$ of degree $D-2$ for $G \neq SO_1(n,1)$ with n odd, and of degree $D-1=2m$ for $G = SO_1(2m+1,1)$, $|C(r)|^{-2}$ is given by

$$|C(r)|^{-2} = \begin{bmatrix} C_G\pi r P(r)\tanh(\pi r) & \text{for } G = SO_1(2m,1), & \\ C_G\pi r P(r)\tanh(\pi r/2) & \text{for } G = SU(n,1), & n \text{ odd} \\ & \text{or } G = SP(n,1), & F_{4(-20)}, \\ C_G\pi r P(r)\coth(\pi r/2) & \text{for } G = SU(n,1), & n \text{ even}, \\ C_G\pi P(r) & \text{for } G = SO_1(2m+1,1) & \end{bmatrix}. \tag{4.7}$$

The coefficients of $P(r)$ will be denoted by a_{2j}:

$$\begin{aligned} P(r) &= \sum_{j=0}^{D/2-1} a_{2j} r^{2j} \qquad \text{for } G \neq SO_1(2m+1,1), \\ &= \sum_{j=0}^{m} a_{2j} r^{2j} \qquad \text{for } G = SO_1(2m+1,1). \end{aligned} \tag{4.8}$$

The constant C_G and the Miatello coefficients a_{2j} of the polynomials $P(r)$ appear in Eqs. (4.7) and (4.8) respectively. C_G and $P(r)$ for the various rank one simple groups G of this paper are given in the following table:

Structure of the Harish-Chandra-Plancherel measure

G	C_G	P(r)
$SO_1(n,1),\ n \geq 2$	$\left[2^{2n-4}\Gamma\left(\frac{n}{2}\right)^2\right]^{-1}$	$\prod_{j=0}^{n-2}\left[r^2 + \frac{(2j+1)^2}{4}\right],\ n = 2m$
		$\prod_{j=0}^{n-1}\left[r^2 + j^2\right],\ n = 2m+1$
$SU(n,1),\ n \geq 2$	$\left[2^{2n-1}\Gamma(n)^2\right]^{-1}$	$\prod_{j=1}^{n-1}\left[\frac{r^2}{4} + \frac{(n-2j)^2}{4}\right]$
$SP(n,1),\ n \geq 2$	$\left[2^{4n+1}\Gamma(2n)^2\right]^{-1}$	$\left[\frac{r^2}{4} + \frac{1}{4}\right]\prod_{j=3}^{n+1}\left[\frac{r^2}{4} + \left(n-j+\frac{3}{2}\right)^2\right]$
		$\times\left[\frac{r^2}{4} + \left(n-j+\frac{5}{2}\right)^2\right]$
$F_{4(-20)}$	$\left[2^{21}\Gamma(8)^2\right]^{-1}$	$\left[\frac{r^2}{4} + \frac{1}{4}\right]\left[\frac{r^2}{4} + \frac{9}{4}\right]$
		$\times\prod_{j=0}^{4}\left[\frac{r^2}{4} + \left(\frac{2j+1}{2}\right)^2\right]$

Table 4.1

4.2 The Spectral Zeta Function

Let $\Gamma \subset G$ be a discrete, co-compact torsion free subgroup. Then $X_\Gamma = \Gamma\backslash\mathcal{M}$ is a compact Riemannian manifold with fundamental group Γ; namely X_Γ is a compact locally symmetric space. Given a finite-dimensional unitary representation χ of Γ, there is the corresponding vector bundle $V_\chi \to X_\Gamma$ over X_Γ, given by $V_\chi = \Gamma\backslash(\mathcal{M} \otimes F_\chi)$, where F_χ (the fiber of V_χ) is the representation space of χ and where Γ acts on $\mathcal{M} \otimes F_\chi$ by the rule $\gamma \cdot (m, f) = (\gamma \cdot m, \chi(\gamma)f)$ for $(\gamma, m, f) \in (\Gamma \otimes \mathcal{M} \otimes F_\chi)$. Let Δ_Γ be the Laplace–Beltrami operator of X acting on smooth sections of V_χ; we obtain Δ_Γ by projecting the Laplace–Beltrami operator of $\mathcal{M}$ (which is G-invariant and thus Γ-invariant) to X_Γ.

As X_Γ is compact, we can consider the spectrum $\{\lambda_j = \lambda_j(\chi),\ n_j = n_j(\chi)\}_{j=0}^{\infty}$ of $L = -\Delta_\Gamma$, where n_j is the (finite) multiplicity of the eigenvalue λ_j. We use the minus preceding Δ_Γ to have the $\lambda_j \geq 0:\ 0 = \lambda_0 < \lambda_1 < \lambda_2\ ...;\ \lim_{j\to\infty} \lambda_j = \infty$.

The spectral zeta function $\zeta_\Gamma(s|L)$ of X_Γ was introduced by Minakshisundaram-Pleijel [S. Minakshisundaram and A. Pleijel (1949)]. We shall write it in the form

$$\zeta_\Gamma(s|L) = \sum_{j=1}^{\infty} n_j(\chi)\lambda_j(\chi)^{-s}\,, \tag{4.9}$$

for $\mathrm{Re}\, s >> 0$. The zeta function is a holomorphic function on the domain $\mathrm{Re}\, s > D/2$, and, as we known, it admits a meromorphic continuation to the full complex plane $\mathbb{C}$. However, since the manifold X_Γ is quite special, it is desirable to have the meromorphic continuation of $\zeta_\Gamma(s|L)$ in an explicit form, for example in terms of the structure of G and Γ. To state these results, we introduce further notation.

As pointed out earlier, the explicit meromorphic structure of the zeta function $\zeta_\Gamma(s|L)$ of (4.9) is worked out in [F.L. Williams (1998)], in terms of the spherical harmonic analysis of G and Γ-structure (see Theorems 4.2, 5.1 there); also compare Theorems 5.2, Eq. (6.1), and Theorem 6.9 of [F.L. Williams (1997)]. In particular, apart from the case $G = SO_1(n, 1)$ with n odd (a case which we treat separately), $\zeta_\Gamma(s|L)$ is holomorphic except for possibly simple poles at $s = 1, 2, ..., D/2$. By Theorem 5.1 of [F.L. Williams (1998)], or by the results stated in [F.L. Williams (1997)], we can compute the residues at these points $s = 1, 2, ..., D/2$. The results

[A.A. Bytsenko and F.L. Williams (1999)] are the following, where we omit the cotangent case.

Theorem 4.1 *Apart from the cases $SO_1(\ell,1)$, $SU(q,1)$ with ℓ odd and q even, the residue of $\zeta_\Gamma(s|L)$ at $s=m$ (for m an integer, $1\le m\le D/2$) equals*

$$\frac{1}{4}\chi(1)\mathrm{Vol}(\Gamma\backslash G)C_G\sum_{j=0}^{D/2-m}(-1)^j\binom{m+j-1}{j}\rho_0^{2j}a_{2(m+j-1)}, \tag{4.10}$$

in the preceding notation. Also for $n=1,2,...$,

$$\zeta_\Gamma(-n|L)=\frac{1}{4}\chi(1)\mathrm{Vol}(\Gamma\backslash G)C_G\left[\sum_{j=0}^{D/2-1}\frac{(-1)^{j+1}j!\rho_0^{2(j+n+1)}a_{2j}}{(n+1)(n+2)\cdots(n+j+1)}\right.$$

$$\left.+2\sum_{j=0}^{D/2-1}\sum_{k=0}^{n}\frac{(-1)^k n!}{(n-k)!}\rho_0^{2(n-k)}b_{k+1}(j)a_{2j}\right], \tag{4.11}$$

where

$$b_r(j)\stackrel{def}{=}\left[2^{1-2(r+j)}-1\right]\left[\frac{\pi}{a(G)}\right]^{2(r+j)}\frac{(-1)^jB_{2(r+j)}}{2(r+j)[(r-1)!]}, \tag{4.12}$$

for $r=1,2,...$, B_m being the m-th Bernoulli number, and for

$$a(G)\stackrel{def}{=}\left[\begin{array}{ll}\pi & \text{if } G=SO_1(\ell,1) \text{ with } \ell \text{ even,}\\ \frac{\pi}{2} & \text{if } G=SU(q,1) \text{ with } q \text{ odd}\\ & \text{or } G=SP(\ell,1), \text{ any } \ell,\ F_{4(-20)}\end{array}\right]. \tag{4.13}$$

For $n=0$ there is an additional term to Eq. (4.11) which is $-n_0(\chi)$.

Now, we consider the case $G=SO_1(\ell,1)$ with ℓ odd. By the results of [F.L. Williams (1998)], for $G=SO_1(2n+1,1)$, $\zeta_\Gamma(s|L)$ has at most a simple pole at the points $s=D/2-k$, $k=0,1,2,....$ Moreover, the following theorem holds [A.A. Bytsenko and F.L. Williams (1999)]

Theorem 4.2 *For $G=SO_1(2n+1,1)$, the residue of $\zeta_\Gamma(s|L)$ at $s=D/2-k$ (where $D/2=n+1/2$, $k=0,1,2,...$) equals*

$$\frac{1}{4}\chi(1)\mathrm{Vol}(\Gamma\backslash G)C_G\sum_{j=0}^{n}\frac{(-1)^{j+n+k}\rho_0^{2(j+k-n)}\Gamma(j+\frac{1}{2})a_{2j}}{(j-n+k)!\Gamma(n+\frac{1}{2}-k)}, \tag{4.14}$$

for $k \geq n$, and equals

$$\frac{1}{4}\chi(1)\mathrm{Vol}(\Gamma\backslash G)C_G\sum_{j=0}^{k}\frac{(-1)^j\rho_0^{2j}\Gamma(n-k+j+\frac{1}{2})a_{2(n-k+j)}}{j!\Gamma(n+\frac{1}{2}-k)}, \tag{4.15}$$

for $0 \leq k < n$. Here $\rho_0 = n$. Also $\zeta_\Gamma(0|L) = -n_0(\chi)$, whereas $\zeta_\Gamma(-k|L) = 0$ for $k = 1, 2, \ldots$.

In Theorems 4.1 and 4.2 the constant C_G is given in Table 4.1.

4.3 Asymptotics of the Heat Kernel

In this Section, we shall investigate the heat-kernel asymptotics. The literature on this issue is very vast. A partial list of references includes [R. Miatello (1976); R. Miatello (1979); R. Miatello (1980); B.S. DeWitt (1965); R.T. Seeley (1967); M. Atiyah et al. (1973a)] for closed Riemannian manifolds and [I.G. Moss and J.S. Dowker (1989); T.P. Branson and P.B. Gilkey (1990)] for manifolds with a smooth boundary.

To begin with, let us consider the Laplace type operator L acting on a general irreducible rank one locally symmetric space X. In [R. Miatello (1976); R. Miatello (1979); R. Miatello (1980)], R. Miatello studied the case of a closed locally symmetric rank one manifold X, using the representation theory of the group of isometries of X. We consider the same case, but we use the spectral zeta function of X. By our approach, we determine the set of Minakshisundaram–Pleijel coefficients $\{A_k(L)\}_{k=0}^{\infty}$ in the short–time asymptotic expansion of the kernel immediately and explicitly (essentially in one step). The object of interest is the heat kernel $K_\Gamma(t|L)$ defined for $t > 0$, by

$$K_\Gamma(t|L) = \sum_{j=0}^{\infty} n_j(\chi)e^{-\lambda_j(\chi)t}. \tag{4.16}$$

If I_Γ is the fundamental solution of the heat equation on $\mathcal{M}$, then I_Γ and $K_\Gamma(t|L)$ are related by the Selberg trace formula

$$K_\Gamma(t|L) = \chi(1)\mathrm{Vol}(\Gamma\backslash G)I_\Gamma(1) + \theta_\Gamma(t|L), \tag{4.17}$$

where the theta function $\theta_\Gamma(t|L)$ is given by Eq. (4.18) of [F.L. Williams

(1998)] (for $b = 0$ there) and Eq. (4.39) below, and where

$$I_\Gamma(1) = \frac{1}{4\pi} e^{-\rho_0^2 t} \int_{\mathbb{R}} dr e^{-r^2 t} |C(r)|^{-2}. \tag{4.18}$$

Our goal is to compute explicitly all coefficients $A_k = A_k(\Gamma, \chi)$ in the asymptotic expansion

$$K_\Gamma(t|L) \simeq (4\pi t)^{-D/2} \sum_{k=0}^{\infty} A_k t^k, \quad \text{as } t \to 0^+. \tag{4.19}$$

Now, we known that the $\zeta_\Gamma(s|L)$ and $K_\Gamma(t|L)$ are related by the Mellin transform and one knows from general considerations (cf. [S. Minakshisundaram and A. Pleijel (1949); V. Voros (1987)] for example) that the coefficients A_k are related to residues and special values of $\zeta_\Gamma(s|L)$. Namely, for m an integer with $1 \leq m \leq D/2$, for D even

$$A_{\frac{D}{2}-m} = (4\pi)^{D/2} \Gamma(m) \times [\text{residue of } \zeta_\Gamma(s|\mathcal{A}) \text{ at } s = m]. \tag{4.20}$$

Also, for a positive integer n

$$A_{\frac{D}{2}+n} = \frac{(-1)^n (4\pi)^{D/2}}{n!} \zeta_\Gamma(-n|L), \tag{4.21}$$

whereas

$$A_{\frac{D}{2}} = (4\pi)^{D/2} \left[n_0(\chi) + \zeta_\Gamma(0|L) \right]. \tag{4.22}$$

For $G = SO_1(2n+1, 1)$ (the only case in which D is odd), we have for $k = 0, 1, 2, \ldots$

$$A_k = (4\pi)^{D/2} \Gamma\left(\frac{D}{2} - k\right) \times \left[\text{residue of } \zeta_\Gamma(s|L) \text{ at } s = \frac{D}{2} - k\right]; \tag{4.23}$$

$D/2 = n + 1/2$. Hence, by Eqs. (4.10) – (4.13), and Theorems 4.1, 4.2 we obtain the following main result [A.A. Bytsenko and F.L. Williams (1999); A.A. Bytsenko (2002b)]:

Theorem 4.3 *The heat kernel $K_\Gamma(t|L)$ in (4.17) admits an asymptotic expansion (4.19). More precisely, given any non-negative integer N, one has*

$$\lim_{t \to 0^+} \left[(4\pi t)^{D/2} K_\Gamma(t|L) - \sum_{k=0}^{N} A_k(\Gamma, \chi) t^k \right] t^{-N} = 0, \tag{4.24}$$

where, apart from the cotangent case in (4.7) (i.e. the case $G = SU(q,1)$ with q even), the coefficients $A_k(\Gamma,\chi) = A_k(L)$ are given as follows.

For all G, except $G = SO_1(\ell,1)$, $SU(q,1)$ with ℓ odd and q even

$$A_k(\Gamma,\chi) = (4\pi)^{D/2-1}\chi(1)\mathrm{Vol}(\Gamma\backslash G)C_G\pi$$

$$\times\sum_{\ell=0}^{k}\frac{(-\rho_0^2)^{k-\ell}}{(k-\ell)!}[\frac{D}{2}-(\ell+1)]!a_{2[\frac{D}{2}-(\ell+1)]} \quad \text{for } 0\leq k\leq\frac{D}{2}-1, \tag{4.25}$$

$$A_{\frac{D}{2}+n}(\Gamma,\chi) = (-1)^n(4\pi)^{\frac{D}{2}-1}\chi(1)\mathrm{Vol}(\Gamma\backslash G)C_G\pi$$

$$\times\left[\sum_{j=0}^{\frac{D}{2}-1}(-1)^{j+1}\frac{\rho_0^{2(n+1+j)}}{(n+1+j)!}j!a_{2j}+2\sum_{j=0}^{\frac{D}{2}-1}\sum_{\ell=0}^{n}(-1)^{\ell}\frac{\rho_0^{2(n-\ell)}}{(n-\ell)!}b_{\ell+1}(j)a_{2j}\right], \tag{4.26}$$

for $n = 0,1,2,\ldots,$, where $b_r(j)$ $(r = 1,2,\ldots)$ and $a(G)$ are given by (4.12) and (4.13). For $G = SO_1(2n+1,1)$, $k = 0,1,2,\ldots$

$$A_k(\Gamma,\chi) = \pi(4\pi)^{n-\frac{1}{2}}\chi(1)\mathrm{Vol}(\Gamma\backslash G)C_G$$

$$\times\sum_{\ell=0}^{\min(k,n)}\frac{(-n^2)^{k-\ell}\Gamma\left(n-\ell+\frac{1}{2}\right)a_{2(n-\ell)}}{(k-\ell)!}, \tag{4.27}$$

or

$$A_k(\Gamma,\chi) = \pi^{3/2}(4\pi)^{n-\frac{1}{2}}\chi(1)\mathrm{Vol}(\Gamma\backslash G)C_G$$

$$\times\sum_{\ell=0}^{\min(k,n)}\frac{(-\rho_0^2)^{k-\ell}[2(n-\ell)]!a_{2(n-\ell)}}{(k-\ell)!(n-l)!2^{2(n-\ell)}}, \tag{4.28}$$

using that $\Gamma(m+1/2) = \pi^{1/2}(2m)!\left[2^{2m}m!\right]^{-1}$.

Using results [F.L. Williams (1998)] on the meromorphic structure of the zeta function of a rank one locally symmetric space X, we have obtained in a quick computation all the Minakshisundaram–Pleijel coefficients (in closed form) in the short–time asymptotic expansion of the heat kernel on X. Our method differs markedly from that of [R. Miatello (1976); R. Miatello (1979); R. Miatello (1980)].

4.4 Product of Einstein Manifolds

Among the most important geometric structures on manifolds [K. Yano (1965); S.S. Chern (1966); S. Kobayashi (1972)] almost–product structures play an essential role. Structures of this kind appear in a natural way from a variational principle based on a general class of Lagrangians depending on the Ricci square invariant constructed out of a metric and a symmetric connection. The reader can found geometric properties of (pseudo–) Riemannian almost–product structures, for example, in Refs. [K. Yano (1965); B.L. Reinhart (1983)]. In this class of structures an almost-product structure on Einstein manifold is associated with an Einstein metric g (i.e. $\mathrm{Ric}(g) = \gamma g$ holds, where $\mathrm{Ric}(g)$ is the Ricci tensor and γ is a constant).

The purpose of the present section is to investigate the global additive properties of invertible elliptic operators of Laplace type acting on Einstein manifolds, especially on a non-compact symmetric space, and related zeta functions. We consider the problem of the global existence of zeta functions on (pseudo–) Riemannian product manifolds, a product of two Einstein manifolds [K. Yano (1965); K. Yano and M. Kon (1984)]

$$(\mathcal{M}, g, \mathcal{P}) = (\mathcal{M}_1, g_1, \mathcal{P}_1) \otimes (\mathcal{M}_2, g_2, \mathcal{P}_2), \tag{4.29}$$

where $g = g_1 \otimes g_2$ and the metric g separates the variables, i.e.

$$ds^2 = g_{\alpha\beta}(x)dx^\alpha \otimes dx^\beta + g_{\mu\nu}(y)dy^\mu \otimes dy^\nu. \tag{4.30}$$

The tangent bundle splits as $T\mathcal{M} = T\mathcal{M}_1 \oplus T\mathcal{M}_2$ and $\mathcal{P} = \mathcal{P}_1 + \mathcal{P}_2$, where $\mathcal{P}_j$ $(j = 1, 2)$ are the corresponding projections on $T\mathcal{M}_j$,

$$\mathcal{P}^2 = Id, \quad g(\mathcal{P}\mathcal{X}, \mathcal{P}\mathcal{Y}) = g(\mathcal{X}, \mathcal{Y}), \quad \mathcal{X}, \mathcal{Y} \in \chi(\mathcal{M}), \tag{4.31}$$

and $\chi(M)$ being the Lie algebra of vector fields $\mathcal{X}$ and $\mathcal{Y}$ on $\mathcal{M}$. The trivial examples of an almost–product structure are given by the choices $\mathcal{P} = \pm Id$ ($\pm$ identity).

We recall some facts about Einstein manifolds necessary for the next considerations. We start with an almost-product (pseudo-) Riemannian structure $(\mathcal{M}, g, \mathcal{P})$ which is integrable iff $\Delta_g \mathcal{P} = 0$ for the Levi–Civita connection Δ_g of g. The two integrable complementary subbundles, i.e. both foliations, are totally geodesic [K. Yano (1965); K. Yano and M. Kon (1984)]. Let $\mathcal{M}$ be a pseudo-Kähler manifold; the metric g splits as $g_1 \otimes g_2$.

Such a manifold is an Einstein manifold iff in any adapted coordinates (x^α, y^α) both metrics g_1 and g_2 are Einstein metrics for the same constant γ [K. Yano (1965); K. Yano and M. Kon (1984); A.L. Besse (1987)], i.e.

$$R_{\alpha_j\beta_j} = \gamma g_{\alpha_j\beta_j}, \quad j = 1, 2. \tag{4.32}$$

Our consideration will be restricted to only locally decomposable manifolds. A wide class of (pseudo–) Riemannian manifolds includes non-locally decomposable manifolds as well, which are given by warped product spacetimes ([B. O'Neil (1983)]). Note that many exact solutions of Einstein equations (associated with Schwarzschild, Robertson–Walker, Reissner–Nordström, de Sitter spacetimes) and brane solutions are, in fact, warped product spacetimes [A.L. Besse (1987)].

4.4.1 *The Krönecker sum of Laplace operators*

Like before let us consider irreducible rank one symmetric spaces $\mathcal{M}_j \equiv X_j = K_j\backslash G_j$ of non-compact type. Thus G_j will be connected non-compact simple split rank one Lie groups with finite center and $K_j \subset G_j$ will be maximal compact subgroups. Let $\Gamma_j \subset G_j$ be discrete, co-compact, torsion free subgroups. Let $L_j : C^\infty(V(\mathcal{M}_j)) \to C^\infty(V(\mathcal{M}_j))$ be partial differential operators acting on smooth sections of vector bundles $V(\mathcal{M}_j)$. Let χ_j be a finite-dimensional unitary representation of Γ_j, let $\{\lambda_\ell(j)\}_{\ell=0}^\infty$ be the set of eigenvalues of the second-order operator of Laplace type $L_j = -\Delta_{\Gamma_j}$ acting on smooth sections of the vector bundle over $\Gamma_j\backslash X_j$ induced by χ_j, and let $n_\ell(\chi_j)$ denote the multiplicity of $\lambda_\ell(j)$.

We shall need further a suitable regularization of the determinant of an elliptic differential operator, since the naive definition of the product of eigenvalues gives rise to a badly divergent quantity. We shall make the choice of zeta-function regularization. The zeta function associated with the operators $\mathcal{L}_j \equiv L_j + b_j$ has the form

$$\zeta(s|\mathcal{L}_j) = \sum_\ell n_\ell(\chi_j)\{\lambda_\ell(j) + b_j\}^{-s}, \tag{4.33}$$

here b_j are arbitrary constants (endomorphisms of the vector bundle $V(\mathcal{M}_j)$), in the physical literature potential terms. $\zeta(s|\mathcal{L}_j)$ is a well-defined analytic function for $\mathrm{Re}\, s > \dim(\mathcal{M}_j)/2$, and can be analytically continued to a meromorphic function on the complex plane $\mathbb{C}$, regular at $s = 0$. One can define the heat kernel of the elliptic operator $\mathcal{L}_j$ by

$$K_{\Gamma_j}(t|\mathcal{L}_j) \equiv \mathrm{Tr}\left(e^{-t\mathcal{L}_j}\right) = \frac{-1}{2\pi i}\mathrm{Tr}\int_{\mathcal{C}_0} e^{-zt}(z-\mathcal{L}_j)^{-1}dz, \tag{4.34}$$

where $\mathcal{C}_0$ is an arc in the complex plane $\mathbb{C}$. From standard results in operator theory, there exist $\varepsilon, \delta > 0$ such that, for $0 < t < \delta$, the following heat-kernel expansion holds

$$K_{\Gamma_j}(t|\mathcal{L}_j) = \sum_{\ell=0}^{\infty} n_\ell(\chi_j)e^{-(\lambda_\ell(j)+b_j)t} = \sum_{0\leq\ell\leq\ell_0} A_\ell(\mathcal{L}_j)t^{-\ell} + \mathcal{O}(t^\varepsilon), \tag{4.35}$$

$\zeta(s|\mathcal{L}_j)$ and $K_{\Gamma_j}(t|\mathcal{L}_j)$ are related by Eq. (1.13). Eventually, we would also like to take $b_j = 0$, but for now we consider only non-zero modes: $b_j + \lambda_\ell(j) > 0$, $\forall \ell : \lambda_0(j) = 0$, $b_j > 0$.

We study the zeta function

$$\zeta(s|\oplus\mathcal{L}_j) = \zeta_{\Gamma_1\backslash X_1\otimes\Gamma_2\backslash X_2}(s|\mathcal{L}_{1,2}) = \frac{1}{\Gamma(s)}\int_0^\infty dt\, t^{s-1}\prod_j K_{\Gamma_j}(t|\mathcal{L}_j), \tag{4.36}$$

where $\mathrm{Re}\, s > (D_1+D_2)/2$, $D_j = \dim X_j$.

Before starting the main theorem of this section (Theorem 4.6 below), we shall need further notation and few more definitions, some of which arise in connection with Selberg's trace formula – a formula of crucial importance which we now review. Let the data (G, K, Γ) be as before, G being one of the four groups of Table 4.1. The trace formula holds [N. Wallach (1976); F.L. Williams (1990)]

$$K_\Gamma(t|\mathcal{L}) = V\int_{\mathbb{R}} dr\, |C(r)|^{-2}e^{-(r^2+b+\rho_0^2)t} + \theta_\Gamma(t|\mathcal{L}), \tag{4.37}$$

where

$$V \stackrel{def}{=} \frac{1}{4\pi}\chi(1)\mathrm{Vol}(\Gamma\backslash G), \tag{4.38}$$

the number ρ_0 is associated with the positive restricted (real) roots of G (with multiplicity) with respect to a nilpotent factor N of G in an Iwasawa decomposition $G = KAN$. One has, $\rho_0 = (n-1)/2, n, 2n+1, 11$ in the cases $(I)-(IV)$, respectively, in Eq. (4.1). Finally, the function $\theta_\Gamma(t|\mathcal{L})$ is defined as follows

$$\theta_\Gamma(t|\mathcal{L}) \stackrel{def}{=} \frac{1}{(4\pi t)^{1/2}} \sum_{\gamma\in C_\Gamma-\{1\}} \chi(\gamma)t_\gamma j(\gamma)^{-1}C(\gamma)e^{-(bt+\rho_0^2 t+t_\gamma^2/(4t))}, \quad (4.39)$$

for a function $C(\gamma),\ \gamma\in\Gamma$, defined on $\Gamma-\{1\}$ by

$$C(\gamma) \stackrel{def}{=} e^{-\rho_0 t_\gamma}|\det{}_{n_0}\left(\mathrm{Ad}(m_\gamma e^{t_\gamma H_0})^{-1}-1\right)|^{-1}. \quad (4.40)$$

The notation used in Eqs. (4.39) and (4.40) is the following. Let a_0, n_0 denote the Lie algebras of A, N. Since the rank of G is 1, $\dim a_0 = 1$ by definition, say $a_0 = \mathbb{R}H_0$ for a suitable basis vector H_0. One can normalize the choice of H_0 by $\beta(H_0) = 1$, where $\beta : a_0 \mapsto \mathbb{R}$ is the positive root which defines n_0; for more detail see Ref. [F.L. Williams (1997)]. Since Γ is torsion free, each $\gamma\in\Gamma-\{1\}$ can be represented uniquely as some power of a primitive element $\delta : \gamma = \delta^{j(\gamma)}$, where $j(\gamma)\geq 1$ is an integer and δ cannot be written as γ_1^j for $\gamma_1\in\Gamma$. Taking $\gamma\in\Gamma$, $\gamma\neq 1$, one can find $t_\gamma>0$ and $m_\gamma\in K$ satisfying $m_\gamma a = a m_\gamma$ for every $a\in A$ such that γ is G conjugate to $m_\gamma\exp(t_\gamma H_0)$, namely for some $\mathrm{g}\in G$, $\mathrm{g}\gamma\mathrm{g}^{-1} = m_\gamma\exp(t_\gamma H_0)$. With Ad denoting the adjoint representation of G on its complexified Lie algebra, one can compute t_γ as follows [N. Wallach (1976)]

$$e^{t_\gamma} = \max\{|c||c = \text{an eigenvalue of } \mathrm{Ad}(\gamma)\}, \quad (4.41)$$

in the case $G = SO_1(m,1)$, with $|c|$ replaced by $|c|^{1/2}$ in the other cases of Eq. (4.1). We will write V_j for $G = G_j$ in (4.37), $j = 1,2$, G_1, G_2 being one of the groups in (4.37). Similarly we write $\rho_0(j), a_{2k}(j)$, C_{G_j}, etc., in place of ρ_0, a_{2k}, C_G for $G = G_j$. Let

$$v \stackrel{def}{=} \rho_0^2(1)+\rho_0^2+b_1+b_2, \quad v_j \stackrel{def}{=} b_j+\rho_0^2(j)\,, \quad j=1,2,$$

$$y_j(s;z) \stackrel{def}{=} \frac{s}{2}+(-1)^{j-1}z, \quad \text{for } s,z\in\mathbb{C}. \quad (4.42)$$

Also for $a,\delta>0$, $n\in\mathbb{N}$, $s\in\mathbb{C}$, define functions $\mathcal{K}_n(s;\delta,a)$ of s by

$$\mathcal{K}_n(s;\delta,a) \stackrel{def}{=} \int_{\mathbb{R}} \frac{r^{2n}\mathrm{sech}^2(ar)dr}{(\delta+r^2)^s}. \quad (4.43)$$

4.4.2 *The Selberg zeta function. Factorization formula*

More statements are needed for the analysis of the Selberg zeta function $Z_\Gamma(s,\chi)$ and its factorization attached to X_Γ; cf. [R. Gangolli (1977); A. Selberg (1956)]. Let χ be a finite-dimensional unitary representation of Γ (χ also denote the corresponding character). The induced unitary representation $\pi_\chi = \mathrm{ind}_{\Gamma\uparrow G}\chi$ of G decomposes directly: $\pi_\chi = \sum_{\pi\in\widehat{G}} m_\pi(\Gamma)\pi$, where $\widehat{G}$ denoting the unitary dual of G, where each multiplicity $m_\pi(\Gamma) \geq 0$ is finite. Let $\{\pi_j\}_{j\geq 0} \subset \widehat{G}$ be the subset of all class one representations (with respect to a maximal compact subgroup K) such that $n_j(\chi) \stackrel{def}{=} m_{\pi_j}(\Gamma) > 0$. Attached to the class one spectrum $\{\pi_j, n_j(\chi)\}_{j\geq 0}$ of π_χ is a spectral zeta series $\zeta_\Gamma(s|\mathcal{L})$. First, we discuss a meromorphic continuation to the whole complex plane of $s \mapsto \zeta_\Gamma(s|\mathcal{L})$ in terms of the harmonic analysis of X. Then, these results will be used to a general factorization of Selberg's zeta function attached to X_Γ.

For $b > 0$, let

$$\Theta_\Gamma(s|\mathcal{L}) = \int_0^\infty dt\, t^{s-1}\theta_\Gamma(t|\mathcal{L}), \tag{4.44}$$

which is an entire function of s. There is the theorem where the cotangent case in (4.7) has been excluded. For the latter case see remarks in the end of that Section.

Theorem 4.4 ([F.L. Williams (1981b)]) *For $s \in \mathbb{C}$ define*

$$\Phi_\Gamma(s;b,\chi) \stackrel{def}{=} \frac{\Theta_\Gamma(s|\mathcal{L})}{\Gamma(s)} + \frac{a(G)}{8}C_G\chi(1)\mathrm{Vol}(\Gamma\backslash G)$$

$$\times \sum_{j=0}^{(D-2)/2} \frac{m_j(b)}{s-j-1}\mathcal{K}_0(s-j-1, b+\rho_0^2, a(G)), \tag{4.45}$$

$$\Phi_\Gamma^{(n)}(s;b,\chi) \stackrel{def}{=} \frac{\Theta_\Gamma(s|\mathcal{L})}{\Gamma(s)} + \frac{\pi^{1/2}}{4}C_G\chi(1)\mathrm{Vol}(\Gamma\backslash G)(b+n^2)^{-s+1/2}$$

$$\times \sum_{j=0}^{n} m_j(b+n^2)^j \frac{\Gamma(s-j-1/2)}{\Gamma(s-j)}, \tag{4.46}$$

where the coefficients $m_j(b)$ are defined by: $\prod(r; b+\rho_0^2) = \sum_j m_j(br^j)$, $\prod(r^2+z; z)$ is an even polynomial in r (see Eq. (4.8)), independent of z, of degree $D-2$ for $G \neq SO_1(2n+1,1)$, and of degree $D-1 = 2n$ for $G = SO_1(2n+1,1)$. Then,

$$\zeta_\Gamma(s|\mathcal{L})_{\text{Re } s > D/2} = \Phi_\Gamma(s; b, \chi)\,, \quad \text{for } G = SO_1(2n,1),\ SP(n,1),\ F_{4(-20)},$$

$$\text{or } SU(\ell,1) \text{ with } \ell \text{ odd (up to local isomorphisms)}, \tag{4.47}$$

$$\zeta_\Gamma(s|\mathcal{L})_{\text{Re } s > D/2 = n+1/2} = \Phi_\Gamma^{(n)}(s; b, \chi)\,, \quad \text{for } G = SO_1(2n+1,1). \tag{4.48}$$

Therefore, the functions $\Phi_\Gamma(s; b, \chi)$, $\Phi_\Gamma^{(n)}(s; b, \chi)$ provide the meromorphic continuation of $\zeta_\Gamma(s|\mathcal{L})$.

Theorem 4.4 can be applied to give a factorization of the Selberg zeta function $Z_\Gamma(s;\chi)$. Let $B_\Gamma(\chi) = (\pi/2)a(G)C_G V$. For $a, b > 0$, $c \in \mathbb{C}$, $n = 0, 1, 2, 3, \ldots$, define

$$I^{(n)}(a; b, c) \stackrel{def}{=} \int_{\mathbb{R}} dt\, t^{2n}(\text{sech}^2 at)(b^2+t^2)[\log(b^2+t^2) - c].$$

These integrals are evaluated in [F.L. Williams (1993)], where it is shown that $b \to I^{(n)}(a; b, c)$ admits a holomorphic continuation to the domain $\mathbb{C} - \{x+iy | y = 0, x \leq -a/2\pi\}$. The following result holds:

Theorem 4.5 ([F.L. Williams (1981b)]) *For* Re $s > 2\rho_0$, *and* $G \neq SO_1(2n+1,1)$, $SU(\ell,1)$, *with ℓ even,* $\exists\, c(\chi) > 0$ *such that*

$$Z_\Gamma(s,\chi) = c(\chi)\det\left(L + s(s-2\rho_0)1\right) e^{(m_0(s(s-2\rho_0))B_\Gamma(\chi)I^{(0)}(a(G); s-\rho_0, 1)}$$

$$\times \prod_{j=1}^{D/2-1}\prod_{\ell=0}^{j} e^{\frac{m_j(s(s-2\rho_0))}{j+1}\binom{j}{\ell}(s-\rho_0)^{2(j-\ell)}B_\Gamma(\chi)I^{(\ell)}(a(G); s-\rho_0, (j+1)^{-1})}\,, \tag{4.49}$$

where $L \equiv -\Delta_\Gamma$ is the Laplace operator and

$$\det(L + b1) \stackrel{def}{=} \exp\left(-\frac{\partial}{\partial s}\Phi_\Gamma(s; b, \chi)|_{s=0}\right). \tag{4.50}$$

Theorem 4.5 was obtained for group $G = SL(2,\mathbb{R})$ in Refs. [P. Sarnak (1987); V. Voros (1987)] (for the case $\chi = 1$ Theorem 4.5 has been proven in [F.L. Williams (1992)]).

Remark. For G locally isomorphic to $SU(\ell,1)$ with ℓ even (the cotangent case in (4.7)) the definition of $\Phi_\Gamma(s;b,\chi)$ in (4.45) has to be modified by adding to it the following extra term, namely:

$$\widehat{\Phi}_\Gamma(s;b,\chi) = \Phi_\Gamma(s;b,\chi) + \frac{1}{4}\chi(1)\mathrm{Vol}(\Gamma\backslash G)C_G$$

$$\times \sum_{j=0}^{(D-2)/2} m_j(b)I_0(s-j;a(G)=\pi/2,b), \tag{4.51}$$

where

$$I_0(s;a,b) = \int_{\mathbb{R}} \frac{dx\, x\, \mathrm{csch}\, ax \cdot \mathrm{sech}\, ax}{(b+\rho_0^2+x^2)^s}.$$

The integral $I_0(s;a,b)$ converges uniformly on compact subset of the plane and is thus an entire function of s. Then, Theorem 4.5 holds for $\widehat{\Phi}_\Gamma(s;b,\chi)$ (see [F.L. Williams (1981b)] for detail).

4.4.3 *Meromorphic continuation*

We denote by $\Psi_\Gamma(s;\chi) \equiv (d/ds)\log Z_\Gamma(s;\chi)$ the logarithmic derivative of Selberg's zeta function $Z_\Gamma(s;\chi)$, meromorphic suitably normalized, attached to G, K, Γ, χ [A. Selberg (1956)] (see also [D. Fried (1977a); R. Gangolli (1977); R. Gangolli and G. Warner (1980); D. Scott (1980); M. Wakayama (1985); F.L. Williams (1990); A.A. Bytsenko et al. (1996)]. For convenience we have assumed (recall) that if G_1 or G_2 is $SO_1(m,1)$ or $SU(q,1)$, then m is even and q is odd. We can now state the following.

Theorem 4.6 ([A.A. Bytsenko and F.L. Williams (1998)]) *The function $\zeta(s|\oplus\mathcal{L}_j)$ admits an explicit meromorphic continuation to $\mathbb{C}$ with at most a simple pole at $s = 1,2,...,(D_1+D_2)/2$. In particular, on the domain* $\mathrm{Re}\, s < 1$,

$$\zeta(s|\oplus\mathcal{L}_j) = \frac{\pi^2}{2}\prod_j^2 V_j a(G_j)C_{G_j} \sum_{k=0}^{\frac{D_1}{2}-1}\sum_{\ell=0}^{j}\sum_{\mu=0}^{\frac{D_2}{2}-1}\sum_{\nu=0}^{\mu} \frac{a_{2k}(1)a_{2\mu}(2)k!\mu!}{(k-\ell)!(\mu-\nu)!}$$

$$\times\frac{\int_0^\infty dr\, r^{2(k-\ell)}\mathrm{sech}^2(a(G_1)r)\mathcal{K}_{\mu-\nu}(s-\ell-\nu-1;r^2+v,a(G_1))}{(s-1)(s-2)...(s-(\ell+1))(s-(\ell+2))...(s-(\ell+1+\nu+1))}$$

$$+C_{G_1}V_1\sin(\pi s)\sum_{k=0}^{D_1/2-1}a_{2k}(1)\int_{\mathbb{R}}dr\, r^{2k+1}\tanh(a(G_1)r)$$

$$\times\left[\int_0^\infty dt\,\Psi_{\Gamma_2}\left(\rho_0(2)+t+\sqrt{r^2+v};\chi_2\right)\left(2t\sqrt{r^2+v}+t^2\right)^{-s}\right]$$

$$+C_{G_2}V_2\sin(\pi s)\sum_{k=0}^{D_2/2-1}a_{2k}(2)\int_{\mathbb{R}}dr\, r^{2k+1}\tanh(a(G_2)r)$$

$$\times\left[\int_0^\infty dt\,\Psi_{\Gamma_1}\left(\rho_0(1)+t+\sqrt{r^2+v};\chi_1\right)\left(2t\sqrt{r^2+v}+t^2\right)^{-s}\right]$$

$$+\frac{1}{2\pi^3 i\Gamma(s)}\int_{\mathrm{Re}\,z=\varepsilon}dz\,\sin[\pi(z+\frac{s}{2})]\sin[\pi(\frac{s}{2}-z)]\Gamma(z+\frac{s}{2})\Gamma(\frac{s}{2}-z)$$

$$\times\prod_j^2\left[\int_0^\infty dt\,\Psi_{\Gamma_j}\left(\rho_0(j)+t+v_j^{\frac{1}{2}};\chi_1\right)\left(2tv_j^{\frac{1}{2}}+t^2\right)^{-y_j(s,z)}\right], \qquad (4.52)$$

for any $-1/2\leq\varepsilon\leq 1/2$. *All of the integrals are entire functions of* s.

Here $\zeta(s|\oplus\mathcal{L}_j)$ is defined by (4.36), which involves the parameters b_1,b_2. In Theorem 4.6 we require that $b_1,b_2>0$. One can, however, consider zero modes. More precisely assume $b_1,b_2\geq 0$, but at least either $b_1>0$ or $b_2>0$. Then, Theorem 4.6 holds on the domain Re $s<0$ when the condition $-1/2\leq\varepsilon\leq 1/2$ is replaced by the condition $\mathrm{Re}(s/2)<\varepsilon<-\mathrm{Re}(s/2)$.

The proof of Theorem 4.6 involves quite a number of steps. One of the steps, among many, is an analysis of the Mellin transform $\Theta_\Gamma(s|\mathcal{L})$ of the holomorphic function $\theta_\Gamma(t|\mathcal{L})$ on the domain Re $s<0$ in Eq. (4.39). Let

$$K_\nu(s)=\frac{1}{2}\int_0^\infty dt\, e^{-(s/2)(t+1/t)}t^{\nu-1} \qquad (4.53)$$

be the modified Bessel function. Then using the result of [A.A. Bytsenko et al. (1996); F.L. Williams (1997)] one can obtain for $\mathrm{Re}\, s < 0$,

$$\Theta_\Gamma(s|L) = \sum_{\gamma\in C_\Gamma-\{1\}} \chi(\gamma)t_\gamma j(\gamma)^{-1}C(\gamma)\int_0^\infty dt\, t^{s-1}\frac{e^{-(\rho_0^2 t+t_\gamma^2/(4t))}}{(4\pi t)^{1/2}}$$

$$= \frac{(2\rho_0)^{\frac{1}{2}-s}}{\pi^{1/2}} \sum_{\gamma\in C_\Gamma-\{1\}} \chi(\gamma)t_\gamma j(\gamma)^{-1}C(\gamma)t_\gamma^{s+\frac{1}{2}}K_{\frac{1}{2}-s}(t_\gamma\rho), \tag{4.54}$$

and

$$\Theta_\Gamma(s|\mathcal{L}) = \frac{\sin(\pi s)}{2\pi}\Gamma(s)\int_0^\infty dt\, \Psi_\Gamma(\rho_0+t+a(b);\chi)(2a(b)t+t^2)^{-s}\,, \tag{4.55}$$

$\mathrm{Re}\ s < 1$, for $a(b) \stackrel{def}{=} \sqrt{b+\rho_0^2}$, $b>0$. Equation (4.55) holds for $b=0$ (the zero modes case) in the domain $\mathrm{Re}\ s<0$. Given the notation in Eq. (4.55), the following statement holds [A.A. Bytsenko and F.L. Williams (1998)]:

Proposition 4.1 *The integral of the product of the functions Θ_{Γ_j} is an entire function of s and allows the form*

$$\int_0^\infty dt\, t^{s-1}\prod_j \theta_{\Gamma_j}(t|\mathcal{L}_j) = \frac{1}{2\pi i}\int_{\mathrm{Re}\, z=c} dz\prod_j \Theta_{\Gamma_j}(y_j(s,z)|\mathcal{L}_j)$$

$$= \frac{1}{2i\pi^3}\int_{\mathrm{Re}\, z=c} dz\prod_\ell^2 \Gamma(y_\ell(s;z))\sin[\pi(y_\ell(s,z))]$$

$$\times\prod_j\int_0^\infty dt\, \Psi_{\Gamma_j}(\rho_0(j)+t+v_j^{\frac{1}{2}};\chi_j)(2v_j^{\frac{1}{2}}t+t^2)^{-y_j(s;z)}, \tag{4.56}$$

for $c\in\mathbb{R}$ with $\mathrm{Re}(s/2)<c<\mathrm{Re}(-s/2)$, and $b_j\geq 0$ with at least $b_1>0$ or $b_2>0$.

The proposition is a consequence of the first integral transformation of Eq. (4.56) as an integral in the complex plane with the help of the Mellin–Parseval identity

$$\int_0^\infty dt\, h(t)g(t) = \frac{1}{2\pi i}\int_{\mathrm{Re}\, z=c} dz\,\widehat{h}(z)\widehat{g}(1-z), \quad c\in\mathbb{R}. \tag{4.57}$$

To this end, one can choose $h(t) = \theta_{\Gamma_1}(t|\mathcal{L}_1)t^{s/2}$, and $g(t) = \theta_{\Gamma_2}(t|\mathcal{L}_2)t^{s/2-1}$.

The meromorphic continuation of $\zeta(s| \oplus \mathcal{L}_j)$ to the full complex plane, whereby Theorem 4.6 follows in particular, can be expressed as follows. For $a, b > 0$, $\ell = 0, 1, 2, 3, ...$, define

$$f^{(\ell)}(t; a, G) \stackrel{def}{=} \int_0^\infty dr\, r^{2\ell+1} e^{-(r^2+a)t} \tanh(a(G)r), \tag{4.58}$$

$$H_\Gamma^{(\ell)}(s; a, b, \chi, G) \stackrel{def}{=} \int_0^\infty dt\, f^{(\ell)}(t; a, G)\theta_\Gamma(t|\mathcal{L})t^{s-1}, \tag{4.59}$$

for $t > 0$ and $s \in \mathbb{C}$, $H_\gamma^{(\ell)}(s; a, b, \chi, G)$ is an entire function of s. Then for $b_1, b_2 \geq 0$ with either $b_1 > 0$ or $b_2 > 0$, we have the following theorem.

Theorem 4.7 ([A.A. Bytsenko and F.L. Williams (1998)]) *Suppose $G_p \neq SO_1(m, 1)$, $SU(q, 1)$ with m odd and q even. Then, for* $\mathrm{Re}\, s > (D_1 + D_2)/2$,

$$\zeta(s| \oplus \mathcal{L}_j) \stackrel{def}{=} \frac{1}{\Gamma(s)} \int_0^\infty dt \prod_j K_{\Gamma_j}(t|\mathcal{L}_j)t^{s-1}$$

$$= \frac{\pi^2}{2} \prod_j V_j a(G_j) C_{G_j} \sum_{k=0}^{\frac{D_1}{2}-1} \sum_{\ell=0}^{j} \sum_{\mu=0}^{\frac{D_2}{2}-1} \sum_{\nu=0}^{\mu} \frac{a_{2k}(1)a_{2\mu}(2)k!\mu!}{(k-\ell)!(\mu-\nu)!}$$

$$\times \frac{\int_0^\infty dr\, r^{2(k-\ell)} \mathrm{sech}^2(a(G_1)r)\mathcal{K}_{\mu-\nu}(s-\ell-\nu-1; r^2+v, a(G_1))}{(s-1)(s-2)...(s-(\ell+1))(s-(\ell+2))...(s-(\ell+1+\nu+1))}$$

$$+ \frac{2\pi}{\Gamma(s)} \left\{ V_1 C_{G_1} \sum_{k=0}^{\frac{D_1}{2}-1} a_{2k}(1) H_{\Gamma_2}^{(k)}(s; b_1+v_1, b_2, \chi_2, G_1) \right.$$

$$\left. + V_2 C_{G_2} \sum_{k=0}^{\frac{D_2}{2}-1} a_{2k}(2) H_{\Gamma_1}^{(k)}(s; b_2+v_2, b_1, \chi_1, G_2) \right\}$$

$$+\frac{1}{\Gamma(s)}\int_0^\infty dt\, t^{s-1}\prod_j^2 \theta_{\Gamma_j}(t|\mathcal{L}_j). \tag{4.60}$$

The formulae (4.56) and (4.60) give the merormorphic continuation of the zeta function (4.36).

4.5 Real Hyperbolic Manifolds

Up to local isomorphism, we choose $X = SO_1(D,1)/SO(D)$. Thus the isotropy group K of the base point $(1,0,...0)$ is $SO(D)$; X can be identified with the hyperbolic $D-$space $\mathbb{H}^D$, $D = \dim X$. It is possible to view $\mathbb{H}^D$, for example, as one sheet of the hyperboloid of two sheets in $\mathbb{R}^{D+1}$ given by $q(x) = -x_0^2 + x_1^2 + ... + x_D^2 = -1$, $x_0 > 0$, with the metric induced by the quadratic form $q(x)$.

Going back to Eq. (4.60), note that the simplest case is, for example, $G_1 = G_2 = G = SO_1(2,\ell) \simeq SL(2,\mathbb{R})$; besides $X_1 = X_2 = \mathbb{H}^2$ is a two-dimensional real hyperbolic space. Then we have $\rho_0^2 = \rho_0^2(1) + \rho_0^2(2) = \frac{1}{4}+\frac{1}{4} = \frac{1}{2}$, $a_{20} = 1$, $C_G = 1$, $a(G) = \pi$, $\Gamma_1 = \Gamma_2 = \Gamma$, and finally $|C(r)|^{-2} = \pi r\tanh(\pi r)$. Using Eq. (4.52) of Theorem 4.6 for $\mathrm{Re}\, s < 1$ we have

$$\zeta(s|\oplus\mathcal{L}_j) = \frac{\pi^3 V_1^2}{2(s-1)(s-2)}\int_0^\infty dr\,\mathrm{sech}^2(\pi r)\mathcal{K}_0\left(s-2; r^2+v,\pi\right)$$

$$+2V_1\sin(\pi s)\int_{\mathbb{R}} dr\, r\tanh(\pi r)\int_0^\infty dt\,\frac{\Psi_\Gamma(\frac{1}{2}+t+\sqrt{r^2+v};\chi)}{\left(2t\sqrt{r^2+v}+t^2\right)^s}$$

$$+\frac{1}{2\pi^3 i\Gamma(s)}\int_{\mathrm{Re}\, z=\epsilon} dz\,\sin[\pi(z+\frac{s}{2})]\sin[\pi(\frac{s}{2}-z)]\Gamma\left(z+\frac{s}{2}\right)\Gamma\left(\frac{s}{2}-z\right)$$

$$\times\int_0^\infty dt\,\Psi_\Gamma\left(\frac{1}{2}+t+v_1^{\frac{1}{2}};\chi\right)\left(2tv_1^{\frac{1}{2}}+t^2\right)^{-y_1(s;z)}$$

$$\times\int_0^\infty dt\,\Psi_\Gamma\left(\frac{1}{2}+t+v_2^{\frac{1}{2}};\chi\right)\left(2tv_2^{\frac{1}{2}}+t^2\right)^{-y_2(s;z)}, \tag{4.61}$$

where $v = 1/2 + b_1 + b_2$ and $v_j = 1/4 + b_j$ $(j = 1,2)$.

4.5.1 *Laplacian on forms*

Let $\Gamma \subset G$ be discrete, co-compact, torsion free subgroups, and let $\chi(\gamma) = \mathrm{trace}(\chi(\gamma))$ be the character of a finite-dimensional unitary representation of Γ for $\gamma \in \Gamma$. Let $L^{(p)}$ be a Laplacian on p-forms acting on the vector bundle $V(X_\Gamma)$ over $X_\Gamma = \Gamma\backslash G/K$ induced by χ. Note that the non-twisted p-forms on X_Γ are obtained by taking $\chi = 1$. One can define the heat kernel $\mathrm{Tr}\exp(-t\mathcal{L}^{(p)})$ of the elliptic operator $\mathcal{L}^{(p)} = L^{(p)} + b^{(p)}$. Let $\{\lambda_\ell^{(p)}\}_{\ell=0}^{\infty}$ be the set of eigenvalues of the operator $L^{(p)}$ and $n_\ell(\chi)$ denote the multiplicity of $\lambda_\ell^{(p)}$. Eventually, we would also like to take $b^{(p)} = 0$, but for now we consider only non zero modes: $b^{(p)} + \lambda_\ell^{(p)} > 0, \forall \ell$ with $\lambda_0^{(p)} = 0, b^{(p)} > 0$.

Theorem 4.8 (Fried's trace formula [D. Fried (1977b)]) *For* $0 \leq p \leq D-1$,

$$\mathrm{Tr}\left(e^{-t\mathcal{L}^{(p)}}\right) = I^{(p)}(t, \mathcal{L}^{(p)}) + I^{(p-1)}(t, \mathcal{L}^{(p-1)})$$

$$+H^{(p)}(t, \mathcal{L}^{(p)}) + H^{(p-1)}(t, \mathcal{L}^{(p-1)}), \tag{4.62}$$

where

$$I^{(p)}(t, \mathcal{L}^{(p)}) \stackrel{def}{=} \frac{\chi(1)\mathrm{Vol}(\Gamma\backslash G)}{4\pi} \int_{\mathbb{R}} dr\, \mu_{\sigma_p}(r) e^{-t[r^2+b^{(p)}+(\rho_0-p)^2]}, \tag{4.63}$$

$$H^{(p)}(t, \mathcal{L}^{(p)}) \stackrel{def}{=} \frac{1}{(4\pi t)^{1/2}} \sum_{\gamma \in C_\Gamma - \{1\}} \chi(\gamma) t_\gamma j(\gamma)^{-1} C(\gamma) \chi_{\sigma_p}(m_\gamma)$$

$$\times \exp\left\{-\left[b^{(p)}t + (\rho_0 - p)^2 t + \frac{t_\gamma^2}{4t}\right]\right\}. \tag{4.64}$$

Here C_Γ is a complete set of representatives in Γ of its conjugacy classes, the Haar measure on G being suitably normalized.

In our case $K \simeq SO(D)$. For $p = 0$ (i.e. for smooth functions or smooth vector bundle sections) the measure $\mu_0(r)$ corresponds to the trivial representation of $SO(D-1)$. For $p \geq 1$ there is a measure $\mu_\sigma(r)$ corresponding to a general irreducible representation σ of $SO(D-1)$. Let σ_p is the standard representation of $SO(D-1)$ on $\Lambda^p\mathbb{C}^{(D-1)}$. If $D = 2n$ is even then σ_p $(0 \leq p \leq D-1)$ is always irreducible; if $D = 2n+1$ then every σ_p is irreducible except for $p = (D-1)/2 = n$, in which case σ_n is the direct sum of two (1/2)-spin representations $\sigma^\pm : \sigma_n = \sigma^+ \oplus \sigma^-$. For $p = n$ the representation τ_n of $K = SO(2n)$ on $\Lambda^n \mathbb{C}^{2n}$ is not irreducible: $\tau_n = \tau_n^+ \oplus \tau_n^-$ is the direct sum of two spin-(1/2) representations. The measures $\mu_{\sigma_p}(r)$ are given by the following theorem.

Theorem 4.9 *Let the group $G = SO_1(2n,1)$. Then,*

$$\mu_{\sigma_p}(r) = \binom{2n-1}{p} \frac{\pi r}{2^{4n-4}\Gamma(n)^2} \prod_{j=2}^{p+1} \left[r^2 + (n + \frac{3}{2} - j)^2\right]$$

$$\times \prod_{j=p+2}^{n} \left[r^2 + (n + \frac{1}{2} - j)^2\right] \tanh(\pi r) \quad for \quad 0 \leq p \leq n-1, \tag{4.65}$$

$$\mu_{\sigma_p}(r) = \binom{2n-1}{p} \frac{\pi r}{2^{4n-4}\Gamma(n)^2} \prod_{j=2}^{2n-p} \left[r^2 + (n + \frac{3}{2} - j)^2\right]$$

$$\times \prod_{j=2n-p+1}^{n} \left[r^2 + (n + \frac{1}{2} - j)^2\right] \tanh(\pi r) \quad for \quad n \leq p \leq 2n-1, \tag{4.66}$$

and $\mu_{\sigma_p}(r) = \mu_{\sigma_{2n-p-1}}(r)$.

For the group $G = SO_1(2n+1,1)$, one has

$$\mu_{\sigma_p}(r) = \binom{2n}{p} \frac{\pi}{2^{4n-2}\Gamma(n+\frac{1}{2})^2} \prod_{j=1}^{p+1} \left[r^2 + (n+1-j)^2\right]$$

$$\times \prod_{j=p+2}^{n} \left[r^2 + (n-j)^2\right] \quad for \quad 0 \leq p < n, \tag{4.67}$$

and

$$\mu_{\sigma_p}(r) = \binom{2n}{p} \frac{\pi}{2^{4n-2}\Gamma(n+\frac{1}{2})^2} \prod_{j=1}^{2n-p+1} \left[r^2 + (n+1-j)^2\right]$$

$$\times \prod_{j=2n-p+2}^{n} \left[r^2 + (n-j)^2\right] \quad for \quad n+1 \leq p \leq 2n-1. \tag{4.68}$$

We should note that the reason for the pair of terms $\{I^{(p)}, I^{(p-1)}\}$, $\{H^{(p)}, H^{(p-1)}\}$ in the trace formula, Eq. (3.64), is that τ_p satisfies $\tau_p = \sigma_p \oplus \sigma_{p-1}$. Finally, using the result of Theorem 4.9 we have

$$\mu_{\sigma_p}(r) = \pi C^{(p)}(D)P(r)\tanh(\pi r) = \pi C^{(p)}(d)\sum_{\ell=0}^{n-1} a_{2\ell}^{(p)} r^{2\ell+1}\tanh(\pi r), \tag{4.69}$$

$$C^{(p)}(D) = \left(\begin{array}{c} D-1 \\ p \end{array}\right)\frac{\pi}{2^{4n-2}\Gamma(D/2)^2}\left\{\begin{array}{ll} 4, & \text{for } D=2n \\ 1, & \text{for } D=2n+1 \end{array}\right., \tag{4.70}$$

where the $P(r)$ are even polynomials (with suitable coefficients $a_{2\ell}^{(p)}$ of degree $D-2$ for $G \neq SO(2n+1,1)$, and of degree $D-1 = 2n$ for $G = SO_1(2n+1,1)$ [A.A. Bytsenko et al. (1996)].

Case of the trivial representation. For $p=0$ we take $I^{(-1)} = H^{(-1)} = 0$. Since σ_0 is a trivial representation, then $\chi_{\sigma_0}(m_\gamma) = 1$. In this case, Fried's formula Eq. (4.62) reduces exactly to the trace formula for $p=0$ [N. Wallach (1976); F.L. Williams (1990)]:

$$K_\Gamma(t,\mathcal{L}) = \frac{\chi(1)\mathrm{vol}(\Gamma\backslash G)}{4\pi}\int_{\mathbb{R}} dr\, \mu_{\sigma_0}(r)e^{-(r^2+b+\rho_0^2)t} + H^{(0)}(t,\mathcal{L}), \tag{4.71}$$

where $\rho_0 = (n-1)/2$ is associated with the positive restricted (real) roots of G (with multiplicity) with respect to a nilpotent factor N of G in an Iwasawa decomposition $G = KAN$. The function $H^{(0)}(t,\mathcal{L})$ has the form

$$H^{(0)}(t,\mathcal{L}) = \frac{1}{(4\pi t)^{1/2}}\sum_{\gamma\in C_\Gamma-\{1\}} \chi(\gamma)t_\gamma j(\gamma)^{-1}C(\gamma)e^{-[bt+\rho_0^2 t+t_\gamma^2/(4t)]}. \tag{4.72}$$

Case of the zero modes. It can be shown [F.L. Williams (1998)] that the Mellin transform of $H^{(0)}(t,L)$ ($b=0$, i.e. the zero modes case)

$$\mathfrak{H}^{(0)}(s,L) = \int_0^\infty dt\, t^{s-1}H^{(0)}(t,L) \tag{4.73}$$

is a holomorphic function on the domain $\mathrm{Re}\, s < 0$. Then using the result of [A.A. Bytsenko et al. (1996); F.L. Williams (1997)] one can obtain on

$\mathrm{Re}\, s < 0$,

$$\mathfrak{H}^{(0)}(s, L) = \sum_{\gamma \in C_\Gamma - \{1\}} \chi(\gamma) t_\gamma j(\gamma)^{-1} C(\gamma) \int_0^\infty dt\, t^{s-1} \frac{e^{-(\rho_0^2 t + t_\gamma^2/(4t))}}{(4\pi t)^{1/2}}$$

$$= \frac{(2\rho_0)^{\frac{1}{2}-s}}{\pi^{1/2}} \sum_{\gamma \in C_\Gamma - \{1\}} \chi(\gamma) t_\gamma j(\gamma)^{-1} C(\gamma) t_\gamma^{s+\frac{1}{2}} K_{\frac{1}{2}-s}(t_\gamma \rho), \tag{4.74}$$

where $K_\nu(s)$ is the modified Bessel function, and finally

$$\mathfrak{H}^{(0)}(s, L) = \frac{\sin(\pi s)}{\pi} \Gamma(s) \int_0^\infty dt\, \Psi_\Gamma(t + 2\rho_0; \chi)(2\rho_0 t + t^2)^{-s}. \tag{4.75}$$

4.5.2 *Simple complex Lie group*

In the following, we would like to illustrate the geometry and local isometry associated with a simple 3-dimensional complex Lie group. The discrete group can be chosen in the form $\Gamma \subset PSL(2, \mathbb{C}) \equiv SL(2, \mathbb{C})/\{\pm Id\}$, where Id is the $2 \otimes 2$ identity matrix and is an isolated element of Γ. The group Γ acts discontinuously at point $z \in \bar{\mathbb{C}}$, $\bar{\mathbb{C}}$ being the extended complex plane. We recall that a transformation $\gamma \neq Id$, $\gamma \in \Gamma$, with

$$\gamma z = \frac{az + b}{cz + d}, \qquad ad - bc = 1, \qquad (\mathrm{Tr}\gamma)^2 = (a + d)^2, \quad a, b, c, d \in \mathbb{C}, \tag{4.76}$$

is called elliptic if $(\mathrm{Tr}\gamma)^2$ satisfies $0 \leq (\mathrm{Tr}\gamma)^2 < 4$, hyperbolic if $(\mathrm{Tr}\gamma)^2 > 4$, parabolic if $(\mathrm{Tr}\gamma)^2 = 4$ and loxodromic if $(\mathrm{Tr}\gamma)^2 \in \mathbb{C}\backslash[0, 4]$. The classification of these transformations can also be based on the properties of their fixed points, the number of which is one for the parabolic transformations and two for all other cases. The element $\gamma \in SL(2, \mathbb{C})$ acts on $z = (y, w) \in \mathbb{H}^3$, $w = x_1 + ix_2$, by means of the following linear-fractional transformation:

$$\gamma z = \left(\frac{y}{|cw + d|^2 + |c|^2 y^2}, \frac{(aw + b)(\bar{c}\bar{w} + \bar{d}) + a\bar{c}y^2}{|cw + d|^2 + |c|^2 y^2} \right). \tag{4.77}$$

The isometric circle of a transformation $\gamma \in SL(2, \mathbb{C})/\{\pm Id\}$ for which ∞ is not a fixed point is defined to be the circle

$$I(\gamma) = \{z : \ |\gamma z| = 1\}, \quad \text{or} \quad I(\gamma) = \{z : \ |z + d/c| = |c|^{-1}\},\ c \neq 0. \tag{4.78}$$

A transformation γ carries its isometric circle $I(\gamma)$ into $I(\gamma^{-1})$. The isometric fundamental domain of a Fuchsian group (Kleinian group without loxodromic elements) has the following structure: it is bounded by arcs of circles orthogonal to the invariant circle and consists either of two symmetric components or of a single component, while the mappings connecting its equivalent sides, generate the whole group. In many cases, it is more convenient to deal with other fundamental regions. For example, the so-called normal fundamental Dirichlet's polygons are often used for Fuchsian groups and we shall follow this approach here.

We consider a special discrete group $SL(2, \mathbb{Z} + i\mathbb{Z})/\{\pm Id\}$, where $\mathbb{Z}$ is the ring of integers. The element $\gamma \in \Gamma$ will be identified with $-\gamma$. The group Γ has, within a conjugation, one maximal parabolic subgroup Γ_∞ $(c = 0)$. Thus, the fundamental domain related to Γ has one parabolic vertex and can be taken in the form [A.B. Venkov (1987); S. Lang (1985)]

$$F(\Gamma) = \left\{(y, w) : x_1^2 + x_2^2 + y^2 > 1, \ -\frac{1}{2} < x_2 < x_1 < \frac{1}{2}\right\}. \tag{4.79}$$

The Selberg trace formula. Now we are ready to start the discussion of the Selberg trace formula, which can be constructed as an expansion in eigenfunctions of the automorphic Laplacian. To begin with, we assume that the group Γ is generated by parabolic mappings. Since the discrete group Γ has a cusp at ∞ $(c = 0)$, each element of the stabiliser Γ_∞ is a translation. Computing the conjugacy class $\{\gamma\}_\Gamma$, $\gamma \in \Gamma_\infty$ with γ different from identity, one easily gets

Proposition 4.2 *Let*

$$\gamma = \begin{pmatrix} 1 & n_1 + in_2 \\ 0 & 1 \end{pmatrix}, \qquad n_1, n_2 \in \mathbb{Z}. \tag{4.80}$$

The conjugacy class with representative γ consists in element γ and γ^{-1}, where

$$\gamma^{-1} = \begin{pmatrix} 1 & -n_1 - in_2 \\ 0 & 1 \end{pmatrix}. \tag{4.81}$$

The remaining conjugacy classes have the representatives in Γ_∞ of the *form*

$$\gamma_1 = \begin{pmatrix} i & 0 \\ 0 & -i \end{pmatrix}, \ \gamma_2 = \begin{pmatrix} i & 1 \\ 0 & -i \end{pmatrix}, \ \gamma_3 = \begin{pmatrix} i & -i \\ 0 & -i \end{pmatrix}, \ \gamma_4 = \begin{pmatrix} i & 1-i \\ 0 & -i \end{pmatrix}. \tag{4.82}$$

The centralisers related to these representations read

$$\Gamma^{\gamma} = \begin{pmatrix} 1 & m_1 + im_2 \\ 0 & 1 \end{pmatrix}, \qquad m_1, m_2 \in \mathbb{Z}, \tag{4.83}$$

$$\Gamma^{\gamma_1} = \left\{ \begin{pmatrix} 1 & 0 \\ 0 & 1 \end{pmatrix}, \begin{pmatrix} i & 1 \\ 0 & -i \end{pmatrix}, \begin{pmatrix} 0 & 1 \\ -1 & 0 \end{pmatrix}, \begin{pmatrix} 0 & i \\ i & 0 \end{pmatrix} \right\}, \tag{4.84}$$

$$\Gamma^{\gamma_2} = \left\{ \begin{pmatrix} 1 & 0 \\ 0 & 1 \end{pmatrix}, \begin{pmatrix} i & 1 \\ 0 & -i \end{pmatrix}, \begin{pmatrix} i & 0 \\ 2 & -i \end{pmatrix}, \begin{pmatrix} -1 & i \\ 2i & 1 \end{pmatrix} \right\}, \tag{4.85}$$

$$\Gamma^{\gamma_3} = \left\{ \begin{pmatrix} 1 & 0 \\ 0 & 1 \end{pmatrix}, \begin{pmatrix} i & -i \\ 0 & -i \end{pmatrix}, \begin{pmatrix} 1 & -1 \\ 2 & -1 \end{pmatrix}, \begin{pmatrix} i & 0 \\ 2i & -i \end{pmatrix} \right\}, \tag{4.86}$$

$$\Gamma^{\gamma_4} = \left\{ \begin{pmatrix} 1 & 0 \\ 0 & 1 \end{pmatrix}, \begin{pmatrix} i & 1-i \\ 0 & -i \end{pmatrix}, \begin{pmatrix} i & 0 \\ 1+i & -i \end{pmatrix}, \begin{pmatrix} 1 & -1-i \\ 1-i & -1 \end{pmatrix} \right\}. \tag{4.87}$$

Let us consider an arbitrary integral operator with kernel $k(z, z')$. Invariance of the operator is equivalent to fulfillment of the condition $k(\gamma z, \gamma z') = k(z, z')$ for any $z, z' \in \mathbb{H}^3$ and $\gamma \in PSL(2, \mathbb{C})$. So the kernel of the invariant operator is a function of the geodesic distance between z and z'. It is convenient to replace such a distance with the fundamental invariant of a pair of points $u(z, z') = |z - z'|^2/yy'$, thus $k(z, z') = k(u(z, z'))$. Let λ_i be the isolated eigenvalues of the self-adjoint extension of the Laplace operator and let us introduce a suitable analytic function $h(r)$ and $r_j^2 = \lambda_j - 1$. It can be shown that $h(r)$ is related to the quantity $k(u(z, \gamma z))$ by means of the Selberg transform. Let us denote by $g(u)$ the Fourier transform of $h(r)$, namely $g(u) = (2\pi)^{-1} \int_{\mathbb{R}} dr\, h(r) \exp(-iru)$. For one parabolic vertex let us introduce a subdomain F_Y of the fundamental region $F(\Gamma)$ by means

$$F_Y = \{z \in F(\Gamma),\, z = \{y, \mathbf{x}\} | y \le Y\},$$

where Y is a sufficiently large positive number.

Lemma 4.1 ([A.B. Venkov (1987)]) *Suppose $h(r)$ to be an even analytic function in the strip $|\operatorname{Im} r| < 1+\varepsilon$ $(\varepsilon > 0)$ and $h(r) = \mathcal{O}(1+|r|^2)^{-2}$. Then the following formula holds:*

$$\sum_j h(r_j) = \lim_{Y\to\infty} \left\{ \int_{F_Y} d\mu(z) \sum_{\{\gamma\}_\Gamma} k(u(z,\gamma z)) \right.$$

$$\left. -\frac{1}{2\pi}\int_0^\infty dr\, h(r) \int_{F_Y} d\mu(z) |E(z,1+ir)|^2 \right\}, \tag{4.88}$$

where $d\mu(z) = y^{-3}dydx_1dx_2$ is the invariant measure on $\mathbb{H}^3$ and $E(z,s)$ is the Eisenstein–Maass series associated with one cusp, namely

$$E(z,s) = \sum_{\gamma\in(\Gamma/\Gamma_\infty)} y^s(\gamma z)\,, \quad x_2(z) = \operatorname{Im} z. \tag{4.89}$$

The series (4.89) converges absolutely for $\operatorname{Re} s > 1$ and uniformly in z on compact subset of $\mathbb{H}^3$. All poles of $E(z,s)$ are contained in the union of the region $\operatorname{Re} s < 1/2$ and the interval $[1/2, 1]$ and those contained in such an interval are simple. Furthermore, for each s, the series $E(z,s)$ is a real analytic function on $\mathbb{H}^3$, automorphic relative to the group Γ and satisfies the eigenvalues equation

$$LE(z,s) = s(s-1)E(z,s),$$

where L is the Laplace operator. The asymptotic expansion of the second integral in Eq. (4.88) can be found with the help of Maass–Selberg relation [A.B. Venkov (1987)]. For $Y \to \infty$ one has

$$\frac{1}{2\pi}\int_0^\infty dr\, h(r)\int_{F_Y} d\mu(z)\,|E(z,1+ir)|^2 = g(0)\log Y + \frac{h(0)}{4}S(1)$$

$$-\frac{1}{4\pi}\int_{\mathbb{R}} dr\, h(r)\frac{d}{ds}\log S(s)\Big|_{s=1+ir} + \mathcal{O}(1). \tag{4.90}$$

The function $S(s)$ (in the general case it is the S-matrix) is given by a generalized Dirichlet series, convergent for $\operatorname{Re} s > 1$,

$$S(s) = \frac{\pi^{1/2}\Gamma(s-1/2)}{\Gamma(s)} \sum_{c\neq 0}\sum_{0\leq d<|c|} |c|^{-2s},$$

where the sums are taken over all pairs c, d of the matrix $\begin{pmatrix} * & * \\ c & d \end{pmatrix} \subset \Gamma_\infty \backslash \Gamma / \Gamma_\infty$. Also the poles of the meromorphic function $S(s)$ are contained in the region $\mathrm{Re}\, s < 1/2$ and in the interval $[1/2, 1]$. The functions $E(z,s)$ and $S(s)$ can be analytically extended on the whole complex s-plane, where they satisfy the functional equations

$$S(s)S(1-s) = Id\,, \quad \overline{S(s)} = S(\bar{s})\,, \quad E(z,s) = S(s)E(z, 1-s).$$

It should be noted that the terms of the trace formula associated with the elements γ and γ^{-1} coincide. Then the contribution to the first integral in Eq. (4.88), which comes from all conjugacy classes of the γ-type ($\gamma \in \Gamma^\gamma$), for $Y \to \infty$ can be written as follows

$$\int_{F_Y} d\mu(z) \sum_{\{\gamma\}_{\Gamma_\infty}} k(u(z, \gamma z)) = (\log Y + C)g(0) + \frac{h(0)}{4}$$

$$-\frac{1}{4\pi} \int_{\mathbb{R}} dr\, h(r) \psi(1 + ir/2) + \mathcal{O}(1), \qquad (4.91)$$

where $\psi(s)$ is the logarithmic derivative of the Euler Γ-function and C a computable constant which reads [A.B. Venkov (1987)]

$$C = \frac{5\log 2}{16} - \frac{\gamma}{2} + C_0,$$

$$C_0 = \lim_{N \to \infty} \frac{1}{4\pi} \sum_{i=1}^{N} \left[|\xi^i|^{-2} - 2\pi \log \frac{|\xi^{i+1}|}{|\xi^i|} \right] - \frac{1}{2} \log |\xi^1|.$$

In the latter equation γ is the Euler–Mascheroni constant and ξ is a two-dimensional vector, such that $\gamma z = \{y, \omega + \xi\}$, $\xi \neq 0$, $|\xi^{i+1}| \geq |\xi^i|$.

For the derivation of the Selberg trace formula, one has to consider the contributions coming from the identity and the non-parabolic elements in Γ, the normalized Eisenstein–Maass series, and all γ-type conjugacy classes, Eq. (4.91). The final result we state should be considered as an explicit addition to Lemma 4.1.

Theorem 4.10 *For the special discrete group $SL(2, \mathbb{Z} + i\mathbb{Z})/\{\pm Id\}$ and $h(r)$ satisfying the conditions of Lemma 4.1, we have the Selberg's trace* *formula*

$$\sum_j h(r_j) - \sum_{\substack{\{\gamma\}_\Gamma, \gamma \neq Id, \\ \gamma - non-parabolic}} \int d\mu(z)\, k(u(z,\gamma z))$$

$$-\frac{1}{4\pi}\int_{\mathbb{R}} dr\, h(r)\frac{d}{ds}\log S(s)\big|_{s=1+ir} + \frac{h(0)}{4}(S(1)-1) - Cg(0)$$

$$= \mathrm{Vol}(F)\int_0^\infty \frac{dr\, r^2}{2\pi^2}\, h(r) - \frac{1}{4\pi}\int_{\mathbb{R}} dr\, h(r)\psi(1+ir/2). \qquad (4.92)$$

The first term in the right-hand side of Eq. (4.92) is the contribution of the identity element, while $\mathrm{Vol}(F)$ is the (finite) volume of the fundamental domain F with respect to the measure $d\mu$.

4.5.3 *An example of functional determinant evaluation*

Here we would like to present a simple application of the formalism developed so far to evaluate the functional determinant of a Laplace-like operator defined on $\Gamma\backslash\mathbb{H}^3$. It is convenient to introduce the operator $L(\delta) = -\Delta_3 + \delta^2 + \kappa$, where $\kappa = -1$ is the curvature of H^3 and δ a suitable parameter. Here $\Delta_3 = \Delta_\Gamma$. We know that the meromorphic structure of the analytically continued zeta function, is related to the asymptotic properties of the heat-kernel trace. For the rank one symmetric space $\Gamma\backslash\mathbb{H}^3$ the trace of the operator $\exp(-tL(\delta))$ may be computed by using Theorem 4.10 (Eq. (4.92)) with the choice $h(r) = \exp(-t(r^2+\delta^2))$ (we use units in which the curvature of $\mathbb{H}^3$ is equal to -1). We have

$$g(u) = (4\pi t)^{-1/2}\exp(-t\delta^2 - u^2/4t), \qquad h(0) = \exp(-t\delta^2).$$

In this and next sections we shall consider additive terms of the zeta function associated with identity and parabolic elements of group Γ only (the heat kernel and the zeta function analysis for co-compact discrete group Γ has been done, for example, in [E. Elizalde et al. (1994); A.A. Bytsenko et al. (1996)]). As a result

$$\mathrm{Tr}e^{-tL(\delta)} = e^{-t\delta^2}\left[\frac{\mathrm{Vol}(F)}{(4\pi t)^{3/2}} + \frac{C}{(4\pi t)^{1/2}} + \frac{1}{4} - \frac{1}{4\pi}\int_{\mathbb{R}} dr\, \psi(1+ir/2)e^{-tr^2}\right]. \qquad (4.93)$$

The asymptotic behavior of the last integral for $t \to 0^+$ can be easily evaluated. In fact we may write

$$\mathrm{Tr}e^{-tL(\delta)} = e^{-t\delta^2}\left[\frac{\log t}{8(\pi t)^{1/2}} + \frac{\mathrm{Vol}(F)}{(4\pi t)^{3/2}} + \frac{C+\log 2+\gamma/4}{(4\pi t)^{1/2}}\right.$$

$$\left. -\frac{t}{\pi i}\int_{\mathbb{R}} dr\, e^{-tr^2} f(ir/2)\right]. \tag{4.94}$$

The function $f(z)$ is defined by (see [I.S. Gradshteyn and I.M. Ryzhik (1980)])

$$f(z) = \frac{1}{2}\sum_{k=1}^{\infty}\frac{k}{(k+1)(k+2)}\sum_{n=1}^{\infty}\frac{1}{(n+z)^{k+1}},$$

and has an asymptotic expansion for large $|z|$ in terms of the Bernoulli numbers B_k given by

$$f(z) \sim \sum_{k=1}^{\infty}\frac{B_{2k}}{2k(2k-1)z^{2k-1}}.$$

The contribution for short t comes from this asymptotics. Thus we have

Proposition 4.3 *The asymptotic behavior of the heat kernel for $t \to 0^+$ reads*

$$\mathrm{Tr}e^{-tL(\delta)} \simeq e^{-t\delta^2}\left[\frac{\log t}{8(\pi t)^{1/2}} + \sum_{n=0}^{\infty}K_n t^{n-3/2}\right] = \sum_{r=0}^{\infty}(A_r + P_r\log t)t^{r-3/2}, \tag{4.95}$$

where the first K_n coefficients are given by

$$K_0 = \frac{\mathrm{Vol}(F)}{(4\pi)^{3/2}}, \quad K_1 = \frac{C+\log 2+\gamma/4}{(4\pi)^{1/2}}, \quad K_2 = (36\pi)^{-1/2}, \tag{4.96}$$

and

$$A_r = \sum_{n=0}^{r}(-1)^n\frac{B_{r-n}\delta^{2n}}{n!}, \qquad P_0 = 0, \qquad P_r = (-1)^{r-1}\frac{\delta^{2(r-1)}}{8\pi^{1/2}(r-1)!}. \tag{4.97}$$

It should be noted that due to the presence of parabolic elements in Γ besides the usual terms one has terms with logarithmic factors. These terms are absent for co-compact group Γ (compact hyperbolic manifolds).

Furthermore, the contribution of hyperbolic elements is exponentially small in t. Thus, in general, the result of Proposition 4.3 still holds true.

Let us analyze the consequences of the presence of logarithmic terms in the latter expansion. As usual, we may introduce the zeta function associated with the elliptic operator $L(\delta)$ by means of the Mellin transform. In order to get the meromorphic structure of the function (5.14), we splits the integration range into two intervals $[0,1)$ and $[1,\infty)$ obtaining in this way two integrals. The last one is regular for $s \to 0^+$, while the structure of the first follows from the asymptotics, Eq. (4.95). Thus, we have

Proposition 4.4 *The meromorphic structure of the zeta function reads*

$$\zeta(s|L(\delta)) = \frac{1}{\Gamma(s)}\sum_{r=0}^{\infty}\left[\frac{A_r}{s+r-\frac{3}{2}} - \frac{P_r}{(s+r-\frac{3}{2})^2}\right] + \frac{J(s)}{\Gamma(s)}, \tag{4.98}$$

where $J(s)$ is an analytic function.

It follows that the analytic continuation of the zeta function is regular at $s=0$. It has to be noted also the presence of double poles, caused by the logarithmic terms.

Asymptotic behavior. The asymptotic behavior for small t gives [V. Voros (1987)]

Proposition 4.5

$$\frac{d}{ds}\zeta(s|L(\delta))|_{s=0} = \frac{\mathrm{Vol}(F)\delta^3}{6\pi} + \frac{\delta}{2}\log\delta - \delta\left(C + \frac{1}{2}\log 2 + \frac{1}{2}\right) + \mathcal{O}(\delta^{-1}). \tag{4.99}$$

The zeta function for $\mathrm{Re}\, s$ sufficiently large, can be rewritten in the form

$$\zeta(s|L(\delta)) = \sum_{\sigma}\rho_\sigma\left(\lambda_\sigma + \delta^2 - 1\right)^{-s} = \sum_j \left(\lambda_j + \delta^2 - 1\right)^{-s}$$

$$+ \int_0^\infty d\lambda\, \rho_\lambda \left(\lambda + \delta^2 - 1\right)^{-s}, \tag{4.100}$$

where the sum over j run over the discrete spectrum, λ_j being the eigenvalues. For the continuous spectrum, ρ_λ is proportional to the logarithmic derivative of the S-matrix $S(s)$. One has

$$\frac{d}{ds}\zeta(s|L(\delta)) = -\sum_{\sigma}\rho_\sigma\left(\lambda_\sigma+\delta^2-1\right)^{-s}\log(\lambda_\sigma+\delta^2-1). \tag{4.101}$$

From the latter equation one gets

$$\frac{d}{d\delta}\left(\frac{1}{2\delta}\frac{d}{d\delta}\frac{d}{ds}\zeta(s|L(\delta))\right) = 2\delta\sum_{\sigma}\rho_\sigma\left(\lambda_\sigma+\delta^2-1\right)^{-s-2}+\mathcal{O}(s). \tag{4.102}$$

A standard Tauberian argument and Eq. (4.95) gives a Weyl's estimate for large σ, namely $(\lambda_\sigma+\delta^2-1)^{-1}\simeq\sigma^{-2/3}$. As a consequence, in the limit $s\to 0^+$, the right hand side of Eq. (4.102) is finite. This works for $D=2$ as well as for $D=3$ dimensions. In higher dimensions it is necessary to take further derivatives with respect to δ [V. Voros (1987)]. On the other hand, we may rewrite the formula Eq. (4.92) as (here $r^2+1=\lambda$ and γ is the identity or parabolic element in Γ)

$$G(\delta)=\sum_{\sigma}\rho_\sigma h(\delta,r_\sigma)=\sum_{j}h(\delta,\lambda_j)-\frac{1}{4}h(\delta,0)S(1)$$

$$-\frac{1}{4\pi}\int_{\mathbb{R}}dr\;h(\delta,r)\frac{d}{ds}\log S(s)|_{(s=1+ir)}. \tag{4.103}$$

$G(\delta)$ denoting the "geometrical" part

$$G(\delta)=\mathrm{Vol}(F)\int_0^\infty\frac{dr\,r^2}{2\pi^2}\,h(\delta,r)+Cg(\delta,0)+\frac{1}{4}h(\delta,0)$$

$$-\frac{1}{4\pi}\int_{\mathbb{R}}dr\;h(\delta,r)\psi(1+ir/2). \tag{4.104}$$

Let us choose the function $h(\delta,r)$ as

$$h(\delta,r)=\frac{1}{r^2+\delta^2}-\frac{1}{r^2+a^2}\,, \tag{4.105}$$

with a a non-vanishing constant. Its Fourier transform at the origin reads

$$g(\delta,0)=\frac{1}{2\delta}-\frac{1}{2a}.$$

Thus,

$$G(\delta) = \sum_{\sigma} \rho_\sigma \left(\frac{1}{r^2 + \delta^2} - \frac{1}{r^2 + a^2} \right) .$$

Taking the derivative with respect to δ we have

$$\frac{d}{d\delta} G(\delta) = -2\delta \sum_{\sigma} \rho_\sigma \left(\lambda_\sigma + \delta^2 - 1\right)^{-2} = -\frac{d}{d\delta} \left(\frac{1}{2\delta} \frac{d}{d\delta} \frac{d}{ds} [\zeta(s|L(\delta))]|_{s=0} \right) . \tag{4.106}$$

Making the comparison between Eqs. (4.102) and (4.106), integrating twice in the variable δ, we obtain

$$\frac{d}{ds} \zeta(s|L(\delta))|_{s=0} = -2 \int d\delta\, \delta G(\delta) + c_1 \delta^2 + c_2 , \tag{4.107}$$

where the constants c_1 and c_2 will be computed making use of the asymptotics for large δ. The Selberg trace formula with the choice (4.105) leads to

$$G(\delta) = -\frac{\mathrm{Vol}(F)}{4\pi} (\delta - a) + \frac{C}{2} \left(\frac{1}{\delta} - \frac{1}{a} \right) + \frac{1}{4} \left(\frac{1}{\delta^2} - \frac{1}{a^2} \right)$$

$$- \frac{1}{4} \left(\frac{\psi(1 + \delta/2)}{\delta} - \frac{\psi(1 + a/2)}{a} \right) . \tag{4.108}$$

As a consequence,

$$\frac{d}{ds} \zeta(s|L(\delta))|_{s=0} = \frac{\mathrm{Vol}(F)\delta^3}{6\pi} + [c_1 + Q(a)]\delta^2 - C\delta$$

$$- \frac{1}{2} \log\delta + \log\Gamma(1 + \delta/2) + c_2 , \tag{4.109}$$

where

$$Q(a) = \frac{1}{4a^2} - \frac{\mathrm{Vol}(F)a}{4\pi} + \frac{C}{2a} - \frac{\psi(1 + a/2)}{4a} .$$

The inclusion of the contribution related to the hyperbolic elements is almost straightforward and can be found in [E. Elizalde et al. (1994); A.A. Bytsenko et al. (1996)]. It is additive and reads simply $\log Z_\Gamma(1+\delta, \chi)$,

$Z_\Gamma(s,\chi)$ being the Selberg zeta function. In the large δ-limit, $\log Z_\Gamma(1+\delta,\chi)$ is vanishing and one has

$$\frac{d}{ds}\zeta(0|L(\delta))|_{s=0} \simeq \frac{\mathrm{Vol}(F)\delta^3}{6\pi} + [c_1 + Q(a)]\delta^2 + \frac{1}{2}\delta\log\delta$$

$$-\delta\left(C + \frac{1}{2}\log 2 + \frac{1}{2}\right) + \frac{1}{2}\log\pi + c_2. \qquad (4.110)$$

Making the comparison, we get $c_1 = -Q(a)$, $c_2 = -(1/2)\log\pi$.

Summarizing we have proved the following theorem.

Theorem 4.11 ([A.A. Bytsenko et al. (1997a)]) *One has the identity*

$$\det L(\delta) = \frac{2}{(\pi\delta)^{1/2}\Gamma(\delta/2)} \exp\left(-\frac{\mathrm{Vol}(F)\delta^3}{6\pi} + C\delta\right) Z_\Gamma(1+\delta,\chi). \quad (4.111)$$

4.5.4 *Scalar fields in spacetime with spatial section of the form* $\Gamma\backslash\mathbb{H}^3$

As physical application, here we consider a self-interacting scalar field in an ultrastatic spacetime $\mathcal{M}$ with the spatial section of the form $\Gamma\backslash\mathbb{H}^3$ of finite volume of the fundamental domain (the discrete group of isometries Γ contains hyperbolic and parabolic elements only). We start with the classical Euclidean action for the scalar field in $\mathbb{R}\times\Gamma\backslash\mathbb{H}^3$

$$S[\phi,g] = \int_{\mathcal{M}} d^4x\,\sqrt{g}\left[-\frac{1}{2}\phi\Delta\phi + V_c(\phi)\right], \qquad (4.112)$$

where the classical potential reads

$$V_c(\phi) = \frac{1}{2}(m^2+\xi R)\phi^2 + \frac{\lambda}{4!}\phi^4 ,$$

where Δ is the Laplace–Beltrami operator, R is the scalar curvature, m is the mass, λ and ξ are arbitrary constants. The small disturbance operator associated with the action (4.112) may be written in the form

$$L_4 = -\Delta_4 + V''(\phi_c) = -\Delta_4 + \delta^2 - 1 , \qquad (4.113)$$

where $\delta^2 - 1 = m^2 + \xi R + \lambda\phi_c^2/2$ is a constant and the prime stands for the derivative with respect to the classical (constant) field ϕ_c. Since we are dealing with an ultrastatic spacetime, we have

$$L_4 = -\partial_\tau^2 - \Delta_3 + \delta^2 + \kappa = -\partial_\tau^2 + L(\delta)\,, \tag{4.114}$$

Δ_3 being the Laplace–Beltrami operator acting on $\Gamma\backslash\mathbb{H}^3$.

The one-loop effective potential is defined by

$$V_{eff}(\phi_c) = V(\phi_c) + \frac{1}{2V_1 V(F)}\log\det(\ell^2 L_4)\,, \tag{4.115}$$

where ℓ is a dimensional renormalization parameter, V_1 the (infinite) interval of imaginary time, and $V(F)$ is the volume of the fundamental domain of $\Gamma\backslash\mathbb{H}^3$.

Our interest in this example is the computation of the above functional determinant. It requires a generalized zeta function regularization, since the zeta function $\zeta(s|L_4)$ turns out to be non regular at the origin, and a simple pole being present there. To show that it is the case, let us compute the asymptotic properties of the heat-kernel trace, which in our case reads

$$\mathrm{Tr}\exp(-tL_4) = \frac{V_1}{\sqrt{4\pi t}}\,\mathrm{Tr}\exp(-tL(\delta))\,. \tag{4.116}$$

The asymptotics related to $\mathrm{Tr}\exp(-tL(\delta))$ has been evaluated in the previous subsection (see Proposition 4.3). As a consequence, the short t-expansion for the heat-kernel trace related to the operator L_4 reads

$$\mathrm{Tr}\, e^{-tL_4} \simeq \frac{V_1}{\sqrt{4\pi}}\sum_{r=0}^{\infty}(A_r + P_r\log t)\, t^{r-2}\,. \tag{4.117}$$

As we known, the presence of logarithmic terms in the expansion Eq. (4.117) leads to additional poles in the meromorphic structure of zeta function, which reads

$$\zeta(s|L_4) = \frac{V_1}{\sqrt{4\pi}\Gamma(s)}\sum_{r=0}^{\infty}\left[\frac{A_r}{s+r-2} - \frac{P_r}{(s+r-2)^2}\right] + J_4(s)\,, \tag{4.118}$$

where $J_4(s)$ is an analytic function. We see that in contrast with the usual cases, the analytic continuation of zeta function is not regular at $s = 0$, but it has a simple pole. It has to be noted also the presence of a double pole at $s = 1$ and new simple poles at $s = -n$ ($n \geq 2$). With regard to the computation of the effective action, we have to use the expression (1.7) of Section 1. The term whose presence requires the use of the generalized zeta-function regularization depends on

$$P_2 = -\frac{1}{8\sqrt{\pi|\kappa|}}\left[m^2 + \frac{\lambda\phi_c^2}{2} + |R|\left(\frac{1}{6} - \xi\right)\right], \tag{4.119}$$

where κ has been recovered for dimensional reasons. Furthermore, in the case of massless free scalar field with the conformal coupling $\xi = 1/6$, we have $P_2 = 0$ and thus, in this particular case the usual evaluation of the effective action by means of usual zeta function methods works without any modification.

Chapter 5

Spinor Fields

Within the Euclidean formulation, we have seen that second order, elliptic, non-negative differential operators play a role in the quantum dynamics of Boson fields (scalar, vector or tensor fields). Heat-kernel techniques are at disposal and the related zeta function may be defined via the Mellin transform, Eq. (1.13). The situation is quite different in the case of fermion fields, since one is dealing with a first order differential operator $\mathfrak{D}$ (the Dirac or the Rarita - Schwinger operator).

In this Chapter, we shall mainly deal with spinor fields. To begin with, let us consider the self-adjoint case. The spectrum is unbounded over the whole real axis, thus the heat-kernel techniques are no longer useful and one has to make use of the general theory of complex power of pseudo-differential operators (ΨDO) [R.T. Seeley (1967)]. Here we recall some definitions on ΨDOs. More details can be found in Appendix B. Let us consider a D-dimensional, smooth (compact) manifold without boundary, $\mathcal{M}$, and a (classical) ΨDO, Q, acting on sections of vector bundles on $\mathcal{M}$. To any classical ΨDO, it corresponds to a complete symbol $\sigma(Q) = Q(x,k) = e^{ikx}Qe^{-ikx}$, such that, modulo infinitely smoothing operators, one has

$$(Qf)(x) \sim \int_{\mathbb{R}^D} \frac{dk}{(2\pi)^D} \int_{\mathbb{R}^D} dy \; e^{i(x-y)k} Q(x,k) f(y) \,. \tag{5.1}$$

The complete symbol of Q admits an asymptotic expansion for $|k| \to \infty$, given by

$$Q(x,k) \sim \sum_{j=0}^{\infty} Q_{q-j}(x,k) \,, \tag{5.2}$$

where the coefficients fulfill the homogeneity property

$$Q_{q-j}(x, tk) = t^{q-j} Q_{q-j}(x, k) \,, \tag{5.3}$$

for $t > 0$, being $Q_q(x, k) \neq 0$. The number q is called the order of Q and $Q_q(x, k)$ its principal symbol. Recall that Q is elliptic when $Q_q(x, k)$ is invertible.

The complex powers Q_θ^{-s} related to a ΨDO, Q, can be defined as soon as there exists a conical neighborhood C_θ of a ray $L_\theta = \{\lambda \in \mathbb{C}, \arg\lambda = \theta\}$ such that it contains, at most, only a finite number of eigenvalues of Q. Under this condition, the zeta function can be defined for $\operatorname{Re} s$ larger than an abscissa of convergence, given by D/q, as $\zeta_\theta(s|Q) = \operatorname{Tr} Q_\theta^{-s} = \sum_i \lambda_i^{-s}$, where the sum is over the whole spectrum of Q and λ^{-s} is defined with respect to the admissible cut L_θ, namely

$$\lambda^{-s} = \exp\left(-s\log_\theta\lambda\right), \qquad \theta - 2\pi \leq \operatorname{Im}\log_\theta\lambda < \theta \,. \tag{5.4}$$

As in the case of differential operators, it is possible to show that $\zeta_\theta(s|Q)$ possesses a meromorphic continuation to the whole complex plane and that it is regular at the origin [R.T. Seeley (1967)]. Thus, one can define the functional determinant related to Q by means of the equation

$$\log\det Q = -\zeta'_\theta(0|Q) \,. \tag{5.5}$$

In general, both the zeta function as well as its derivative may depend on the spectral cut L_θ. Following Wodzicki, one is led to study the function

$$\rho(s|Q) = \zeta_\theta(s|Q) - \zeta_{\theta_1}(s|Q) \,, \tag{5.6}$$

where L_{θ_1} is another admissible cut. The general properties of $\rho(s|Q)$ have been investigated in [M. Wodzicki (1982); M. Wodzicki (1984)] and they can be expressed by means of the non-commutative residue (see Appendix B). We shall mainly be interested in $\rho'(0|Q)$, since it is related to the ambiguity in the definition of the functional determinant of Q, and we will discuss it, in an elementary way. Here we only recall the case when the ambiguity may be present.

Let us introduce the two sectors of the complex plane associated with the two admissible cuts, namely

$$S = \{\lambda, \theta < \arg\lambda < \theta_1\} \,, \qquad S_1 = \{\lambda, \theta_1 - 2\pi < \arg\lambda < \theta\} \,. \tag{5.7}$$

Now, it is possible to show that if there is an infinite number of eigenvalues of Q in both sectors S and S_1, then $\rho'(0|Q)$ is non-vanishing and there exists an ambiguity in the definition of the functional determinant of Q [M. Wodzicki (1982); M. Wodzicki (1984)] . As shown by Wodzicki, a new spectral function plays a role in relation to this ambiguity. It is the eta function [M. Atiyah et al. (1973b)], which, for Re s sufficiently large, is defined by

$$\eta(s|Q) = \sum_i \text{sign}\lambda_i |\lambda_i|^{-s} \,. \tag{5.8}$$

It can be proven that it has a meromorphic extension, which is regular at the origin [M. Wodzicki (1982); M. Wodzicki (1984)] .

If we apply these general results to the Dirac operator, it follows that, in general, there is an ambiguity in the definition of the effective action, already noticed in [R.E. Gamboa Saravi et al. (1996); S. Deser et al. (1998)], and the eta invariant $\eta(0|Q)$ will enter the game. This happens only in odd dimension and it is known as the appearance of an intrinsic parity anomaly term, the so-called induced Chern - Simons term [A. Niemi and G. Semenoff (1983); A. Niemi and G. Semenoff (1983a); A.J. Niemi and G.W. Semenoff (1986); A.N. Redlich (1984); A.N. Redlich (1984a)]. In even dimension the ambiguity is still present, but it depends on a Seeley - DeWitt coefficient and may be absorbed by a redefinition of the scale renormalization (see, for example [S. Deser et al. (1998)]).

5.1 The Dirac Operator and Spectral Invariants

As mentioned above, the general theory of complex powers of elliptic operators works very well even in cases when the elliptic operator is *not* positive definite. In such situations, besides the zeta function, a very important role is played by the eta function (see for example [J. Park and K.P. Wojciechowski (2002)] and references quoted therein and [E. Elizalde et al. (2003a)]).

For simplicity, we start with a self-adjoint operator, which for the moment we assume to be the massless Euclidean Dirac operator on the compact (curved) manifold $\mathcal{M}$ without boundary. It has the form

$$\mathfrak{D} = i\gamma^\mu \nabla_\mu \,, \qquad \gamma_\mu = e^\mu_k \gamma^k \,, \tag{5.9}$$

where e_k^μ are "viel-bein" fields, γ^k are the Dirac matrices in D-dimensions and ∇_μ is the covariant derivative in the spin connection. Since $\mathfrak{D}^+ = \mathfrak{D}$, the eigenvalues of $\mathfrak{D}$ are real. Let λ_i and μ_i be the positive and negative eigenvalues, respectively. We have essentially two possible definitions of the zeta function, because there exist two essentially different cuts, one in the lower and other in the upper half-planes respectively. In both cases, the associated sectors contain an *infinite* number of eigenvalues and related zeta functions read (for $\mathrm{Re}\, s$ sufficiently large)

$$\zeta_+(s|\mathfrak{D}) = \sum_i \lambda_i^{-s} + e^{-i\pi s} \sum_i (-\mu_i)^{-s} \tag{5.10}$$

and

$$\zeta_-(s|\mathfrak{D}) = \sum_i \lambda_i^{-s} + e^{i\pi s} \sum_i (-\mu_i)^{-s} . \tag{5.11}$$

On the other hand, the eta function may also be defined as

$$\eta(s|\mathfrak{D}) = \sum_i \lambda_i^{-s} - \sum_i (-\mu_i)^{-s} = \mathrm{Tr}\left(\frac{\mathfrak{D}}{|\mathfrak{D}|} (\mathfrak{D}^2)^{-\frac{s}{2}} \right) . \tag{5.12}$$

Making use of the above definitions, a direct computation gives [M. Wodzicki (1982)]

$$\eta(s|\mathfrak{D}) = \frac{1+e^{i\pi s}}{2i \sin \pi s} \zeta_+(s|\mathfrak{D}) - \frac{1+e^{-i\pi s}}{2i \sin \pi s} \zeta_-(s|\mathfrak{D}) \tag{5.13}$$

and

$$\zeta\left(\frac{s}{2}|\mathfrak{D}^2\right) = \frac{1-e^{-i\pi s}}{2i \sin \pi s} \zeta_-(s|\mathfrak{D}) - \frac{1-e^{i\pi s}}{2i \sin \pi s} \zeta_+(s|\mathfrak{D}) , \tag{5.14}$$

where $\zeta(z|\mathfrak{D}^2)$ is the zeta function associated with the spinor Laplacian-like operator $\mathfrak{D}^2$, which is a second order non-negative self-adjoint elliptic operator.

In order to study the Wodzicki function,

$$\rho(s|\mathfrak{D}) = \zeta_+(s|\mathfrak{D}) - \zeta_-(s|\mathfrak{D}), \tag{5.15}$$

we may express the two zeta functions in terms of the eta function and $\zeta(z|\mathfrak{D}^2)$. From Eqs. (5.13) and (5.14), one gets

$$\zeta_\pm(s|\mathfrak{D}) = \frac{1}{2}\left[1 + e^{\mp i\pi s}\right] \zeta\left(\frac{s}{2}|\mathfrak{D}^2\right) + \frac{1}{2}\left[1 - e^{\mp i\pi s}\right] \eta(s|\mathfrak{D}) . \tag{5.16}$$

As a consequence, we obtain the result

$$\rho(s|\mathfrak{D}) = i\sin(\pi s)\left[\eta(s|\mathfrak{D}) - \zeta(\tfrac{s}{2}|\mathfrak{D}^2)\right]. \tag{5.17}$$

For physical applications, it is also useful to give the first derivative with respect to s, i.e.

$$\rho'(s|\mathfrak{D}) = i\sin(\pi s)\left[\eta'(s|\mathfrak{D}) - \frac{1}{2}\zeta'\left(\frac{s}{2}|\mathfrak{D}^2\right)\right] + i\pi\cos(\pi s)\left[\eta(s|\mathfrak{D}) - \zeta\left(\frac{s}{2}|\mathfrak{D}^2\right)\right].$$

Then, the ambiguity is described by the value at $s = 0$ of the Wodzicki function and its derivative. These values depend on the regularity at $s = 0$ of $\eta(s|\mathfrak{D})$ and $\zeta(\frac{s}{2}|\mathfrak{D}^2)$. The latter is known to be regular (the Seeley theorem). With regard to the eta function, Wodzicki has proven its analyticity at the origin. However, in the particular but important case of the Dirac operator, a direct approach is known and we will use it.

First, let us consider the even dimensional case, $D = 2p$. Here, the 'γ^5' matrix exists, namely: $\gamma^{D+1} = \gamma^1 \cdots \gamma^D$, and, as is well known, since γ^{D+1} anticommutes with $\mathfrak{D}$, both λ_i and $-\lambda_i$ are eigenvalues for $\mathfrak{D}$ with eigenfunctions $\psi_i^\dagger$ and $\psi_i^- = \gamma^{D+1}\psi_i^\dagger$ respectively. In these conditions, the eta function identically vanishes. We have $\rho(0|\mathfrak{D}) = 0$ and

$$\zeta_\pm(s|\mathfrak{D}) = \frac{1}{2}\left[1 + e^{\mp i\pi s}\right]\zeta\left(\frac{s}{2}|\mathfrak{D}^2\right), \tag{5.18}$$

$$\zeta_+(0|\mathfrak{D}) = \zeta_-(0|\mathfrak{D}) = A_p(\mathfrak{D}^2), \tag{5.19}$$

$$\zeta'_\pm(0|\mathfrak{D}) = \frac{1}{2}\zeta'(0|\mathfrak{D}^2) \mp \frac{i\pi}{2}A_p(\mathfrak{D}^2), \tag{5.20}$$

and

$$\zeta'_+(0|\mathfrak{D}) - \zeta'_-(0|\mathfrak{D}) = -i\pi A_p(\mathfrak{D}^2). \tag{5.21}$$

Here the ambiguity depends on the local functional of the external gauge field, the Seeley - DeWitt coefficient $A_p(\mathfrak{D}^2)$, and since the one-loop effective action is defined modulo counterterms of the same nature, it can be reabsorbed by the renormalization procedure.

5.1.1 *The eta invariant*

In the odd dimensional case, $D = 2p+1$, however there is no symmetry in the spectrum and one has a non-trivial eta function. Furthermore, starting from (5.12), one may also use the Mellin representation of the eta function, namely for $\mathrm{Re}\, s$ sufficiently large:

$$\eta(s|\mathfrak{D}) = \frac{1}{\Gamma(s)} \int_0^\infty dt\ t^{s-1}\, \mathrm{Tr}\left(\frac{\mathfrak{D}}{|\mathfrak{D}|} e^{-t|\mathfrak{D}|}\right), \tag{5.22}$$

which can be conveniently rewritten in the form of

$$\eta(s|\mathfrak{D}) = \frac{1}{\Gamma(\frac{s+1}{2})} \int_0^\infty dt\ t^{\frac{s+1}{2}-1}\, \mathrm{Tr}\left(\mathfrak{D} e^{-t\mathfrak{D}^2}\right). \tag{5.23}$$

With standard heat-kernel techniques, we can try to find the analytic continuation of the eta invariant. To this aim, one needs the short t asymptotics of the integrand. This involves a highly non-trivial calculation and the answer is provided by a theorem due to Bismut and Freed [J.M. Bismut and D.S. Freed (1986b)], which states that, for $t \to 0$,

$$\mathrm{Tr}\left(\mathfrak{D} e^{-t\mathfrak{D}^2}\right) \simeq \sum_{l=0}^{\infty} C_l t^{l+\frac{1}{2}}, \tag{5.24}$$

where the C_l are suitable coefficients. Then, the meromorphic extension reads

$$\eta(s|\mathfrak{D}) = \frac{1}{\Gamma(\frac{s+1}{2})} \sum_{l=0}^{\infty} \frac{C_l}{s/2+l+1} + \eta_0(s|\mathfrak{D}), \tag{5.25}$$

where $\eta_0(s|\mathfrak{D})$ is an analytic function of s. Thus, the regularity of the eta function for $\mathrm{Re}\, s > -2$ follows and $\eta(s|\mathfrak{D})$ is analytic at the origin. As a result,

$$\zeta_+(0|\mathfrak{D}) = \zeta_-(0|\mathfrak{D}) = 0, \tag{5.26}$$

$$\zeta'_\pm(0|\mathfrak{D}) = \frac{1}{2}\zeta'(0|\mathfrak{D}^2) \pm \frac{i\pi}{2}\eta(0|\mathfrak{D}), \tag{5.27}$$

and

$$\zeta'_+(0|\mathfrak{D}) - \zeta'_-(0|\mathfrak{D}) = i\pi\eta(0|\mathfrak{D}). \tag{5.28}$$

Here, the ambiguity is non-trivial, since, as it is clear from (5.25), $\eta(0|\mathfrak{D})$ is a non-local functional of the external gauge field and it cannot be removed by addition counterterms. Note that, in fact, all our results are particular cases of Wodzicki's main theorem and its corollaries and we need no new proofs.

Using the theorem of Bismut and Freed [J.M. Bismut and D.S. Freed (1986b)], and making a Mellin transform inversion (see Appendix A, Section A.2), we also get

$$\mathrm{Tr}\left(\frac{\mathfrak{D}}{|\mathfrak{D}|}e^{-t|\mathfrak{D}|}\right) = \frac{1}{2\pi i}\int_{\mathrm{Re}\, z>0} dz t^{-z}\Gamma(z)\eta(z|\mathfrak{D})\,. \tag{5.29}$$

For $z = 0$ one has a simple pole, while for $z = -2, -4, ..,$ one has double poles. Shifting the vertical line to the left, one obtains the small-t asymptotics:

$$\mathrm{Tr}\left(\frac{\mathfrak{D}}{|\mathfrak{D}|}e^{-t|\mathfrak{D}|}\right) \simeq \eta(0|\mathfrak{D}) + \sum_{l=0}^{\infty}\frac{2C_{l+2+p}}{(2l+2)!\Gamma(-l-\frac{1}{2})} \times\left[\gamma - \Psi(-l-\frac{1}{2}) - \mathrm{log} t\right]t^{2l+2}\,, \tag{5.30}$$

where $\Psi(z)$ is the di-gamma function. It should be noted the presence of logarithmic terms in this short-t expansion, due to the fact that one is dealing with a ΨDO. From the latter equation we have the following result

$$\lim_{t\to 0}\mathrm{Tr}\left(\frac{\mathfrak{D}}{|\mathfrak{D}|}e^{-t|\mathfrak{D}|}\right) = \eta(0|\mathfrak{D})\,, \tag{5.31}$$

which is useful in the evaluation of the eta invariant. Other two representations of the eta invariant can be easily obtained. The first can be obtained integrating Eq. (5.22) and analytically continue the eta function at $s = 0$,

$$\eta(0|\mathfrak{D}) = \int_0^{\infty} dt\ \mathrm{Tr}\left(\mathfrak{D}e^{-t|\mathfrak{D}|}\right)\,. \tag{5.32}$$

The second one can be obtained from Eq. (5.23) evaluated at $s = 0$, namely

$$\eta(0|\mathfrak{D}) = \frac{1}{\sqrt{\pi}}\int_0^{\infty} dt\ t^{-1/2}\,\mathrm{Tr}\left(\mathfrak{D}e^{-t\mathfrak{D}^2}\right)\,. \tag{5.33}$$

We present now an argument for the regularity of the eta function at the origin. Let us start from Eq. (5.12), $\eta(s|\mathfrak{D}) = \mathrm{Tr}\left(Q(\mathfrak{D}^2)^{-\frac{s}{2}}\right)$, with $Q =$

$\frac{\mathfrak{D}}{|\mathfrak{D}|}$, $|\mathfrak{D}| = \sqrt{\mathfrak{D}^2}$. Q is a ΨDO of zeroth order. By making use of Lemma B.1 of Appendix B, near the origin, we have

$$\eta(s|\mathfrak{D}) = \frac{\mathrm{res}(Q)}{s} + \frac{\gamma}{2}\mathrm{res}(Q) - \alpha'_{D/2+1}(0) + \mathcal{O}(s)\,. \tag{5.34}$$

Thus, we have arrived at the following statement:

Proposition 5.1 *The eta function related to a non-positive elliptic operator $\mathfrak{D}$ is regular at $s = 0$, if and only if* $\mathrm{res}(\frac{\mathfrak{D}}{|\mathfrak{D}|}) = 0$, *where* $|\mathfrak{D}| = \sqrt{\mathfrak{D}^2}$.

In fact, if $\mathrm{res}(\frac{\mathfrak{D}}{|\mathfrak{D}|}) = 0$, then from Eq. (5.34), the eta function is regular at the origin. Vice versa, if the eta invariant is finite, one has an asymptotic in t for $\mathrm{Tr}\left(\frac{\mathfrak{D}}{|\mathfrak{D}|}e^{-t|\mathfrak{D}|}\right)$ similar to Eq. (5.29) but without $\log t$ term. As a consequence, according to the result of Appendix B, $\mathrm{res}(\frac{\mathfrak{D}}{|\mathfrak{D}|}) = 0$. We illustrate Proposition 5.1 in the following simple example.

Example 5.1 On a compact flat manifold $\mathcal{M}$ with odd dimension D, the massless Dirac operator has the form $\mathfrak{D} = i\gamma^\mu\nabla_\mu$, γ^μ being the Dirac matrices in $D-1$ dimensions. For the sake of simplicity, let $D = 3$ and assume a constant gauge potential V_μ. Thus, $\mathfrak{D}^2 = -\Box$ and $|\mathfrak{D}| = \sqrt{-\Box}$. The local symbol reads

$$\sigma\left(\frac{\mathfrak{D}}{|\mathfrak{D}|}\right)(x,k) = \frac{\not{k}\not{V}}{\sqrt{k^2+V^2+2\not{k}\not{V}}}\,, \tag{5.35}$$

where, as usual, for any vector V, we indicate by $\not{V} = \gamma^\mu V_\mu$ the contraction with the Dirac matrices. In order to select the homogeneous component of degree -3, we may consider $k \to \lambda k$, with λ very large and pick up the term proportional to $1/\lambda^3$. One gets, with $|k| = 1$,

$$\sigma\left(\frac{\mathfrak{D}}{|\mathfrak{D}|}\right)(x,\lambda k) = ... + \left(C_1\not{V} + C_2(\not{V})^3\right)\lambda^{-3} + ..., \tag{5.36}$$

where the C_1 and C_2 are simple functions of the gauge potential. In the evaluation of the Wodzicki residue one has to take the internal trace tr and this gives a vanishing contribution because, in $D = 3$, not only $\mathrm{tr}\,\not{V} = 0$, but also $\mathrm{tr}(\not{V}^3) = 0$, since $\mathrm{tr}\,\gamma_\mu\gamma_\nu\gamma_\rho = 2\varepsilon_{\mu\nu\rho}$.

5.1.2 *Induced Chern-Simons terms by quantum effects*

As an application of Eq. (5.31), we will present a simple proof of the well known result obtained in [A.N. Redlich (1984); A.N. Redlich (1984a)], where it has been shown that quantum massless fermions coupled to a classical gauge field in $2+1$ dimensions generate a Chern-Simons term. The same result has been obtained in [R.E. Gamboa Saravi et al. (1985)] within a local zeta-function approach. The idea is to compute the first variation of the eta invariant. Let us work in $D = 2p+1$ dimension and consider the massless Dirac operator in the form $\mathfrak{D} = i\gamma^\mu\partial_\mu - e\gamma^\mu V_\mu$, V_μ being the Abelian gauge vector field and e the electric charge. The Euclidean matrices are γ_μ with γ_{2p+1} which plays the role of chirality matrix in $2p$ dimensions. Starting from Eq. (5.31), one gets

$$\delta\eta(0|\mathfrak{D}) = -\lim_{t\to 0}\left[t\,\mathrm{Tr}\left(\delta\mathfrak{D}e^{-t|\mathfrak{D}|}\right)\right]. \tag{5.37}$$

Here $\delta\mathfrak{D} = \delta(i\gamma^\mu\partial_\mu - e\gamma^\mu V_\mu) = -e\gamma^\mu\delta V_\mu$. Thus,

$$\mathrm{Tr}\left(\delta\mathfrak{D}e^{-t|\mathfrak{D}|}\right) = -e\int dx\delta V_\mu\,\mathrm{tr}\left(\gamma^\mu e^{-t|\mathfrak{D}|}(x,x)\right). \tag{5.38}$$

We may write

$$\frac{\delta\eta(0|\mathfrak{D})}{\delta V_\mu(x)} = -e\lim_{t\to 0}\left[t\,\mathrm{tr}(\gamma^\mu e^{-t|\mathfrak{D}|}(x,x))\right]. \tag{5.39}$$

If we start from (5.33), we obtain

$$\delta\eta(0|\mathfrak{D}) = \frac{1}{\sqrt{\pi}}\int_0^\infty dt\,t^{-1/2}\,\mathrm{Tr}\left(\delta\mathfrak{D}(1-2t\mathfrak{D}^2)e^{-t\mathfrak{D}^2}\right). \tag{5.40}$$

Due to the fact that

$$2\frac{d}{dt}\left[t^{1/2}\,\mathrm{Tr}\left(\delta\mathfrak{D}e^{-t\mathfrak{D}^2}\right)\right] = t^{-1/2}\,\mathrm{Tr}\left(\delta\mathfrak{D}e^{-t\mathfrak{D}^2}\right) - t^{1/2}\,\mathrm{Tr}\left(\delta\mathfrak{D}\mathfrak{D}^2e^{-t\mathfrak{D}^2}\right),$$

one gets

$$\delta\eta(0|\mathfrak{D}) = -\frac{2}{\sqrt{\pi}}\lim_{t\to 0}\left[\sqrt{t}\,\mathrm{Tr}\left(\delta\mathfrak{D}e^{-t\mathfrak{D}^2}\right)\right]. \tag{5.41}$$

As a result,

$$\frac{\delta\eta(0|\mathfrak{D})}{\delta V_\mu(x)} = -\frac{2e}{\sqrt{\pi}}\lim_{t\to 0}\left[\sqrt{t}\mathrm{tr}(\gamma^\mu e^{-t\mathfrak{D}^2}(x,x))\right]. \tag{5.42}$$

The operator $\mathfrak{D}^2$ is self-adjoint, thus, the related spectral theorem gives

$$e^{-t|\mathfrak{D}|}(x,x)=\int_0^\infty d\lambda e^{-t\sqrt{\lambda}}\rho_\lambda(\mathfrak{D}^2)(x,x)\,, \tag{5.43}$$

and

$$e^{-t\mathfrak{D}^2}(x,x)=\int_0^\infty d\lambda e^{-t\lambda}\rho_\lambda(\mathfrak{D}^2)(x,x)\,. \tag{5.44}$$

In the $t\to 0$ limit, only the asymptotics for $\lambda\to\infty$ of the local spectral density is relevant. For $D=2p+1$, when $\lambda\to\infty$ one has (see Eq. (1.9) in Chapter 1)

$$\rho_\lambda(\mathfrak{D}^2)(x,x)\simeq\sum_{r=0}^{p}a_r(x)\frac{\lambda^{p-\frac{1}{2}-r}}{\Gamma(p+\frac{1}{2}-r)}+\ldots\,, \tag{5.45}$$

where $a_r(x)$ are the heat-kernel coefficients associated with the spinor Laplacian $\mathfrak{D}^2$. Thus, one has the usual heat-kernel expansion for $e^{-t\mathfrak{D}^2}$ and

$$e^{-t|\mathfrak{D}|}(x,x)\simeq\sum_{r=0}^{p}a_r(x)\frac{\Gamma(2p+1-2r)}{\Gamma(p+\frac{1}{2}-r)}t^{2r-2p-1}\,. \tag{5.46}$$

As a result we get

$$\frac{\delta\eta(0|\mathfrak{D})}{\delta V_\mu(x)}=-\frac{2e}{\sqrt{\pi}}\mathrm{tr}(\gamma^\mu a_p(x))-2e\lim_{t\to 0}\sum_{r<p}\frac{\Gamma(2p+1-2r)}{\Gamma(p+1/2-r)}\mathrm{tr}(\gamma^\mu a_r(x))t^{2r-2p}\,.$$

Due to the trace properties of the Dirac matrices in $2p$ dimensions, it is possible to show that

$$\mathrm{tr}(\gamma^\mu a_r(x))=0\,,\;\; r=0,1,...,p-1\,. \tag{5.47}$$

We will give a proof of this fact later. Thus,

$$\frac{\delta\eta(0|\mathfrak{D})}{\delta V_\mu(x)}=\frac{2e}{\sqrt{\pi}}\mathrm{tr}(\gamma^\mu a_p(x))\,. \tag{5.48}$$

With respect to the related $2p$-dimensional theory, it turns out that the above variation is proportional to the $2p$-dimensional chiral anomaly [C.D. Fosco et al. (1999)]. As an example, let us consider the three-dimensional case. We have

$$\rho_\lambda(\mathfrak{D}^2)(x,x)\simeq\frac{\lambda^{\frac{1}{2}}}{\Gamma(\frac{3}{2})}+a_1(x)\frac{\lambda^{-\frac{1}{2}}}{\Gamma(\frac{1}{2})}+\mathcal{O}(\lambda^{-\frac{3}{2}})\,, \tag{5.49}$$

where $a_1(x) = (4\pi)^{-3/2}\gamma^\nu F^*_\nu$, with $F_\mu = \varepsilon_{\mu\nu\rho}F^{\nu\rho}$, is the first non-trivial Seeley–DeWitt coefficient related to the spinor Laplacian. The leading term gives no contribution and taking the internal traces, one has for the non-trivial ground state fermion current contribution

$$\frac{\delta\eta(0|\mathfrak{D})}{\delta V_\mu(x)} = -\frac{e}{2\pi^2}F^*_\mu\,, \tag{5.50}$$

which is the Redlich result. We note that the sign ambiguity is always present in the massless case, because the total effective action is given by (5.27) and has nothing to do with the Pauli-Villars regularization, as often claimed in the literature.

5.1.3 *Another form for the eta invariant variation*

Another general formula can be obtained along the line suggested in [J. Park and K.P. Wojciechowski (2002)]. Starting from (5.23) we get

$$\delta\eta(s|\mathfrak{D}) = \frac{1}{\Gamma(\frac{s+1}{2})}\int_0^\infty dt\; t^{\frac{s+1}{2}-1}\,\mathrm{Tr}\left(\delta\mathfrak{D}e^{-t\mathfrak{D}^2}(1-2t\mathfrak{D}^2)\right). \tag{5.51}$$

Since

$$\begin{aligned}2t^{\frac{s+1}{2}}\,\mathrm{Tr}\left(\delta\mathfrak{D}e^{-t\mathfrak{D}^2}\mathfrak{D}^2\right) &= t^{\frac{s-1}{2}}(s+1)\,\mathrm{Tr}\left(\delta\mathfrak{D}e^{-t\mathfrak{D}^2}\right)\\ &\quad -2\frac{d}{dt}\left(t^{\frac{s+1}{2}}\,\mathrm{Tr}(\delta\mathfrak{D}e^{-t\mathfrak{D}^2})\right),\end{aligned}$$

we have

$$\begin{aligned}\delta\eta(s|\mathfrak{D}) &= -\frac{s}{\Gamma(\frac{s+1}{2})}\int_0^\infty dt\; t^{\frac{s-1}{2}}\,\mathrm{Tr}\left(\delta\mathfrak{D}e^{-t\mathfrak{D}^2})\right)\\ &\quad+\frac{1}{\Gamma(\frac{s+1}{2})}\left[t^{\frac{s+1}{2}}\,\mathrm{Tr}(\delta\mathfrak{D}e^{-t\mathfrak{D}^2})\right]_0^\infty.\end{aligned} \tag{5.52}$$

Now heat-kernel techniques can profitably be used. In fact, we have the following small-t asymptotics [P.B. Gilkey and G. Grubb (1998)]:

$$\mathrm{Tr}(\delta\mathfrak{D}e^{-t\mathfrak{D}^2}) = \sum_{r=0}^\infty t^{\frac{r-a-1}{2}-p}b_r(\delta\mathfrak{D})\,, \tag{5.53}$$

where a is the (integer) order of $\delta\mathfrak{D}$ and b_r are evaluable expansion coefficients. For manifold without boundary, all b_r coefficients are vanishing as soon as $r+a$ is odd. As a consequence, for large $\mathrm{Re}\, s$, the boundary term is vanishing and we are left with

$$\delta\eta(s|\mathfrak{D}) = -\frac{s}{\Gamma(\frac{s+1}{2})}\int_0^\infty dt\; t^{\frac{s-1}{2}}\,\mathrm{Tr}\left(\delta\mathfrak{D}e^{-t\mathfrak{D}^2}\right). \tag{5.54}$$

Making use of the asymptotic expansion (5.53), we can perform the analytic continuation in s of the above expression, separating the small values of t, namely

$$\begin{aligned}\delta\eta(s|\mathfrak{D}) &= -\frac{s}{\Gamma(\frac{s+1}{2})}\int_0^1 dt\; t^{\frac{s-1}{2}}\sum_{r=0}^{\infty} t^{\frac{s+r-a}{2}-p-1} b_r(\delta\mathfrak{D}) + \mathcal{O}(s)\\ = -\frac{s}{\Gamma(\frac{s+1}{2})}\sum_{r=0}^{\infty}\frac{2b_r(\delta\mathfrak{D})}{s+r-a-2p} &+ \mathcal{O}(s) = -\frac{2b_{2p+a}(\delta\mathfrak{D})}{\Gamma(\frac{s+1}{2})} + \mathcal{O}(s)\,.\end{aligned} \tag{5.55}$$

As a result we obtain the general formula

$$\delta\eta(0|\mathfrak{D}) = -\frac{2b_{2p+a}(\delta\mathfrak{D})}{\sqrt{\pi}}\,. \tag{5.56}$$

Let us check the above formula in the case of a deformation of the classical external gauge field, $\delta\mathfrak{D} = -e\gamma^\mu\delta V_\mu$. In fact, the order of $\delta\mathfrak{D}$ is zero, i.e. $a=0$. Thus, we have to compute the coefficient $b_{2p}(\delta\mathfrak{D})$. We have

$$\begin{aligned}\mathrm{Tr}(\delta\mathfrak{D}e^{-t\mathfrak{D}^2}) &= \int dx\delta V_\mu(x)\mathrm{tr}\left(\gamma^\mu e^{-t\mathfrak{D}^2}(x,x)\right)\\ &= \sum_{r=0}^{\infty} t^{r-1/2-p}\int dx\delta V_\mu(x)\mathrm{tr}\left(\gamma^\mu a_r(x)\right),\end{aligned} \tag{5.57}$$

here we have used the heat-kernel asymptotic expansion of $e^{-t\mathfrak{D}^2}$. It follows that

$$b_{2p}(\delta\mathfrak{D}) = -\int dx\delta V_\mu(x)\mathrm{tr}\left(\gamma^\mu a_p(x)\right). \tag{5.58}$$

Thus, Eq. (5.56) gives again Eq. (5.48) directly.

5.2 The Massive Dirac Operator

Let us consider now the case of massive spinors. Denoting the massive operator by $K = \mathfrak{D} + im$, we have

$$K\psi_i = (\mathfrak{D} + im)\psi_i = (\lambda_i + im)\psi_i \,, \tag{5.59}$$

where, as above, λ_i and ψ_i are the eigenvalues and eigenfunctions of $\mathfrak{D}$. K is non-Hermitian and its eigenvalues are now complex numbers. As in the massless case, we also have here two possible spectral cuts, and the two zeta functions may in principle be defined as

$$\zeta_\pm(s|K) = \sum_{\lambda_i>0} (\lambda_i + im)^{-s} + \sum_{\mu_i<0} \left[e^{\mp i\theta}(-\mu_i) + im\right]^{-s} \Bigg|_{\theta=\pi} . \tag{5.60}$$

In order to avoid a singularity in the complex powers of the spectrum of $K = \mathfrak{D} + im$ (mass m is positive) when we perform the analytical continuation of the spectral values (from positive $-\mu_i$ to negative μ_i, along a path $\Gamma_\pm$: $-\mu_i e^{\pm i\theta}, 0 \leq \theta \leq \pi$), it is compulsory to take $\zeta_+(s|K)$ — e.g., to perform the analytic continuation in the upper half plane. In fact, a singular point is lying in the lower half complex μ-plane, namely at $-im$, that gives rise to a cut in the complex s-plane when performing the analytic continuation in s. Note that we have to perform two different analytical continuations, in a compatible way, and this is actually guaranteed in our method. On the contrary, when we will consider the adjoint operator $K^\dagger = \mathfrak{D} - im$, the analytic continuation will have to be performed on the lower half-plane, the relevant zeta function will be $\zeta_-(s|K^\dagger)$, in order to avoid the singularity. Notice also that our definition of the zeta function for the operators K and $K^\dagger$ slightly differs from the one adopted in [S. Deser et al. (1998)], where the imaginary part of the eigenvalues was also rotated.* We believe that our definition is the most natural one in the massive case, being the minimal extension of the corresponding definition for self-adjoint operators with (in general) positive and negative spectrum — that was on its turn obtained from the rigorous mathematical theory for elliptic self-adjoint operators, with no negative eigenvalues.

As we have argued, the choice of the cut in the complex λ-plane depends on the meromorphic structure of $R_\lambda(x,k)$. If K is Hermitian, the poles are

*To compare, the definition in [S. Deser et al. (1998)] corresponds to rotating $(-\mu_i - im)e^{\pm i\theta}, 0 \leq \theta \leq \pi$, from $\theta = 0$ to $\theta = \pi$.

on the real axis and two possible inequivalent cuts can be chosen in the lower or upper complex plane, as discussed in the Introduction. If a mass is present or, in general, the operator K is non-Hermitian, the situation changes.

In the following, we shall consider a convenient algorithm, which is equivalent to the calculus of ΨDOs, in order to construct a parametrix for the resolvents of Dirac-like operators and its related meromorphic continuation. We follow the classical [M. Atiyah et al. (1973a)] (see also [G. Cognola et al. (1990b)]). For the sake of simplicity, let us consider now the massive Dirac operator on a flat (compact) manifold. Let us introduce the ε-symbol,

$$\sigma_\varepsilon(K) = e^{i\frac{kx}{\varepsilon}} K e^{-i\frac{kx}{\varepsilon}} = K(x, p + \frac{k}{\varepsilon}) = \not{p} - \mathfrak{D} + \frac{\not{k}}{\varepsilon} + im, \tag{5.61}$$

where ε is a real parameter, defined in order to determine recursively the parametrix. The ε-symbol for the resolvent is given by

$$\left(\sigma_\varepsilon(K) - \frac{\lambda}{\varepsilon}\right) R_\lambda(\varepsilon; x, k) = I\,. \tag{5.62}$$

Here, I is the identity operator and λ a complex number. The parametrix for the symbol of the complex power is obtained putting $\varepsilon = 1$. It is given by

$$\sigma(K^{-s})(x,k) = \frac{1}{2\pi i}\int_{\Gamma_\theta} d\lambda \lambda^{-s} R_\lambda(x,k)\,, \tag{5.63}$$

where Γ_θ is a suitable cut in the complex λ-plane. Recall that the Fourier transform of (5.63) gives the parametrix for the kernel of the complex power K^{-s}. One may solve Eq. (5.62) with the ansatz

$$R_\lambda(\varepsilon; x, k) = \sum_{l=1}^{\infty} \varepsilon^l R_l(x,k)\,. \tag{5.64}$$

As a result one has the recurrence relations

$$(\not{k} - \lambda)R_1 = I\,, \qquad (\not{k} - \lambda)R_l + KR_{l-1} = 0\,, \qquad l = 2, 3, \ldots. \tag{5.65}$$

Thus, formally one obtains

$$R_\lambda(x,k) = \frac{1}{\not{k}-\lambda}\left(I - K\frac{1}{\not{k}-\lambda} + K\frac{1}{\not{k}-\lambda}K\frac{1}{\not{k}-\lambda} \right.$$
$$\left. -K\frac{1}{\not{k}-\lambda}K\frac{1}{\not{k}-\lambda}K\frac{1}{\not{k}-\lambda} + \ldots\right)\,. \tag{5.66}$$

As an illustration, we can consider the free case, namely $K^{(0)} = \not{p} + im$. A formal resummation gives

$$R_\lambda^{(0)}(x,k) = \frac{1}{\not{k} + im - \lambda} = \frac{\not{k} - im + \lambda}{k^2 - (im - \lambda)^2} \,. \tag{5.67}$$

Here the poles are located in the upper complex plane and one is forced to choose the lower cut in Eq. (5.63). This argument is also valid in the interacting case, because the "free" contribution can be additively separated in Eq. (5.66).

On the other hand, in the case of the adjoint operator $K^\dagger$, the only change is $im \to -im$ and, in this situation, the cut must be chosen in the upper complex λ-plane.

If the mass is sufficiently small, binomial expansion gives

$$\zeta_+(s|K) = \sum_{k=0}^{\infty} \frac{(-1)^k \Gamma(s+k)\,(im)^k}{\Gamma(s)k!}\,\zeta_+(s+k|\mathfrak{D})\,, \tag{5.68}$$

and analogously for $\zeta_-(s|K^\dagger)$. Making use of Eq. (5.16), one arrives at

$$\begin{aligned} \zeta_+(s|K) \;&=\; \frac{1}{2}[1+e^{-i\pi s}]\,\zeta\left(\frac{s}{2}|\mathfrak{D}^2\right) + \frac{1}{2}[1-e^{-i\pi s}]\,\eta(s|\mathfrak{D}) \\ &\quad + \sum_{k=1}^{\infty} \frac{(-1)^k\Gamma(s+k)\,(im)^k}{2\Gamma(s)k!}\left\{[1+e^{-i\pi(s+k)}]\zeta\left(\frac{s+k}{2}|\mathfrak{D}^2\right)\right. \\ &\qquad \left. +[1-e^{-i\pi(s+k)}]\eta(s+k|\mathfrak{D})\right\}. \end{aligned} \tag{5.69}$$

Thus we have obtained, in a direct way, the meromorphic extension of the zeta function associated with the massive Dirac operator, starting from the knowledge of the massless one.

As an example, if $\mathcal{M}$ is a flat even-dimensional manifold, then $\zeta(s|\mathfrak{D}^2)$ has only a simple pole at $s = D/2$ with residue equal to $(4\pi)^{-D/2} D \not{V}$, $\not{V}$ being the volume of $\mathcal{M}$. This means that $\zeta(s|K)$ has a finite number of simple poles at the points $s = D - k > 0$ $(k = 0, 1, ..., D-1)$, whose residues read

$$\operatorname{Res}\zeta_+(s|K)|_{s=D-k>0} = \frac{2\not{V}\Gamma(D+1)\;(-im)^k}{(4\pi)^{D/2}\Gamma(D/2)\Gamma(D-k)k!} \,. \tag{5.70}$$

Observe that this result is independent of the choice of definition for the zeta function $\zeta_\pm$. Furthermore, consider the free massive Dirac operator in a flat manifold. For the symbol, we have

$$\sigma(i\mathfrak{D}+im)=\not{p}+im\,,\quad \sigma(i\mathfrak{D}+im)^{-\alpha}=\frac{1}{\not{p}^{\alpha}}\left[\sum_{k=0}^{\infty}\left(\frac{-im}{\not{p}}\right)^{k}\right]^{\alpha} \tag{5.71}$$

and, in order to compute the non-commutative residue, we must pick up the coefficient of $\not{p}^{-D}$ in the previous expansion. We see that such a coefficient is non-vanishing only when $\alpha = n$ is an integer and $\alpha \leq D$. Under such condition, we get

$$\mathrm{res}([\mathfrak{D}+im]^{-n})=\frac{\Gamma(D+1)\ (-im)^{D-n}}{(4\pi)^{D/2}\Gamma(D/2)\Gamma(n)(D-n)!}\,. \tag{5.72}$$

Using this result in Eq. (B.13) we again obtain Eq. (5.70).

5.3 One-Dimensional Example

As a simple example of exact computation, let us consider the first-order operator $\mathcal{D}=P-a$, with $P=-i\partial_\tau$, acting on $\mathbb{S}^1$, and a, $a\neq 0,-1,-2,..$ is a constant which can be interpreted as the mean value of a one-dimensional periodic vector potential. We may define the domain of $\mathcal{D}$ as consisting of periodic or antiperiodic functions with period 2π. Let us consider periodic (untwisted) spinors. The antiperiodic case is recovered through the replacement $a\to a+1/2$.

To begin with, the zeta functions related to the massless case read

$$\zeta_{\pm}(s|\mathcal{D})=\sum_n(n-a)^{-s}=\zeta_H(s,-a)+e^{\mp i\pi s}\zeta_H(s,a)-(-a)^{-s}\,, \tag{5.73}$$

where $\zeta_H(s,b)$ is the Hurwitz zeta function. Thus,

$$\zeta'_{\pm}(0|\mathcal{D})=\zeta'_H(0,-a)+\zeta'_H(0,a)+\log(-a)\mp i\pi\zeta_H(0,a)\,. \tag{5.74}$$

Making use of the known properties of the Hurwitz zeta function and its first derivative, we get

$$\zeta'_H(0,a)=\log\Gamma(a)-\frac{1}{2}\log(2\pi)\,,\qquad \zeta_H(0,a)=\frac{1}{2}-a\,. \tag{5.75}$$

Then,

$$\zeta'_H(0,-a)+\zeta'_H(0,a)+\log(-a)=-\log(2\sin(\pi a))\mp i\pi\zeta_H(0,a)\,, \tag{5.76}$$

where

$$\Gamma(a)\Gamma(-a) = -\frac{\Gamma(a)\Gamma(1-a)}{a} = -\frac{\pi}{a\sin(\pi a)} \tag{5.77}$$

has been used. Thus, we have

$$\zeta'_{\pm}(0|\mathcal{D}) = -\log(2\sin(\pi a)) \mp \frac{i\pi}{2}(2a-1)\,. \tag{5.78}$$

Now, a direct computation gives

$$\zeta(s|\mathcal{D}^2) = \sum_n ((n-a)^2)^{-s} = \zeta_H(2s,-a) + \zeta_H(2s,a) - (-a)^{-2s}\,. \tag{5.79}$$

As a result,

$$\frac{1}{2}\zeta'(0,\mathcal{D}^2) = -\log(2\sin(\pi a))\,. \tag{5.80}$$

As a consequence, we have obtained an explicit formula for $\log\det\mathcal{D}$. Furthermore, in order to check the agreement with the general theory, let us show that indeed, the last term in Eq. (5.78) is the eta invariant. The eta function related to $\mathcal{D}$, for large $\operatorname{Re} s$, is

$$\eta(s|\mathcal{D}) = \sum_n \frac{n-a}{|n-a|}(n-a)^{-\frac{s}{2}}\,. \tag{5.81}$$

The analytic extension to $s = 0$ can be obtained by using the spectral theorem,

$$\operatorname{Tr}\mathcal{D}e^{-t\mathcal{D}^2} = \sum_n (n-a)e^{-t(n-a)^2}\,. \tag{5.82}$$

The Poisson resummation formula yields

$$\operatorname{Tr}\mathcal{D}e^{-t\mathcal{D}^2} = -2\left(\frac{\pi}{t}\right)^{3/2}\sum_{n=1}^{\infty} n e^{-\frac{\pi^2 n^2}{t}}\sin(2\pi n a)\,. \tag{5.83}$$

Direct calculation gives

$$\eta(s|\mathcal{D}) = -2\sqrt{\pi}\frac{\Gamma(1-s)}{\Gamma(\frac{s+1}{2})}\sum_{n=1}^{\infty}(\pi n)^{2s-1}\sin(2\pi n a)\,, \tag{5.84}$$

and, as a consequence, the eta invariant reads (for $0 < a < 1$),

$$\eta(\mathcal{D}) = \eta(0|\mathcal{D}) = -\frac{2}{\pi}\sum_{n=1}^{\infty} n^{-1}\sin(2\pi n a) = 2a-1\,. \tag{5.85}$$

For any a,

$$\eta(\mathcal{D}) = \eta(0|\mathcal{D}) = 2a - 2[a] - 1 \,. \tag{5.86}$$

Note that the result is defined modulo k, k is an integer (i.e., it is a discontinuous and periodic function of a, of period 1).

We may also check the formula (5.31). For the sake of simplicity let us take $0 < a < 1$, but the analysis can be extended to a generic a. The spectral theorem gives

$$\begin{aligned} \mathrm{Tr}\left(\frac{\mathcal{D}}{|\mathcal{D}|}e^{-t|\mathcal{D}|}\right) &= \sum_n \frac{n-a}{|n-a|}e^{-t|n-a|} = -e^{at} + \left(e^{at} - e^{-at}\right)\sum_{n=1}^{\infty} e^{-nt} \\ &= \frac{\sinh[(a-1/2)t]}{\sinh(t/2)} \,. \end{aligned}$$

Thus,

$$\lim_{t\to 0} \mathrm{Tr}\left(\frac{\mathcal{D}}{|\mathcal{D}|}e^{-t|\mathcal{D}|}\right) = 2a - 1 = \eta(0|\mathcal{D}) \,. \tag{5.87}$$

With regard to the massive case, we may proceed in a similar way. Now $K = P - a + im$, and we choose the lower cut, i.e.

$$\begin{aligned} \zeta_+(s|K) = \sum_n (n-a+im)^{-s} &= \zeta_H(s, -a+im) \\ &+ e^{-i\pi s}\zeta_H(s, a+im) - (-a+im)^{-s} \,. \end{aligned} \tag{5.88}$$

As a result,

$$\begin{aligned} \zeta'_+(0|K) &= -\log(2\sin[\pi(a-im)]) + \pi m + \frac{i\pi}{2}(2a-1) \\ &= -\log\left(2\sqrt{\sin^2(\pi a) + \sinh^2(\pi m)}\right) + \pi m + \frac{i\pi}{2}(2a-1) \\ &\quad + i\arctan\left[\tanh(\pi m)\tan(\pi(a-1/2))\right] . \end{aligned} \tag{5.89}$$

This is a direct computation. Let us consider the general formula (5.79). We have, for $\mathrm{Re}\, s > 1/2$,

$$\zeta(s|\mathcal{D}^2 + m^2) = \sum_n \left((n-a)^2 + m^2\right)^{-s} \,. \tag{5.90}$$

The analytical continuation of this zeta function can be done by means of (see, for example [E. Elizalde et al. (1994)])

$$\sum_n \left((n-a)^2+m^2\right)^{-s} = \sqrt{\pi}\frac{\Gamma(s-1/2)}{\Gamma(s)}m^{1-2s}$$
$$+\frac{4\pi^s}{\Gamma(s)}m^{1/2-s}\sum_{j=1}^{\infty} j^{s-1/2}\cos(2\pi ja)K_{s-1/2}(2\pi jm)\,. \tag{5.91}$$

The derivative at the origin reads

$$\zeta'(0|\mathcal{D}^2+m^2) = -2\pi m + 2\sum_{j=1}^{\infty}\frac{\cos(2\pi ja)}{j}e^{-2\pi mj}\,. \tag{5.92}$$

The sum can be easily performed and we get

$$\frac{1}{2}\zeta'(0|\mathcal{D}^2+m^2) = -\log(2\sqrt{\sin^2(\pi a)+\sinh^2(\pi m)}\,. \tag{5.93}$$

With respect to the eta-like contribution, a direct evaluation gives

$$\int_0^\infty dt\frac{\sin mt}{t}\,\mathrm{Tr}\left(\frac{\mathcal{D}}{|\mathcal{D}|}e^{-t|\mathcal{D}|}\right) = \int_0^\infty dt\frac{\sin(mt)}{t}\frac{\sinh[(a-1/2)t]}{\sinh t/2}$$
$$= \arctan\left[\tanh(\pi m)\tan(\pi(a-1/2))\right]. \tag{5.94}$$

As a result, the general formula (5.96) is verified. It is easy to show that Eq. (5.32) is verified too.

Finally, we would like to mention that in [A.A. Bytsenko et al. (1999); A.A. Bytsenko et al. (2000)], explicit examples of computations of the eta invariant, related to three-dimensional compact hyperbolic manifolds, can be found.

5.4 The One-Loop Effective Action

In this Section we will compute the effective action for a massive spinor field in a space $\mathcal{M}$ in terms of the zeta function and the eta function of the related massless operator. The effective action turns out to be proportional to the logarithm of the functional determinant of the Dirac operator, which may be expressed as minus the first derivative of the zeta function at zero. The massive Dirac operator $K = \mathfrak{D} + im$ is non-Hermitian, and one has

$K^\dagger = \mathfrak{D} - im = K - 2im$. However, $K^\dagger K = \mathfrak{D}^2 + m^2$ is a positive self-adjoint second order Laplace-like operator, and binomial expansion gives

$$\zeta(s|K^\dagger K) = \zeta(s|\mathfrak{D}^2 + m^2) = \zeta(s|\mathfrak{D}^2) + \sum_{j=1}^{\infty}(-1)^j \frac{\Gamma(s+j)m^{2j}}{\Gamma(s)j!}\zeta(s+j|\mathfrak{D}^2) . \tag{5.95}$$

It is plain that the functional determinant associated with $\mathfrak{D}^2 + m^2$ will play here an important role.

Let us begin with the even dimensional case, i.e. $D = 2p$. Recall that, in this situation, $\eta(s|\mathfrak{D}) = 0$. Splitting the sum in Eq. (5.69) into even and odd terms and making use of Eqs. (5.16), it is easy to show that the latter gives a contribution of order $\mathcal{O}(s^2)$. We can isolate the singular term by writing

$$\zeta_+(s|K) = \frac{1}{2}[1 + e^{-i\pi s}]\,\zeta\left(\frac{s}{2}|\mathfrak{D}^2\right) + \frac{1}{\Gamma(s)}\left(\frac{G_+(s)}{s} + \phi_+(s)\right) + \mathcal{O}(s^2) ,$$

and

$$\zeta_-(s|K^\dagger) = \frac{1}{2}[1 + e^{i\pi s}]\,\zeta\left(\frac{s}{2}|\mathfrak{D}^2\right) + \frac{1}{\Gamma(s)}\left(\frac{G_-(s)}{s} + \phi_-(s)\right) + \mathcal{O}(s^2) ,$$

where

$$\begin{aligned}\phi_\pm(s) = \frac{1}{2}\sum_{j=1}^{p} \frac{(-1)^j\Gamma(s+2j)\, m^{2j}}{(2j)!}\,[1 + e^{\mp i\pi s}]\left[\zeta\left(\frac{s}{2} + j|\mathfrak{D}^2\right) - \frac{2A_{p-j}(\mathfrak{D}^2)}{s\Gamma(\frac{s}{2}+j)}\right] \\ + \frac{1}{2}\sum_{j=p+1}^{\infty} \frac{(-1)^j\Gamma(s+2j)\, m^{2j}}{(2j)!}\,[1 + e^{\mp i\pi s}]\,\zeta\left(\frac{s}{2} + j|\mathfrak{D}^2\right)\end{aligned}$$

and

$$G_\pm(s) = \sum_{j=1}^{p} \frac{(-1)^j\Gamma(s+2j)\, m^{2j}}{(2j)!}\,[1 + e^{\mp i\pi s}]\,\frac{A_{p-j}(\mathfrak{D}^2)}{\Gamma(\frac{s}{2}+j)} .$$

It is easy to see that both $\phi_\pm(s)$ are regular at $s = 0$. Taking derivatives at $s = 0$, we get

$$\zeta'_+(0|K) = \frac{1}{2}\left[\,\zeta'(0|\mathfrak{D}^2) - i\pi\zeta(0|\mathfrak{D}^2)\right] + \gamma G_+(0) + G'_+(0) + \phi_+(0) , \tag{5.96}$$

and a similar expression for $\zeta'_-(0|K^\dagger)$.

On the other hand, from (5.95), one readily has

$$\frac{1}{2}\zeta'(0|\mathfrak{D}^2+m^2)=\frac{1}{2}\,\zeta'(0|\mathfrak{D}^2)+\gamma G(0)+\phi(0)\,, \tag{5.97}$$

where $G(0) = G_\pm(0)$, and analogously $\phi(0) = \phi_\pm(0)$. Comparing Eqs. (5.96) and (5.97), we finally obtain

$$\zeta'_+(0|K)=\frac{1}{2}\left[\,\zeta'(0|\mathfrak{D}^2+m^2)-i\pi\zeta(0|\mathfrak{D}^2+m^2)\right]+\sum_{j=1}^{p}\frac{(-1)^j c_j m^{2j}A_{p-j}(\mathfrak{D}^2)}{j!}\,, \tag{5.98}$$

and

$$\begin{aligned}\zeta'_-(0|K^\dagger) &= \frac{1}{2}\left[\,\zeta'(0|\mathfrak{D}^2+m^2)+i\pi\zeta(0|\mathfrak{D}^2+m^2)\right]\\ &+ \sum_{j=1}^{p}\frac{(-1)^j c_j m^{2j}A_{p-j}(\mathfrak{D}^2)}{j!}\,,\end{aligned} \tag{5.99}$$

where $c_j = \sum_{l=1}^{j}(2l-1)^{-1}$. The above formula is useful for the actual evaluation of the derivative of the zeta function of the massive Dirac operator. Here, again, the ambiguity depends on the Seeley-DeWitt coefficient, $\zeta(0|\mathfrak{D}^2+m^2)$, only. We should point out that, as it stands, the above equation is valid for arbitrary, but finite, values of the mass m. This observation allows us to obtain a large mass expansion, simply by recalling an asymptotic theorem due to Voros [V. Voros (1987)], which states that, for large m, the asymptotics of the massive Laplacian are determined by the short asymptotics of its heat-kernel trace. As a result we have

$$\begin{aligned}\zeta(s|\mathfrak{D}^2+m^2) &\simeq \sum_{r=0}^{\infty}\frac{A_r(\mathfrak{D}^2)\Gamma(r-p+s)}{\Gamma(s)}m^{2(p-r-s)}\\ &\simeq \sum_{r=0}^{p}\frac{(-1)^{p-r}}{(p-r)!}A_r(\mathfrak{D}^2)m^{2p-2r-2s}+\mathcal{O}(s)\,,\end{aligned} \tag{5.100}$$

and

$$\begin{aligned}\zeta'(0|\mathfrak{D}^2+m^2) \simeq &\sum_{r=0}^{p-1}\frac{(-1)^{p-r}A_r(\mathfrak{D}^2)}{(p-r)!}m^{2(p-r)}\left(-\log m^2+\sum_{j=1}^{p-r}\frac{1}{j}\right)\\ &-A_p(\mathfrak{D}^2)\log m^2+\sum_{r=p+1}^{\infty}A_r(\mathfrak{D}^2)\Gamma(r-p)m^{2(p-r)}.\end{aligned} \tag{5.101}$$

To summarize, we have reduced the problem of the first order massive Dirac operator (in the even dimensional case) to the much more familiar one for the simple Laplacian operator $\mathfrak{D}^2 + m^2$, and traced back the terms coming from the previous ambiguities (fixed for $m \neq 0$) of the zeta function definition, to the term involving the Seeley-DeWitt coefficient $\zeta(0|\mathfrak{D}^2+m^2)$. Our results are in agreement with the ones obtained in [S. Deser et al. (1998)], with a completely different method.

In the odd dimensional case, $D = 2p+1$, the eta invariant is non-vanishing and we have to refer to its meromorphic continuation given by Eq. (5.25). As a consequence,

$$\zeta'_+(0|K) = \frac{1}{2}\,\zeta'(0|\mathfrak{D}^2) + \frac{i\pi}{2}\eta(0|\mathfrak{D}) + \sum_{j=1}^{\infty} \frac{(im)^{2j}}{j}\zeta(j|\mathfrak{D}^2)$$

$$-\sum_{j=0}^{\infty} \frac{(im)^{2j+1}}{2j+1}\eta(2j+1|\mathfrak{D}) + \frac{i\pi}{2}\sum_{j=0}^{p} \frac{(im)^{2j+1}}{\Gamma(j+\frac{3}{2})}A_{p-j}(\mathfrak{D}^2)\,, \qquad (5.102)$$

and

$$\zeta'_-(0|K^\dagger) = \frac{1}{2}\,\zeta'(0|\mathfrak{D}^2) - \frac{i\pi}{2}\eta(0|\mathfrak{D}) + \sum_{j=1}^{\infty} \frac{(im)^{2j}}{2j}\zeta(j|\mathfrak{D}^2)$$

$$-\sum_{j=0}^{\infty} \frac{(-im)^{2j+1}}{2j+1}\eta(2j+1|\mathfrak{D}) - \frac{i\pi}{2}\sum_{j=0}^{p} \frac{(im)^{2j+1}}{\Gamma(j+\frac{3}{2})}A_{p-j}(\mathfrak{D}^2)\,. \qquad (5.103)$$

Recalling Eq. (5.95) one has

$$\zeta'(0|\mathfrak{D}^2) + \sum_{j=1}^{\infty} \frac{(im)^j}{j}\zeta(j|\mathfrak{D}^2) = \zeta'(0|\mathfrak{D}^2 + m^2)\,. \qquad (5.104)$$

If we introduce the eta-like functions

$$\eta(s|\mathfrak{D} \pm im) = \sum_i (\lambda_i \pm im)^{-s} - \sum_i (-\mu_i \pm im)^{-s}\,, \qquad (5.105)$$

for small mass, we have

$$\eta(s|\mathfrak{D} \pm im) = \eta(s|\mathfrak{D}) + \sum_{k=1}^{\infty} \frac{(\mp im)^k\Gamma(s+k)}{\Gamma(s)k!}\eta(s+k|\mathfrak{D})\,. \qquad (5.106)$$

Thus, the meromorphic properties of $\eta(s|\mathfrak{D} \pm im)$ follow from the ones of the eta function. In particular, one obtains

$$\eta'(0|\mathfrak{D}+im) - \eta'(0|\mathfrak{D}-im) = -2\sum_{j=0}^{\infty} \frac{(im)^{2j+1}}{2j+1}\eta(2j+1|\mathfrak{D}) . \quad (5.107)$$

As a result, we have

$$\zeta'_+(0|K) = \frac{1}{2}\,\zeta'(0|\mathfrak{D}^2+m^2) + \frac{i\pi}{2}\eta(0|\mathfrak{D})$$
$$+\frac{1}{2}\left[\eta'(0|\mathfrak{D}+im) - \eta'(0|\mathfrak{D}-im)\right] - \frac{\pi}{2}\sum_{j=0}^{p} \frac{(-1)^j m^{2j+1}}{\Gamma(j+\frac{3}{2})} A_{p-j}(\mathfrak{D}^2), \quad (5.108)$$

and

$$\zeta'_-(0|K^\dagger) = \frac{1}{2}\,\zeta'(0|\mathfrak{D}^2+m^2) - \frac{i\pi}{2}\eta(0|\mathfrak{D})$$
$$-\frac{1}{2}\left[\eta'(0|\mathfrak{D}+im) - \eta'(0|\mathfrak{D}-im)\right] + \frac{\pi}{2}\sum_{j=0}^{p} \frac{(-1)^j m^{2j+1}}{\Gamma(j+\frac{3}{2})} A_{p-j}(\mathfrak{D}^2). \quad (5.109)$$

It should be noted the appearance of other two "eta-like" non-local contributions. This, again, is valid for arbitrary value of the mass m. In fact, we may use Eq. (5.33) and obtain

$$\eta'(0|\mathfrak{D}+im) - \eta'(0|\mathfrak{D}-im) = -2i\int_0^\infty dt\frac{\sin(mt)}{t}\,\mathrm{Tr}\left(\frac{\mathfrak{D}}{|\mathfrak{D}|}e^{-t|\mathfrak{D}|}\right) . \quad (5.110)$$

On the other hand, we also get

$$\eta'(0|\mathfrak{D}+im) - \eta'(0|\mathfrak{D}-im) = -2\sqrt{\pi}\int_0^\infty \frac{dt}{\sqrt{t}}\Phi(m\sqrt{t})\,\mathrm{Tr}\left(\mathfrak{D}e^{-t\mathfrak{D}^2}\right) . \quad (5.111)$$

In the last expression, $\Phi(x)$ is the error function, namely

$$\Phi(x) = \frac{2}{\sqrt{\pi}}\int_0^x dte^{-t^2} . \quad (5.112)$$

The two expressions above vanish for $m \to 0$, as they should. As far as we know, the first expression is new, while the second one is in agreement with the expression obtained in [S. Deser et al. (1998)].

With regard to the large mass limit of the non-local contributions, we observe that, from the Voros theorem, one has

$$\zeta'(0|\mathfrak{D}^2 + m^2) \simeq \sum_{r=0}^{\infty} \Gamma(r - p - \frac{1}{2}) A_r(\mathfrak{D}^2) m^{2p-2r+1}\,, \qquad (5.113)$$

while, from Eqs. (5.31) and (5.110),

$$\begin{aligned} &\lim_{m\to\infty} \left[\eta'(0|\mathfrak{D} + im) - \eta'(0|\mathfrak{D} - im)\right] \\ &= -2i \lim_{m\to\infty} \int_0^\infty dt \frac{\sin t}{t} \operatorname{Tr}\left(\frac{\mathfrak{D}}{|\mathfrak{D}|} e^{-t/m|\mathfrak{D}|}\right) = -i\pi\eta(0|\mathfrak{D}) \end{aligned} \qquad (5.114)$$

easily follows. The same result can be obtained starting from (5.111) and observing that $\lim_{m\to\infty} \Phi(mt^{1/2}) = \pi^{-1/2}$. As a consequence, in the large mass limit the leading eta contributions cancel out and we finally obtain

$$\zeta'_+(0|K) \sim \frac{1}{2} \sum_{j=0}^{p} \left[\Gamma(-j - \frac{1}{2}) - \frac{(-1)^j \pi}{\Gamma(j + \frac{3}{2})}\right] A_{p-j}(\mathfrak{D}^2) m^{2j+1} + \mathcal{O}(1/m)\,. \qquad (5.115)$$

5.5 Dirac Bundle and the Ray–Singer Norm

We recall that the eta invariant was introduced in connection with the index theorem, related to Dirac operator for even-dimensional manifolds with boundary [M.F. Atiyah et al. (1975); M.F. Atiyah et al. (1976)].

We will review in this Section some standard material on Dirac bundles (for details see, for example, [H.B. Lawson and M.L. Michelson (1989)]), and spectral functions. Let $\mathcal{M}$ be an even-dimensional orientable Riemannian manifold with boundary $\partial\mathcal{M}$ and let $\mathbb{E} \to \mathcal{M}$ be a Hermitian vector bundle equipped with a compatible connection. Let $\mathbb{C}\ell(\mathcal{M})$ denote the complexified Clifford bundle. We suppose that there is a bundle map from $\mathbb{C}\ell(\mathcal{M}) \to \operatorname{End}\mathbb{E}$ which is an algebra homomorphism on each fiber and which covers the identity:

$$\begin{array}{ccc} \mathbb{C}\ell(\mathcal{M}) & \longrightarrow & \operatorname{End}\mathbb{E} \\ & \searrow \quad \swarrow & \\ & \mathcal{M} & \end{array}$$

is commutative.

Denote by $\mathcal{S}$ the spin bundles associated with the two half-spin representation of Spin(dim$\mathcal{M}$); then $\mathcal{S} = \mathcal{S}^+ \oplus \mathcal{S}^-$, where $\mathcal{S}^{\pm}$, are half-spin bundles. ∇ induces a connection on $\mathcal{S} \otimes \mathbb{E}$ which is compatible with both the metric and the $\mathbb{Z}_2$-grading. This connection canonically defines a Dirac operator, $\mathcal{D}_{\mathcal{M}} : C^\infty(\mathcal{S}^+ \otimes \mathbb{E}) \to C^\infty(\mathcal{S}^- \otimes \mathbb{E})$, described by

$$\mathcal{D} : C^\infty(\mathbb{S}^+_{\mathbb{E}}) \stackrel{\nabla}{\longrightarrow} C^\infty(T^*\mathcal{M} \otimes \mathbb{S}^+_{\mathbb{E}}) \stackrel{\mathcal{H}}{\longrightarrow} C^\infty(\mathbb{S}^-_{\mathbb{E}}), \tag{5.116}$$

where $\mathcal{H} : T^*\mathcal{M} \longrightarrow \operatorname{Hom}(\mathcal{S}^+_{\mathbb{E}}, \mathcal{S}^-_{\mathbb{E}})$ denotes the Clifford multiplication and $\mathcal{S}^{\pm}_{\mathbb{E}} \stackrel{def}{=} \mathcal{S}^{\pm} \otimes \mathbb{E}$. Then $\dim \operatorname{Ker}^{\dagger}\mathcal{D}_{\mathcal{M}} = \dim \operatorname{Ker}^{-}\mathcal{D}_{\mathcal{M}}$ and the boundary condition is an isometry $I : \operatorname{Ker}^{\dagger}\mathcal{D}_{\mathcal{M}} \to \operatorname{Ker}^{-}\mathcal{D}_{\mathcal{M}}$.

Suppose that the metric on a neighborhood $\mathcal{E} \simeq [0, 1) \times \partial\mathcal{M}$ of the boundary is a product metric. Then, near the boundary $\partial\mathcal{M}$ one has a separation of variables, which leads to the representation $\mathcal{D} = \sigma\left(\frac{\partial}{\partial x} + \mathfrak{D}\right)$, where the decomposition $\phi = dx \wedge \phi_1(x) + \phi_2(x)$ of a smooth form ϕ is used. Thus $\mathcal{D}$ acts on $C^\infty\left((0, 1), \Omega^p(\partial\mathcal{M}) \oplus \Omega^p(\partial\mathcal{M})\right)$, $\Omega^p(\partial\mathcal{M})$ being the space of smooth p-forms on $\partial\mathcal{M}$. A little computation shows that $\mathfrak{D}$ is an elliptic self-adjoint map on $\Omega^p(\partial\mathcal{M})$ [M.F. Atiyah et al. (1975)] defined by

$$\mathfrak{D}\phi = (-1)^{k+p+1}(\varepsilon * d - d*)\phi, \tag{5.117}$$

where the symbol $*$ denotes the duality operator on $\partial\mathcal{M}$, $k = (\dim \partial\mathcal{M})/4$ and ϕ is either a $2p$-form ($\varepsilon = 1$) or a $(2p-1)$-form ($\varepsilon = -1$). The operator $\mathfrak{D}$ maps even (resp. odd) forms to even (resp. odd) forms, and can therefore be decomposed as follows:

$$\mathfrak{D} = \mathfrak{D}^{ev} \oplus \mathfrak{D}^{odd}, \tag{5.118}$$

where $\mathfrak{D}^{ev}$ is isomorphic to $\mathfrak{D}^{odd}$ [M.F. Atiyah et al. (1975)].

We recall now some standard material on the eta invariant of a self-adjoint elliptic differential Dirac-like operator acting on a compact odd–dimensional manifold $\mathcal{M}$. For details we refer the reader to [M.F. Atiyah et al. (1975); M.F. Atiyah et al. (1976)].

We have already seen that, in order to introduce the spectral invariant which measures the asymmetry of the spectrum Spec($\mathfrak{D}$) of the operator $\mathfrak{D}$, one makes use of the proposition:

Proposition 5.2 *The holomorphic function*

$$\eta(s|\mathfrak{D}) \stackrel{def}{=} \sum_{\lambda \in \mathrm{Spec}\mathfrak{D}\setminus\{0\}} \mathrm{sgn}(\lambda)|\lambda|^{-s} = \mathrm{Tr}\left(\mathfrak{D}\left(\mathfrak{D}^2\right)^{-(s+1)/2}\right), \tag{5.119}$$

is well defined for all $\mathrm{Re}\, s \gg 0$ *and extends to a meromorphic function on* $\mathbb{C}$.

For our purposes below, it is useful to sketch the proof again. Indeed, from the asymptotic behavior of the heat operator at $t = 0$ [J.M. Bismut and D.S. Freed (1986b)],

$$\mathrm{Tr}\left(\mathfrak{D}\exp\left(-t\mathfrak{D}^2\right)\right) = \mathcal{O}(t^{1/2}), \tag{5.120}$$

it follows that $\eta(s|\mathfrak{D})$ admits a meromorphic extension to the whole s-plane, with at most simple poles at $s = (\dim\mathcal{M} - k)/(\mathrm{ord}\,\mathfrak{D})$ $(k \in \mathbb{Z}_+)$ and locally computable residues.

It follows that the point $s = 0$ is not a pole (see the discussion in the previous Section), what makes it possible to define the eta invariant of $\mathfrak{D}$ by $\eta(0|\mathfrak{D})$. We also have that $\eta(0|-\mathfrak{D}) = -\eta(0|\mathfrak{D})$ and $\eta(0|\lambda\mathfrak{D}) = \eta(0|\mathfrak{D})$, $\forall\lambda > 0$. One can attach the eta invariant to any operator of Dirac type on a compact Riemannian manifold of odd dimension. As $\mathfrak{D}^{ev}$ is isomorphic to $\mathfrak{D}^{odd}$, we have

$$\eta(s|\mathfrak{D}^{ev}) = \eta(s|\mathfrak{D}^{odd}) = \frac{1}{2}\eta(s|\mathfrak{D}). \tag{5.121}$$

If $\mathfrak{L}_p$ is a self-adjoint Laplacian on p-forms then the following results hold.

Proposition 5.3 *There exist* $\varepsilon, \delta > 0$ *such that for* $0 < t < \delta$ *the heat-kernel expansion for the Laplace operators* $\mathfrak{L}_p$ *on a compact manifold* $\mathcal{M}$ *is* *given by*

$$\mathrm{Tr}\left(e^{-t\mathfrak{L}_p}\right) = \sum_{0\leq\ell\leq\ell_0} a_\ell(\mathfrak{L}_p)t^{-\ell} + \mathcal{O}(t^\varepsilon). \tag{5.122}$$

Proposition 5.4 *The zeta function of $\mathfrak{L}_p$ is the Mellin transform*

$$\zeta(s|\mathfrak{L}_p) = \frac{\mathfrak{M}\left[\mathrm{Tr}e^{-t\mathfrak{L}_p}\right]}{\Gamma(s)} = \frac{1}{\Gamma(s)}\int_0^\infty dt\, t^{s-1}\,\mathrm{Tr}e^{-t\mathfrak{L}_p}. \tag{5.123}$$

This function equals $\mathrm{Tr}\left(\mathfrak{L}_p^{-s}\right)$ *for* $s > \dim(\mathcal{M})/2$.

The Ray–Singer norm. Using the Hodge decomposition, the cohomology $H(\mathcal{M};\xi)$ can be embedded into $\Omega(\mathcal{M};\xi)$ as the space of harmonic forms. This embedding induces a norm $|\cdot|^{RS}$ on the determinant line $\det H(\mathcal{M};\xi)$. The Ray–Singer norm $||\cdot||^{RS}$ on $\det H(\mathcal{M};\xi)$ is defined by [D.B. Ray and I.M. Singer (1973)]

$$||\cdot||^{RS} \stackrel{def}{=} |\cdot|^{RS}\prod_{p=0}^{\dim\mathcal{M}}\left[\exp\left(-\frac{d}{ds}\zeta(s|\mathfrak{L}_p)|_{s=0}\right)\right]^{(-1)^p p/2}, \tag{5.124}$$

where the zeta function $\zeta(s|\mathfrak{L}_p)$ of the Laplacian acting on the space of p-forms orthogonal to the harmonic forms has been used. For a closed connected orientable smooth manifold of odd dimension and for Euler structure $\eta \in \mathrm{Eul}(\mathcal{M})$, the Ray–Singer norm of its cohomological torsion $T_{an}(\mathcal{M};\eta) = T_{an}(\mathcal{M}) \in \det H(\mathcal{M};\xi)$ is equal to the positive square root of the absolute value of the monodromy of ξ along the characteristic class $c(\eta) \in H^1(\mathcal{M})$, namely $||T_{an}(\mathcal{M})||^{RS} = |\det_\xi c(\eta)|^{1/2}$. In the special case where the flat bundle ξ is acyclic ($H^p(\mathcal{M};\xi) = 0$), we have

$$[T_{an}(\mathcal{M})]^2 = |\det_\xi c(\eta)|\prod_{p=0}^{\dim\mathcal{M}}\left[\exp\left(-\frac{d}{ds}\zeta(s|\mathfrak{L}_p)|_{s=0}\right)\right]^{(-1)^{p+1}p}. \tag{5.125}$$

For odd-dimensional manifold the Ray–Singer norm is topological invariant: it does not depend on the choice of metric on $\mathcal{M}$ and ξ, used in the construction. But for even-dimensional $\mathcal{M}$ this is not the case [J.M. Bismut and M. Zhang (1992)].

5.6 The Determinant Line Bundles

The differential geometry of determinant line bundles has been developed in [D. Quillen (1985)], in a special case, and in [J.M. Bismut and D.S. Freed (1986a); J.M. Bismut and D.S. Freed (1986b)] in general. In [X. Dai and D.S. Freed (1994); D.S. Freed (1997)], the results on eta invariants (the exponentiated eta invariant naturally takes values in the determinant line of the boundary) were used to re-prove the holonomy formula for determinant line bundles, known as Witten's global anomaly formula [E. Witten (1985)].

Results which we have considered hold for any Dirac operator on a $\text{Spin}^{\mathbb{C}}$ manifold coupled to a vector bundle with connection (manifolds are assumed spin and its metric is a product near the boundary). On a closed manifold X, the Dirac operator $\mathfrak{D}_X$ is self-adjoint and has discrete spectrum $\text{Spec}\,(\mathfrak{D}_X)$. The eta invariant $\eta(s, \mathfrak{D})$ of Atiyah-Patodi-Singer [M.F. Atiyah et al. (1975)], by general estimates converges for Re s sufficiently large. For Dirac operators, the meromorphic continuation is analytic for Re $s > -2$ [J.M. Bismut and D.S. Freed (1986b)]. At $s = 0$ the eta invariant is regular, and we set

$$\tau_X = \exp\left(\pi i(\eta(0|\mathfrak{D}) + \dim \operatorname{Ker} \mathfrak{D})_X\right) \in \mathbb{C}. \tag{5.126}$$

Generally, the eta invariant $\eta(0|\mathfrak{D})$ is discontinuous but the general theory shows that τ_X varies smoothly in families ($|\tau_X| = 1$). The invariant (5.126) is defined and we have $\tau_X \in \text{Det}^{-1}_{\partial X}$, where $\text{Det}_{\partial X}$ is the determinant line of the Dirac operator $\mathfrak{D}_{\partial X}$ on the boundary: $\text{Det}_{\partial X} = (\text{Det}\,\text{Ker}^{-}\mathfrak{D}_{\partial X}) \otimes (\text{Det}\,\text{Ker}^{\dagger}\mathfrak{D}_{\partial X})^{-1}$.

Let us consider a family of Riemannian manifolds in a smooth fiber bundle $\pi : X \to Z$ together with a metric on the relative tangent bundle $T(X/Z)$ which is endowed with spin structure. Also we assume that the Riemannian metrics on the fibers are products near the boundary. The determinant lines carry the Quillen metric and a canonical connection ∇ [J.M. Bismut and D.S. Freed (1986a)] and the exponentiated eta invariant is a smooth section $\tau_{X/Z} : Z \to \text{Det}^{-1}_{\partial X/Z}$. We mention here two basic results on this invariant, namely variation and curvature formulae.

Theorem 5.1 *([X. Dai and D.S. Freed (1994)]) The covariant derivative* *of the exponentiated eta invariant is*

$$\nabla\tau_{X/Z} = 2\pi i \left[\int_{X/Z} \widehat{A}(\Omega^{X/Z})\right]_{(1)} \cdot \tau_{X/Z}, \tag{5.127}$$

where $\Omega^{X/Z}$ is the Riemannian curvature of $X \to Z$, $\widehat{A}$ is the usual polynomial

$$\widehat{A}(\Omega) = \sqrt{\det\left(\frac{\Omega/4\pi}{\sinh \Omega/4\pi}\right)}, \tag{5.128}$$

$\mathrm{Tr}e^{i\Omega/2\pi} = \mathrm{ch}(\Omega)$ *and the symbol (1) denotes the 1-form piece of the differential form.*

Note that, for a family of closed manifolds, this is the well-known result of Atiyah-Patodi-Singer. Let $KO(X)$ be the group of virtual real vector bundles on X, up to equivalence, and $KO^{-N}(X) \subset KO(\mathbb{S}^N \otimes X)$, $N = \dim(X/Z)$, is the subspace of isomorphism classes of virtual bundles trivial on $\mathbb{S}^N \bigvee X$. The spin structure on the fibers can be determined by a pushforward map $\pi_!^{X/Z} : KO(X) \to KO^{-N}(Z)$. Let $E \to X$ be a real vector bundle; the family of Dirac operators on X/Z coupled to E has an index in $KO^{-N}(Z)$ which equals [M.F. Atiyah et al. (1976)] $\pi_!^{X/Z}([E])$, where $[E] \in KO(X)$ is the isomorphism class of E.

In general, for Dirac operators coupled to complex bundles in K–theory, one can express the Chern character of the index in terms of the Chern character of E by means of the formula [D.S. Freed and E. Witten (1999)]:

$$\mathrm{ch}\,\pi_!^{X/Z}([E]) = \pi_*^{X/Z}(\widehat{A}(X/Z)\mathrm{ch}\,(E)), \tag{5.129}$$

where π_* is the pushforward map in rational cohomology. Let the fibers X/Z be closed. Then, the determinant line bundle $\mathrm{Det}\,\mathcal{D}^{X/Z}(E)$ is well-defined as a smooth line bundle (it carries a canonical metric and connection) [J.M. Bismut and D.S. Freed (1986a)]. The complex Dirac operator for the odd dimensional fibers X/Z is self-adjoint and there is a geometrical invariant $\xi_{X/Z}(E) : Z \to \mathbb{R}/\mathbb{Z}$ defined by Atiyah-Patodi-Singer, $2\xi_{X/Z} = (\eta(0|\mathfrak{D}) + \mathrm{dimKer}\,\mathfrak{D})_{X/Z}$.

Theorem 5.2 *([J.M. Bismut and D.S. Freed (1986b)]) The 2-form cur-*

vature of the determinant line bundle is

$$\Omega^{\mathrm{Det}\,\mathcal{D}^{X/Z}(E)} = \left[2\pi i \int_{X/Z} \widehat{A}(\Omega^{X/Z})\mathrm{ch}\,(\Omega^{E})\right]_{(2)} \in \Omega^2(Z),$$

where $\Omega^{X/Z}$ *and* Ω^E *are the curvature forms.*

Let $X \to \mathbb{S}^1$ be a loop of manifolds in this geometric setup. A metric and spin structure on X could be induced by a metric and boundary spin structure on $\mathbb{S}^1$. The holonomy of the determinant line bundles around a loop takes the form

$$\mathrm{Hol}\,\mathrm{Det}\,\mathfrak{D}^{X/\mathbb{S}^1}(E) = \mathrm{a{-}lim}\,\tau_X^{-1}(E), \tag{5.130}$$

where a-lim is the adiabatic limit, i.e. the limit as the metric on $\mathbb{S}^1$ blows up ($\varepsilon \to 0 : \ g_{\mathbb{S}^1} \to g_{\mathbb{S}^1}\varepsilon^{-2}$). For the flat determinant line bundles, no adiabatic limit is required. As we pointed out before, Eq. (5.130) is the global anomaly formula [E. Witten (1985); J.M. Bismut and D.S. Freed (1986a)].

There are many applications of these results [E. Witten (1985); J.M. Bismut and D.S. Freed (1986a); J.M. Bismut and D.S. Freed (1986b)]. Global anomalies in the worldsheet path integral of string theory in the presence of D-branes have ben studied in [D.S. Freed and E. Witten (1999)]. Recently, agreement between partition functions of IIA string theory and M-theory (a derivation of K-theory from M-theory) was found in [D.E. Diaconescu et al. (2000)]. In all these examples, the holonomy of the determinant line bundles $\mathrm{Hol}(\mathrm{Det}\mathfrak{D}^{X/Z}(\mathbb{E}))$ and the Dirac index play a very important role.

5.7 The Dirac Index of Hyperbolic Manifolds

Note that a topological invariant $\xi_{X/Z}(\mathbb{E})$ is difficult to compute directly unless the metric has some special symmetry. As an example, we mention here topological invariants of the Chern-Simons theory, associated to real hyperbolic 3-manifolds [A.A. Bytsenko et al. (1997c); A.A. Bytsenko et al. (1999); A.A. Bytsenko et al. (2000)]. The semiclassical approximation for the Chern-Simons partition function leads to a series of C^∞-invariants, associated with triplets $\{\mathcal{M}; F; \xi\}$, with $\mathcal{M}$ a smooth homology 3$-$sphere, F a homology class of framings of $\mathcal{M}$, and ξ an acyclic conjugacy class of orthogonal representations of the fundamental group $\pi_1(\mathcal{M})$ [S. Axelrod and I.M. Singer (1994)]. In addition, the cohomology $H(\mathcal{M}; Ad\,\xi)$ of $\mathcal{M}$ with respect to the local system related to $Ad\,\xi$ vanishes.

Vanishing theorems can also be formulated for a complex manifold [F.L. Williams (1981a)]. In that case, if $\mathbb{E} \to \mathcal{M}$ is the induced complex bundle, we write $H^{p,q} \simeq H^{0,q}(\mathcal{M}, \Lambda^{p,0}\mathcal{M} \otimes \mathbb{E})$ for holomorphic vector bundles $\Lambda^{p,0}\mathcal{M} \to \mathcal{M}$ (see [F.L. Williams (1981a)] for detail).

We shall compute the Dirac index for real closed hyperbolic manifolds. Calculating the index of the Dirac operator, we shall follow the lines of [D. Barbasch and H. Moscovici (1983)] and consider the special case $G = SO_1(2n,1)$, $K = SO(2n)$. The complexified Lie algebra $\mathfrak{g} = \mathfrak{g}_0^{\mathbb{C}} = \mathfrak{so}(2n+1, \mathbb{C})$ of G is of Cartan type B_n, with Dynkin diagram

$$\underbrace{\bigcirc - \bigcirc - \bigcirc \cdots \bigcirc - \bigcirc}_{2n\ \text{nodes}} = \bigcirc. \tag{5.131}$$

The standard systems of positive roots Δ^+, Δ_k^+ for $\mathfrak{g}$ and $k = k_0^{\mathbb{C}}$ - the complexified Lie algebra of K, with respect to a Cartan subgroup H of G, $H \subset K$, are given by

$$\Delta^+ = \{\varepsilon_i | 1 \leq i \leq n\} \bigcup \Delta_k^+, \tag{5.132}$$

where

$$\Delta_k^+ = \{\varepsilon_i \pm \varepsilon_j | 1 \leq i < j \leq n\}, \tag{5.133}$$

and

$$\Delta_n^+ \stackrel{def}{=} \{\varepsilon_i | 1 \leq i \leq n\}, \tag{5.134}$$

is the set of positive non-compact roots. Here

$$(\varepsilon_i, \varepsilon_j) = \frac{\delta_{ij}}{2(2n-1)}, \tag{5.135}$$

$(\,,\,)$ being the Killing form of G. Let $\mathfrak{h}_0$ be the Lie algebra of H and let $\mathfrak{h}_{\mathbb{R}}^* = \mathrm{Hom}(i\mathfrak{h}_0, \mathbb{R})$ be the dual space of the real vector space $i\mathfrak{h}_0$, here $i \equiv \sqrt{-1}$. Thus, the $\{\varepsilon_i\}_{i=1}$ are an $\mathbb{R}$-basis of $\mathfrak{h}_{\mathbb{R}}^*$. Of interest are the *integral* elements f of $\mathfrak{h}_{\mathbb{R}}^*$:

$$f \stackrel{def}{=} \{\mu \in \mathfrak{h}_{\mathbb{R}}^* | \frac{2(\mu, \alpha)}{(\alpha, \alpha)} \in \mathbb{Z},\ \forall \alpha \in \Delta^+\}. \tag{5.136}$$

Using Eq. (5.135), we have

$$\frac{2(\mu,\varepsilon_i)}{(\varepsilon_i,\varepsilon_i)} = 2\mu_i\,, \qquad \text{for} \quad 1 \le i \le n,$$

$$\frac{2(\mu,\varepsilon_i \pm \varepsilon_j)}{(\varepsilon_i \pm \varepsilon_j, \varepsilon_i \pm \varepsilon_j)} = \mu_i \pm \mu_j\,, \quad \text{for} \quad 1 \le i < j \le n,$$

where we shall write $\mu = \sum_{j=1}^{n} \mu_j \varepsilon_j$ for $\mu \in \mathfrak{h}_{\mathbb{R}}^*$, $\mu_j \in \mathbb{R}$. Then clearly

$$f = \{\mu \in \mathfrak{h}_R^* | \quad 2\mu_i \ \in \mathbb{Z}\,, \qquad \text{for } 1 \le i \le n\,,$$

$$\mu_i \pm \mu_j \in \mathbb{Z}\,, \qquad \text{for } 1 \le i < j \le n\}\,.$$

Let $\delta_k = \frac{1}{2}\sum_{\alpha\in\Delta_k^+}\alpha$, $\delta_n = \frac{1}{2}\sum_{\alpha\in\Delta_n^+}\alpha$, $\delta = \delta_k + \delta_n = \frac{1}{2}\sum_{\alpha\in\Delta^+}\alpha$. Then $\delta_k = \sum_{i=1}^{n}(n-i)\varepsilon_i$, $\delta_n = \frac{1}{2}\sum_{i=1}^{n}\varepsilon_i$, $\delta = \sum_{i=1}^{n}(n-i-\frac{1}{2})\varepsilon_i$ are all integral. The elements μ of f correspond to characters e^{μ} of H. Following [D. Barbasch and H. Moscovici (1983)] we fix, once and for all, $\mu \in f$, which is Δ_k^+-dominant: $(\mu,\alpha) \ge 0$ for $\alpha \in \Delta_k^+$. For us, in concrete terms, this means the following. Let $\mu = (\mu_1, ..., \mu_n)$ be a sequence of real numbers such that

(**i**) $2\mu_i \in \mathbb{Z}$ for $1 \le i \le n$ and $\mu_i \pm \mu_j \in \mathbb{Z}$ for $1 \le i < j < n$ (i.e. μ is integral).

(**ii**) $|\mu_n| \le \mu_{n-1} \le \mu_{n-2} \le ... \le \mu_2 \le \mu_1$ and either every $\mu_i \in \mathbb{Z}$ or, else, every μ_i is half an odd integer.

Note that, in fact, (**ii**) $\to$ (**i**) so that we may drop condition (**i**) (only for $G = SO_1(2n,1)$ since in general (**ii**) does not imply (**i**)). Then $\mu + \delta_n$ does define a character of H, as has been required in [D. Barbasch and H. Moscovici (1983)] and, by its construction (μ is the highest weight of an irreducible k-module V_μ), there is a twisted Dirac operator $\mathfrak{D}_{\mu,\Gamma}^+$ on a vector bundle over $\Gamma\backslash G/K$ for Γ a discrete subgroup of G for Γ satisfying the conditions of that paper. Thus, $\Gamma\backslash G$ needs not be compact, but one requires of course that $\mathrm{Vol}(\Gamma\backslash G) < \infty$. The paper [D. Barbasch and H. Moscovici (1983)] also requires that a choice ψ of positive root system for $(\mathfrak{g}, \mathfrak{h} = \mathfrak{h}_0^{\mathbb{C}})$ be met such that $\Delta_k^+ \subset \psi$ and such that $\mu + \delta_k$ be ψ-dominant, i.e. $(\mu + \delta_k, \alpha) \ge 0 \;\forall\, \alpha \in \psi$ (condition (1.3.1) of [D. Barbasch and H. Moscovici (1983)] where $\rho_c = \delta_k$, $\Phi = \Delta \overset{def}{=} \Delta^+ \bigcup (-\Delta^+)$, $\Phi_k = \Delta_k^+$, $\Phi_n = \Delta_n^+$, $T = H$). Here, we choose $\psi = \Delta^+$ provided $\mu_n \ge 0$, which we now assume (see (**ii**)).

Twisted Dirac operator. In summary, we assume the following: $\mu = (\mu_1, ..., \mu_n)$ is a sequence of real numbers, and

(**iii**) $0 \leq \mu_n \leq \mu_{n-1} \leq \mu_{n-2} \leq \cdots \leq \mu_2 \leq \mu_1$, where either every $\mu_j \in \mathbb{Z}$ or every μ_j has the form $\mu_j = n_j + \frac{1}{2}$ for some $n_j \in \mathbb{Z}$.

Then we have a twisted Dirac operator $\mathfrak{D}^+_{\mu,\Gamma}$ over $\Gamma \backslash G/K$.

Theorem 5.3 ([D. Barbasch and H. Moscovici (1983)], for the case $G = SO_1(2n, 1), n \geq 2$). *For a suitable normalization of the Haar measure on G, and for μ satisfying the condition* (**iii**), *one has*

$$\operatorname{Index} \mathfrak{D}^+_{\mu,\Gamma} = \operatorname{Vol}(\Gamma \backslash G) \frac{\prod_{\alpha \in \Delta^+} (\mu + \delta_k, \alpha)}{\prod_{\alpha \in \Delta^+_k} (\delta_k, \alpha)}. \tag{5.137}$$

To make this explicit, we must express $P_k \overset{def}{=} \prod_{\alpha \in \Delta^+_k} (\delta_k, \alpha)$ and $P \overset{def}{=} \prod_{\alpha \in \Delta^+} (\mu + \delta, \alpha)$ directly in terms of the real numbers $\mu_1, ..., \mu_n$. We compute $P_k \overset{def}{=} \prod_{\alpha \in \Delta^+_k} (\delta_k, \alpha)$. First, note that the number of pairs (i, j) with $1 \leq i < j \leq n$ is $n(n-1)/2$; $(\delta_k, \delta_j) = \sum_{i=1}^n (n - i)(\varepsilon_i, \varepsilon_j) = [n - j][2(2n-1)]^{-1} \implies (\delta_k, \varepsilon_i \pm \varepsilon_j) = [n - i \pm (n - j)][2(2n-1)]^{-1}$, or

$$\begin{aligned}
P_k &= \prod_{1 \leq i < j \leq n} (\delta_k, \varepsilon_i + \varepsilon_j) \prod_{1 \leq i < j \leq n} (\delta_k, \varepsilon_i - \varepsilon_j) \\
&= \prod_{1 \leq i < j \leq n} \frac{2n - i - j}{2(2n-1)} \prod_{1 \leq i < j \leq n} \frac{-i + j}{2(2n-1)} \\
&= \left[\frac{1}{2(2n-1)}\right]^{\frac{n(n-1)}{2}} \prod_{1 \leq i < j \leq n} (2n - i - j) \left[\frac{1}{2(2n-1)}\right]^{\frac{n(n-1)}{2}} \prod_{1 \leq i < j \leq n} (-i + j) \\
&= \left[\frac{1}{2(2n-1)}\right]^{n(n-1)} \prod_{1 \leq i < j \leq n} (2n - i - j)(-i + j).
\end{aligned} \tag{5.138}$$

We also look at $P \overset{def}{=} \prod_{\alpha \in \Delta^+} (\mu + \delta_k, \alpha)$;

$$\mu + \delta_k = \sum_{j=1}^n (\mu_j + n - j)\varepsilon_j \quad \implies$$

$$(\mu+\delta_k,\varepsilon_i)=\sum_{j=1}^{n}(\mu_j+n-j)(\varepsilon_j,\varepsilon_i)=\frac{\mu_i+n-i}{2(2n-1)}. \tag{5.139}$$

Then, for $i<j$,

$$(\mu+\delta_k,\varepsilon_i+\varepsilon_j)=\frac{\mu_i+\mu_j+2n-i-j}{2(2n-1)}, \tag{5.140}$$

$$(\mu+\delta_k,\varepsilon_i-\varepsilon_j)=\frac{\mu_i-\mu_j+i-j}{2(2n-1)}, \tag{5.141}$$

$$\prod_{\alpha\in\Delta_k^+}(\mu+\delta_k,\alpha)=\prod_{1\leq i\leq j\leq n}\frac{\mu_i+\mu_j+2n-i-j}{2(2n-1)}\prod_{1\leq i\leq j\leq n}\frac{\mu_i-\mu_j-i+j}{2(2n-1)}$$
$$=\left[\frac{1}{2(2n-1)}\right]^{n(n-1)}\prod_{1\leq i\leq j\leq n}(\mu_i+\mu_j+2n-i-j)(\mu_i-\mu_j-i+j).$$

We have

$$\prod_{\alpha\in\Delta_n^+}(\mu+\delta_k,\alpha)=\prod_{i=1}^{n}\frac{(\mu_i+n-i)}{2(2n-1)}=\left[\frac{1}{2(2n-1)}\right]^{n}\prod_{i=1}^{n}(\mu_i+n-i), \tag{5.142}$$

and therefore

$$P=\left[\frac{1}{2(2n-1)}\right]^{n^2}\prod_{i=1}^{n}(\mu_i+n-i)\prod_{1\leq i<j\leq n}(\mu_i+\mu_j+2n-i-j)(\mu_i-\mu_j-i+j). \tag{5.143}$$

We arrive at the final result for $\mu=(\mu_1,...,\mu_n)\in\mathbb{R}^n$, subject to the condition **(iii)** (for a suitable Haar measure on G) [A.A. Bytsenko (2002a)].

Theorem 5.4 *The L^2-index of the twisted Dirac operator $\mathfrak{D}^+_{\mu,\Gamma}$ is equal to*

$$\mathrm{Index}\,\mathfrak{D}^+_{\mu,\Gamma}=\mathrm{Vol}(\Gamma\backslash G)\frac{P}{P_k}=\frac{\mathrm{Vol}(\Gamma\backslash G)}{[2(2n-1)^n]} \tag{5.144}$$
$$\times\ \frac{\prod_{i=1}^{n}(\mu_i+n-i)\prod_{1\leq i<j\leq n}(\mu_i+\mu_j+2n-i-j)(\mu_i-\mu_j-i+j)}{\prod_{1\leq i<j\leq n}(2n-i-j)(-i+j)}.$$

Chapter 6

Field Fluctuations and Related Variances

In this Chapter, making use of zeta-function regularization, we will reconsider the fluctuations of quantum fields. In particular, we shall rederive the vacuum expectation values of some quantities like the stress tensor or its trace (the conformal anomaly) employing a method which is different from the one used in Chapter 2, but that permits a generalization to the evaluation of the variances associated with the fluctuations.

To begin with, we remind that vacuum fluctuations play an important role in many physical processes. The Casimir effect is one of its most interesting physical manifestations and this issue will be considered in detail in Chapter 9. Here we only recall that the Casimir effect is related to the presence of a non-vanishing vacuum energy. This fact mainly occurs when one is dealing with non-trivial spacetimes, where the spatial sections are topologically non-trivial spaces or manifolds with boundaries.

Another interesting issue where quantum vacuum effects are present is quantum field theory in curved spacetime [N.D. Birrell and P.C.W. Davies (1982); S.A. Fulling (1991); R.M. Wald (1994)]. Recall that within the semi-classic approach to quantum gravity, the basic equation reads

$$G_{\mu\nu} + g_{\mu\nu}\Lambda = 8\pi G < T_{\mu\nu} > , \tag{6.1}$$

where $G_{\mu\nu}$ is the Einstein tensor, Λ the cosmological constant and $< T_{\mu\nu} >$ the vacuum expectation value of the matter-stress tensor. As a consequence, fluctuations of the stress tensor can induce fluctuations of the classical gravitational field and, in order to justify the semiclassical approximation, it appears very important to have reliable *a priori* estimates of these fluctuations. Fluctuations of the stress tensor were studied in [L.H. Ford

(1993); C.I. Kuo and L.H. Ford (1993)] making use of canonical methods and the issue concerning the validity of semiclassical gravity has been discussed in [C.I. Kuo and L.H. Ford (1993); B.L. Hu and N.G. Phillips (2000)]. Fluctuations of Casimir forces were investigated in [G. Barton (1991); C. Eberlein (1992)]. Alternatively, other authors [N.G. Phillips and B.L. Hu (1997)] have investigated the same problem by making use of zeta-function regularization.

Regarding the fluctuations, for simplicity here we will mainly consider only two quantities, that is $O = \phi^2$ and the stress tensor trace $O = T^\mu_\mu$ and of course their corresponding variances $\Delta^2 O =< O^2 > - < O >^2$. Within our formalism, it will be convenient to introduce the relative variance

$$\Delta^2 O_{rel} = \frac{< O^2 > - < O >^2}{< O >^2} . \tag{6.2}$$

Due to Einstein equations, the fluctuations of the stress tensor trace induce classical fluctuations of the scalar curvature. In some cases, for example for conformally invariant quantum fields in homogeneous spacetimes, to take into account only the trace of the stress tensor is not really restrictive, since one has

$$< T_{\mu\nu} >= \frac{g_{\mu\nu}}{D} < T^\alpha_\alpha > . \tag{6.3}$$

The fluctuations of $T_{\mu\nu}$ are then proportional to the fluctuations of its trace.

The vacuum expectation values of physical quantities may be obtained from the first variation of the effective action, while the second one is strictly related to the variance. It has to be stressed that the first variation of the effective action is well defined by the use of zeta-function regularization. Unfortunately this is not the case for the second variations. They are intrinsically ill defined, but for the case when they have disjoint supports. In the coincidence limit, the physically interesting case, one has to make use of a further regularization [N.G. Phillips and B.L. Hu (1997)].

We shall show that, modulo regularization problems, the relative variance turns out to be constant, namely $\Delta_{rel} = 2N^{-1}$, where N is the number of scalar fields in some multiplet. This result seems to be general, that is, independent of the quantity one is dealing with. Our results should be compared with the ones recently obtained, for $N = 1$, regarding the vacuum energy density fluctuations via smeared fields and point separation [N.G. Phillips and B.L. Hu (2000)], which give relative variances of the order of unity, albeit dependent on the dimension.

On the other hand, coming back to the case of N neutral scalar non-interacting fields, we recover a well known criterion for neglecting the quantum gravity fluctuations in the large N-limit [E. Toumboulis (1977); C.G. Callan et al. (1992); D. Binosi et al. (1999a)].

6.1 The First Variation of the Effective Action

We shall consider a multiplet of N scalar fields denoted by ϕ in an external field, described through a classical action of the form

$$S = \frac{1}{2}\int dx\, \phi L \phi\,, \tag{6.4}$$

where L is a suitable (matrix valued) differential operator defined on a D-dimensional smooth manifold.

To begin with, let us recall the formal trick that allows one to get the vacuum expectation value of the bilinear $O = \phi K \phi$ [G. Cognola and S. Zerbini (1988)]. We are mainly interested in the two cases, $K = I$, the identity matrix, concerning the fluctuation $O = \phi^2$, and $K = c_1 + c_2 L$, concerning the stress tensor trace $O = T^\mu_\mu$. c_1 and c_2 are known constants. In particular, $c_1 = 0$, in the conformally invariant case.

We denote by $\alpha(x)$ a suitable classical source and consider the Euclidean partition function

$$Z(\alpha) = \int D\phi\, e^{-\frac{1}{2}\int dx\, \phi L(\alpha)\phi} = \left(\det \ell^2 L(\alpha)\right)^{-N/2}\,. \tag{6.5}$$

Here we are assuming that in the massive case, the multiplet has the same common mass and $L(\alpha) = L + \alpha K$ may be regarded as a differential operator.

A formal functional derivation leads to

$$< O(x) >= -2\, \frac{\delta \log Z(\alpha)}{\delta \alpha}\Big|_{\alpha=0}\,. \tag{6.6}$$

Making use of zeta-function regularization, we are confronted with the problem how to compute the first variation involving trace of elliptic operators.

6.1.1 *Deformation of elliptic operators: the first variation*

In this subsection, we present some technical details concerning the deformations related to elliptic operators. In order to compute the variation of the trace of an elliptic invertible operator A, one has to take into account the fact that the variation (deformation) of A does not commute with A. From $A^{-1}A = I$, we have

$$\delta A^{-1} = -A^{-1}\delta A A^{-1} \,. \tag{6.7}$$

For the calculation of complex powers of A we can use the Cauchy-Dunford representation

$$A^{-s} = -\frac{1}{2\pi i}\int_C dz z^{-s}(A-z)^{-1} \,, \tag{6.8}$$

in which C is a suitable contour on the complex z-plane. As a consequence, we have

$$\delta A^{-s} = \frac{1}{2\pi i}\int_C dz z^{-s}(A-z)^{-1}\delta A(A-z)^{-1} \,. \tag{6.9}$$

Making use of the two representations (for $\text{Re}\, z > 0$ and $\text{Re}\, s > 0$):

$$\Gamma(s)(A-z)^{-s} = \int_0^\infty dt t^{s-1} e^{tz} e^{-tA} \,, \tag{6.10}$$

and

$$\frac{1}{\Gamma(z)} = -\frac{1}{2\pi i}\int_C dw(-w)^{-z}e^{-w} \,, \tag{6.11}$$

one gets

$$\delta \text{Tr} A^{-s} = -s\text{Tr}(A^{-s-1}\delta A) \,. \tag{6.12}$$

If the operator A is self-adjoint, we may give another derivation of the above formula. In fact, there exist eigenvalues and eigenvectors, λ_n and Ψ_n, such that $A\Psi_n = \lambda_n \Psi_n$, and we have

$$\begin{aligned} \delta \text{Tr} A^{-s} &= \frac{1}{2\pi i}\int_C dz z^{-s}\text{Tr}((A-z)^{-2}\delta A) \\ &= \frac{1}{2\pi i}\int_C dz z^{-s}\sum_n \frac{1}{(z-\lambda_n)^2} < \Psi_n, \delta A \Psi_n > \\ &= -s\sum_n \lambda_n^{-s-1} < \Psi_n, \delta A\Psi_n >= -s\text{Tr}(A^{-s-1}\delta A) \,. \end{aligned} \tag{6.13}$$

6.1.2 The vacuum expectation values

Making use of (6.12) and zeta-function regularization, we have

$$\delta \log Z(\alpha) = -\frac{N}{2} \lim_{s\to 0} \frac{d}{ds} \left[\ell^{-2s} s \mathrm{Tr}\left(L^{-s-1}(\alpha)\delta L(\alpha)\right)\right] . \tag{6.14}$$

For an arbitrary deformation, we may interpret the above equation in the following way. In general, $L^{-1}(\alpha)\delta L(\alpha)$ is a pseudo-differential operator and L is the elliptic, positive, second-order operator. As a consequence (see Lemma B.1 in Appendix B), we have

$$\begin{aligned} \mathrm{Tr}\left(L^{-s}(\alpha)L^{-1}(\alpha)\delta L(\alpha)\right) &= \frac{1}{2s\Gamma(s+1)}\mathrm{res}\ \left(L^{-1}(\alpha)\delta L(\alpha)\right) \\ &\quad -R_{L^{-1}\delta\alpha}(L(\alpha)) + \mathcal{O}(s)\,, \end{aligned} \tag{6.15}$$

and further

$$\delta \log Z(\alpha) = -\frac{N}{2}\left[\frac{1}{2}(\gamma - \log \ell^2)\mathrm{res}\ \left(L^{-1}(\alpha)\delta L(\alpha)\right) - R_{L^{-1}\delta\alpha}\right] . \tag{6.16}$$

Due to the difficulty in evaluating the quantities written above, in general, one can see that the evaluation of the first variation is not an easy task. For this reason, we shall restrict ourselves to two special cases of deformations.

In the first case, $\delta L(\alpha) = \delta\alpha$, corresponding to $O = \phi^2$. Then one obtains

$$< \phi^2(x) >= N \lim_{s\to 0} \frac{d}{ds}\left[s\ell^{-2s}\zeta(s+1|L)(x)\right] . \tag{6.17}$$

In the latter and in the following equations, $\zeta(s|L)(x)$ is the local zeta function. Making use of the meromorphic expansion (1.14), one gets the result

$$< \phi^2(x) >= N\left(\mathrm{PP}\zeta(1|L)(x) - K_{D-2}(x)\log\ell^2\right) . \tag{6.18}$$

Here PP denotes the finite part of the specified function. When D is odd and the manifold is without a boundary, we simply have

$$< \phi^2(x) >= N\zeta(1|L)(x) . \tag{6.19}$$

In the second case, that is when one is dealing with the stress tensor trace, $\delta L(\alpha) = \delta\alpha(c_1 + c_2 L)$. As a consequence

$$< T^\mu_\mu(x) >= N \lim_{s\to 0} \frac{d}{ds}\left[s\ell^{-2s}(c_1\zeta(s+1|L)(x) + c_2\zeta(s|L)(x))\right] , \tag{6.20}$$

and, as a result,

$$< T_\mu^\mu(x) >= c_1 < \phi^2(x) > + c_2 N\, K_D(x) \,. \tag{6.21}$$

In the conformally coupled case $c_1 = 0$ and one has the usual conformal anomaly, only due to quantum effects. In this particular case, one can also make use of (6.16), because the non-commutative residue contribution is vanishing, since one is dealing with, at most, a multiplicative operator and the R contribution can be evaluated. The result is again the conformal anomaly.

For the sake of completeness, let us consider the case corresponding to a zeta function non-regular at the origin. Zeta-function regularization gives

$$\log Z(\alpha) = \frac{N}{4} \lim_{s\to 0} \frac{d^2}{ds^2} \left[\ell^{-2s} s \mathrm{Tr}\left(L^{-s}(\alpha)\right)\right] . \tag{6.22}$$

Thus, the first variation reads

$$\delta \log Z(\alpha) = -\frac{N}{4} \lim_{s\to 0} \frac{d^2}{ds^2} \left[\ell^{-2s} s^2 \mathrm{Tr}\left(L^{-s-1}(\alpha)\delta L(\alpha)\right)\right] . \tag{6.23}$$

In the case of the field fluctuation, $\delta L(\alpha) = \delta\alpha$. As a result

$$< \phi^2(x) >= N \lim_{s\to 0} \frac{d^2}{ds^2} \left[\ell^{-2s} s^2 \zeta(s+1|L)(x)\right] . \tag{6.24}$$

In the other case, namely stress tensor trace for a conformally coupled massless scalar field, $\delta L(\alpha) = \delta\alpha L$. As a consequence,

$$< T_\mu^\mu(x) >= N \lim_{s\to 0} \frac{d^2}{ds^2} \left[\ell^{-2s} s^2 \zeta(s|L)(x)\right] . \tag{6.25}$$

If we assume the meromorphic extension of the local zeta function of the form

$$\Gamma(z)\zeta(z|L)(x) = \sum_{r=0}^{\infty} \frac{A_r(x)}{s + r - \frac{D}{2}} - \sum_{k=0}^{\infty} \frac{B_k(x)}{(z+k-q)^2} + J(z;x) \,, \tag{6.26}$$

we get

$$\zeta(s+1|L)(x) = \frac{A_{\frac{D}{2}-1}(x)}{s} - \frac{B_{q-1}(x)}{s^2} + \mathrm{PP}\zeta(s+1|L)(x) \tag{6.27}$$

and

$$\zeta(s|L)(x) = A_{D/2}(x) - \frac{B_q(x)}{s} + \mathcal{O}(s) \,. \tag{6.28}$$

Thus, we arrive at

$$< \phi^2(x) >= N\left(\mathrm{PP}\zeta(1|L)(x) - A_{\frac{D}{2}-1}(x)\log\ell^2 - \frac{1}{2}B_{q-1}(x)(\log\ell^2)^2\right),$$

$$< T^\mu_\mu(x) >= N\left(A_{\frac{D}{2}}(x) - B_q(x)(\log\ell^2)\right). \tag{6.29}$$

When the zeta function is regular at the origin, $B_k = 0$ and one recovers Eqs. (6.18) and (6.21). Equation (6.29) may be interpreted as a generalization of the conformal anomaly.

6.2 The Second Variation of the Effective Action

We have seen that the first variation of the effective action is associated with the vacuum expectation value $< O >$ of bilinear quantities in quantum fields. Let us show now that the second variation of the effective action is related to the variance $\Delta^2 O =< O^2 > - < O >^2$.

The second variation of the partition function (6.5) gives

$$< O(x)O(y) >= \frac{4}{Z(\alpha)}\frac{\delta^2 Z(\alpha)}{\delta\alpha(x)\delta\alpha(y)}\bigg|_{\alpha=0}. \tag{6.30}$$

On the other hand, we have the identity

$$\frac{\delta^2\log Z(\alpha)}{\delta\alpha(x)\delta\alpha(y)} = \frac{1}{Z(\alpha)}\frac{\delta^2 Z(\alpha)}{\delta\alpha(x)\delta\alpha(y)} - \frac{1}{Z^2(\alpha)}\frac{\delta Z(\alpha)}{\delta\alpha(x)}\frac{\delta Z(\alpha)}{\delta\alpha(y)}. \tag{6.31}$$

As a consequence, the related variance is given by

$$\begin{aligned}\Delta^2 O(x,y) &= < O(x)O(y) > - < O(x) >< O(y) > \\ &= 4\,\frac{\delta^2\log Z(\alpha)}{\delta\alpha(x)\delta\alpha(y)}\bigg|_{\alpha=0}\end{aligned} \tag{6.32}$$

and the relative variance [N.G. Phillips and B.L. Hu (1997)] has the form

$$\Delta^2 O_r(x,y) = \frac{\Delta^2 O(x,y)}{< O(x) >< O(y) >}. \tag{6.33}$$

The coincidence case, $x = y$, is particularly interesting from the physical point of view. It is formally given by

$$\Delta^2 O_r(x) = \Delta^2 O_{rel}(x,y)|_{x=y} = \frac{\delta^2\log Z(\alpha)}{\delta^2\alpha(x)}\left(\frac{\delta\log Z(\alpha)}{\delta\alpha(x)}\right)^{-2}. \tag{6.34}$$

In this limit, the relative variance $\Delta^2 O_{rel}(x)$ is strictly related to the one used by Kuo and Ford [C.I. Kuo and L.H. Ford (1993)], which reads

$$\Delta^2 O_{(1)} = \frac{\Delta^2 O}{< O^2 >} = \frac{\Delta^2 O_r(x)}{1+\Delta^2 O_r(x)} . \tag{6.35}$$

6.2.1 *Deformation of elliptic operators: the second variation*

In this subsection, we shall compute the second variation of the complex-power trace, again making use of (6.9) and starting from (6.12). The result is

$$\begin{aligned} \delta_2\delta_1 \mathrm{Tr} A^{-s} &= -s\mathrm{Tr}(\delta_2 A^{-s-1}\delta_1 A) \qquad (6.36)\\ &= -s\frac{1}{2\pi i}\int_C dz z^{-s-1}\mathrm{Tr}((A-z)^{-1}\delta_2 A(A-z)^{-1}\delta_1 A) . \end{aligned}$$

This expression is valid only if Supp δ_1 has void intersection with Supp δ_2. Since we are interested in the coincidence limit, $y = x$, we have to introduce an additional analytic regularization. The simplest one is to replace $(A - z)^{-1}$ with $(A-z)^{-1-\varepsilon}$ and employ the representation in terms of the Mellin transform:

$$\Gamma(1+\varepsilon)(A-z)^{-1-\varepsilon} = \int_0^\infty du\, u^\varepsilon\, e^{uz} e^{-uA} , \tag{6.37}$$

in Eq. (6.34) and then use again (6.11). As a result, we arrive at the formula

$$\begin{aligned} &\delta_2\delta_1 \mathrm{Tr} A^{-s}(\varepsilon) = \qquad (6.38)\\ &-\frac{1}{\Gamma^2(1+\varepsilon)}\frac{1}{\Gamma(s)}\int_0^\infty du u^\varepsilon \int_0^\infty dv v^\varepsilon (u+v)^s \mathrm{Tr}\left(e^{-uA}\delta_2 A e^{-vA}\delta_1 A\right) . \end{aligned}$$

This representation for the second variation of the trace of a complex power of an elliptic operator A is in agreement with the one obtained in [N.G. Phillips and B.L. Hu (1997)], making use of a different method based on Schwinger's perturbative expansion.

6.2.2 *The relative variance*

As we already said, in general the second variation of the regularized partition function, in the coincidence limit is an ill-defined quantity and a further regularization is required, as explained in the previous subsection 6.2.1. Making use of (6.38) one has

$$\delta_2\delta_1 \log Z(\alpha)(\varepsilon) = \frac{N\ell^{-2\varepsilon}}{2\Gamma^2(1+\varepsilon)} \lim_{s\to 0} \frac{d}{ds} \left[\frac{\ell^{-2s}}{\Gamma(s)} \times \right.$$

$$\left. \int_0^\infty du\, u^\varepsilon \int_0^\infty dv\, v^\varepsilon (u+v)^s \mathrm{Tr}\left(e^{-uA}\delta_2 A e^{-vA}\delta_1 A\right)\right] . \qquad (6.39)$$

For $\varepsilon > 0$ and sufficiently large, the integrand is regular at $s = 0$, and we have

$$\delta_2\delta_1 \log Z(\alpha)(\varepsilon) =$$

$$\frac{N\ell^{-2\varepsilon}}{2\Gamma^2(1+\varepsilon)} \left[\int_0^\infty du\, u^\varepsilon \int_0^\infty dv\, v^\varepsilon \mathrm{Tr}\left(e^{-uA}\delta_2 A e^{-vA}\delta_1 A\right)\right]$$

$$= N\frac{\ell^{-2\varepsilon}}{2} \mathrm{Tr}\left(A^{-1-\varepsilon}\delta_2 A A^{-1-\varepsilon}\delta_1 A\right) . \qquad (6.40)$$

This general formula for the second variation of the functional determinant is in agreement with a result obtained in [M. Kontsevich and S. Vishik (1993)].

Let us consider the specific variation related to our observables. First, for $O = \phi^2$, $\delta L(\alpha) = \delta\alpha$, and in the coincidence limit, we have

$$\Delta^2 \phi^2(x,\varepsilon) = 2N\ell^{-2\epsilon}\zeta^2(1+\varepsilon|L)(x). \qquad (6.41)$$

For standard zeta functions, when $x \neq y$, $\zeta(1+\varepsilon|L)(x,y)$ is regular at $\varepsilon = 0$, but in the coincidence limit $y \to x$, $\zeta(1|L)(x)$ is well defined only for odd D and boundary free manifolds. In these cases,

$$\Delta^2 \phi^2(x) = 2N\zeta^2(1|L)(x). \qquad (6.42)$$

Recalling that

$$< \phi^2 >= N\zeta(1|L)(x) , \qquad (6.43)$$

we obtain $\Delta^2 < \phi^2 >_{rel} = \frac{2}{N}$.

In the conformally coupled case and for $O = T^\mu_\mu$, one has $\delta L(\alpha) = c_2\delta\alpha L$. As a result, in the standard case,

$$\Delta^2 T^\mu_\mu(x,\varepsilon) = 2c_2^2 N\zeta^2(\varepsilon|L)(x) = 2c_2^2 N\zeta^2(0|L)(x) + \mathcal{O}(\varepsilon) \qquad (6.44)$$

and we can remove the regularization parameter as soon as $\zeta(s|L)(x)$ is regular at the origin. Within this hypothesis, since

$$< T^\mu_\mu(x) >= N c_2 \zeta(0|L)(x)\,, \tag{6.45}$$

we obtain $\Delta^2 < T^\mu_\mu >_{rel} = 2N^{-1}$.

Some remarks are in order. If we limit ourselves to the odd-dimensional case, in a smooth manifold without boundary and L being an elliptic differential operator, the analytic continuation works in a simple manner for both the quantities $O = \phi^2$ and $O = T^\mu_\mu$, and no scale dependence appears in the final expressions. As a consequence, it turns out that the relative variance is always equal to

$$\Delta^2 O_r = \frac{2}{N}\,, \qquad \Delta^2 O_{(1)} = \frac{2}{N+2}\,. \tag{6.46}$$

In the even dimensional case, or/and in the presence of boundary, the situation is more complicated and a further renormalization seems unavoidable. If the manifold is not smooth and L is a pseudo-differential operator, the meromorphic structure of the zeta function may be changed, and one may have a singularity at the origin. In this situation, the variances are always ill defined and the zeta-function regularization, implemented by a further analytic regularization, is not sufficient to regularize the variance.

6.3 Some Examples

6.3.1 *The Casimir slab*

As a simple example, we will revisit the computation of the local zeta function related to the Casimir slab, namely a massless scalar quantum field confined in the x direction between infinite parallel planar Dirichlet boundaries, located at $x = 0$ and $x = a$. In this case, the local zeta function is non-trivial, due to the presence of the planar boundaries (see, for example [A.A Grib et al. (1994); S.A. Fulling (1991); E. Elizalde et al. (1994); E. Elizalde (1995)]).

To start with, recall the form of the local diagonal heat kernel, which depends only on the confining coordinate x, i.e.

$$< x|e^{-tL}|x >= \frac{2}{a(4\pi t)^{\frac{D-1}{2}}} \sum_{n=1}^{\infty} \sin^2\left(\frac{n\pi x}{a}\right) \exp\left(-\frac{n^2\pi^2\, t}{a^2}\right)\,. \tag{6.47}$$

Here L is the Dirichlet Laplacian in the slab. Making use of the trigonometric duplication formula for the sine and Poisson-Jacobi re-summation formula A.1, we may rewrite the above expression in the form

$$< x|e^{-tL}|x >= \frac{1}{(4\pi t)^{\frac{D}{2}}} \left(\sum_{n=-\infty}^{\infty} \exp(-\frac{a^2n^2}{t}) - \sum_{n=-\infty}^{\infty} \exp(-\frac{(na+x)^2}{t}) \right) .$$

All the Seeley-DeWitt coefficients $K_r(x)$ vanish, but the first one $K_0(x) = 1$. As a consequence, we may anticipate that $< T^{\mu}_{\mu}(x) >$ is vanishing. The analytic continuation for the local zeta function may be easily obtained by using the latter expression for the heat kernel in the Mellin representation. To this aim, on has to analytically regularize the integral [I.M. Gelfand and G.E. Chilov (1972)]

$$\int_0^{\infty} dt\ t^z = 0 . \tag{6.48}$$

We present the result in a simple and symmetric form (for another equivalent form corresponding to $D = 4$, see [D. Iellici and V. Moretti (1998)])

$$\zeta(s|L)(x) = \frac{\Gamma(\frac{D}{2} - s)a^{2s-D}}{(4\pi)^{\frac{D}{2}}\Gamma(s)} \times [2\zeta_R(D-2s) - \zeta_H(D-2s, x/a) - \zeta_H(D-2s, 1-x/a)] . \tag{6.49}$$

In the above expression, $\zeta_R(z)$ and $\zeta_H(z,q)$ are the Riemann and Hurwitz zeta functions respectively. This result can also be obtained by making use of the re-summation techniques explained in [A. Actor (1989); E. Elizalde et al. (1994)].

Some comments are in order. The local expression we have obtained is already in the renormalized form. Furthermore, the related trace involves a (infinite, but trivial) volume V_T in the transverse directions and an integration over x. Since

$$\int_0^1 dq\ \zeta_H(z,q) = 0 , \tag{6.50}$$

the resulting zeta function reads

$$\zeta(s|L) = \frac{2V_T\,\Gamma(\frac{D}{2} - s)a^{2s-D+1}}{(4\pi)^{\frac{D}{2}}\Gamma(s)} \zeta_R(D-2s) , \tag{6.51}$$

and this is exactly the zeta function associated with the Casimir slab configuration.

As far as the application to the evaluation of the vacuum expectation value is concerned, first let us consider $D > 2$. The zeta function is regular at $s = 1$ and the result is

$$< \phi^2(x) >= \frac{\Gamma(\frac{D}{2} - 1)a^{2-D}}{(4\pi)^{\frac{D}{2}}} \times [2\zeta_R(D-2) - \zeta_H(D-2, x/a) - \zeta_H(D-2, 1-x/a)] . \quad (6.52)$$

The expression is finite everywhere and gives a vanishing result at $x = 0$ and $x = a$. Due to the simple geometry of the planar boundaries, the boundary divergences are not present.

Things are different in $D = 2$. In this case, the zeta function has a pole at $s = 1$ and the dependence on the scale μ appears. For the sake of completeness, we give here the result. It reads

$$< \phi^2(x) >= \frac{1}{2\pi} \left[\gamma + \log \frac{a\mu}{\pi} + \log \sin \left(\frac{\pi x}{a} \right) \right] . \quad (6.53)$$

In this case, the boundary divergences are present.

6.3.2 *The D-dimensional torus*

In this Section we will consider, as second example, the D-dimensional torus. This is a symmetric flat manifold with finite volume and the local zeta function is simply the ratio of the global zeta function and the torus volume. Thus, we may limit ourselves to the discussion involving the zeta function. This is a general conclusion valid for every symmetric space.

The zeta function for the case of a massive ($N = 1$) scalar field is given by

$$\zeta_D(s|L) = \left(\frac{4\pi^2}{R^2} \right)^{-s} Z_D(s; (mR)^2) \quad (6.54)$$

with [E. Elizalde (2001a)], [M. Bordag et al. (1996a)]

$$Z_D(s; (mR)^2) = (2\pi)^{D/2} (mR)^{D-2s} \frac{\Gamma(s - D/2)}{\Gamma(s)} + \frac{2^{s/2+D/4+1} \pi^s (mR)^{-s+D/2}}{\Gamma(s)} \quad (6.55)$$

$$\times \sum_{\mathbf{n}\in mathbbZ^D}{}' \left(\mathbf{n}^2\right)^{s/2-D/4} K_{D/2-s}\left(2^{3/2}\pi mR\sqrt{\mathbf{n}^2}\right).$$

Here R is the radius of the torus and m the mass of the field. This expansion is exponentially convergent. It is to be seen not just as a big mass expansion (the convergence is then extremely fast), for it is valid in a very wide range of values of mR: $1 \leq mR < \infty$.

In the case of a massless field, the convenient expression to be used is quite different (this is explained in detail in [E. Elizalde (2001a)]):

$$\begin{aligned} Z_D(s) &= \frac{2^{1+s}}{\Gamma(s)} \sum_{j=0}^{D-1} \left[\pi^{j/2}\Gamma(s-j/2)\zeta_R(2s-j) \right. \\ &\left. +4\pi^s \sum_{n=1}^{\infty} \sum_{\mathbf{n}\in\mathbb{Z}^j}{}' n^{j/2-s} \left(\mathbf{n}^2\right)^{s/2-j/4} K_{j/2-s}(2\pi n\sqrt{\mathbf{n}^2}) \right]. \end{aligned} \tag{6.56}$$

On the other hand, for the sake of completeness, it is interesting to have a perturbative expression for the case when the mass of the field is very small but different from zero. This is obtained by means of binomial expansion, in the equation defining the zeta function. The result is:

$$Z_D(s;(mR)^2) = \sum_{k=0}^{\infty}(-1)^k \frac{\Gamma(s+k)}{k!\,\Gamma(s)}(mR)^{2k} Z_D(s+k). \tag{6.57}$$

This expression, combined with the preceding one, Eq. (6.56), yields the desired low-mass expansion. Such expansions for the zeta function are not to be found in the literature. In fact, explicit expressions of the type (6.56) have appeared in the seminal paper [E. Elizalde (2001a)] for the first time. They are convenient, on the one hand, because they exhibit the pole structure of the zeta function explicitly. On the other, they are useful from the computational point of view, because they consist of a term including the main contribution, together with a series that converges extremely fast: only a few first terms need to be computed in order to obtain results with great accuracy.

From these expressions, the first and second variations of the effective action are immediate to compute. Essentially, what we get are equations of the following kind: (i) in the massless case (p is a natural number, a small one for any of the operators here considered)

$$\frac{Z_D(p)}{[Z_D(p-1)]^2} = \frac{\Gamma(p-1)^2}{2^{p-2}\Gamma(p)} \tag{6.58}$$
$$\times \frac{\sum_{j=0}^{D-1}\left[\pi^{j/2}\Gamma(p-j/2)\zeta_R(2p-j)+4\pi^2 S_{D,j}(p)\right]}{\left\{\sum_{j=0}^{D-1}\left[\pi^{j/2}\Gamma(p-1-j/2)\zeta_R(2p-2-j)+4\pi^2 S_{D,j}(p-1)\right]\right\}^2},$$

where the $S_{D,j}(p)$ are fast convergent series providing only corrections to the main terms, (ii)

$$\frac{Z_D(p;(mR)^2)}{[Z_D(p-1;(mR)^2)]^2} = \frac{Z_D(p)}{[Z_D(p-1)]^2} \tag{6.59}$$
$$\times \left[1+2p\left(\frac{Z_D(p)}{[Z_D(p-1)]^2}-\frac{Z_D(p+1)}{[Z_D(p)]^2}\right)(mR)^2+\cdots\right],$$

in the case of a field of very small mass (after doing a small-mass expansion as described above), and (iii)

$$\frac{Z_D(p;(mR)^2)}{[Z_D(p-1;(mR)^2)]^2} = \frac{(mR)^{2p-4}\Gamma(p-1)^2}{(2\pi)^{D/2}\Gamma(p)} \tag{6.60}$$
$$\times \frac{\left[\Gamma(p-D/2)+2^{p/2-D/4+1}\pi^{p-D/2}(mR)^{p-D/2}+\Sigma_D(p)\right]}{\left[\Gamma(p-1-D/2)+2^{p/2-D/4+1/2}\pi^{p-1-D/2}(mR)^{p-1-D/2}+\Sigma_D(p-1)\right]^2},$$

in the massive case, where again $\Sigma_D(p)$ is a very fast convergent series. These expressions simplify very much when poles of the gamma function appear (for even dimension D). They are quite easy to deal with. As advanced before, no singularity appears for D odd, and the whole expression has to be considered in that case (although the series is well approximated by a couple of first terms).

6.4 Remarks

We have revisited in this Chapter the use of zeta-function regularization approach to the evaluation of expectation values of physical quantities and their related quantum variances. The former are associated with the evaluation of the first variation of the effective action, while the latter are related to the second variation. We have shown that the first variation can be regularized by use of zeta-function techniques and explicit expressions for the vacuum expectation values have been exhibited. For issues concerning the second variation, in general, zeta-function regularization works well only when one is dealing with off-diagonal terms, since the coincidence limit is highly singular. A further analytic regularization has to be introduced to treat this coincidence limit.

We may summarize the results as follows. For a multiplet of N scalar fields and $O = T_\mu^\mu$, in the conformally coupled case, analytic regularization gives a relative variance exactly equal to $\frac{2}{N}$, independently of the dimension of the boundary-free manifold. For $O = \phi^2$, we have obtained again a relative variance exactly equal to $\frac{2}{N}$, but now limited to the odd-dimensional case without boundary. These restrictions can be removed in some particular situations, like the case of the Casimir slab, where one is dealing with a flat manifold with flat boundaries. The even dimensional case requires however further renormalization.

The formalism can be directly applied to the expectation value of the stress tensor (see [N.G. Phillips and B.L. Hu (1997)]), but only after much more work and effort. Again, the first variation of the effective action with respect to the metric tensor gives $< T_{\mu\nu} >$.

Chapter 7
The Multiplicative Anomaly

7.1 Introduction

Anomalies, as processes where classical symmetries are broken at the quantum level, have been studied by physicists for a long time in the past, in the framework of perturbative analysis based on Feynman diagrams [S. Adler (1969); J. Bell and R. Jackiw (1969); W.A. Bardeen (1969)] and also in the context of the path-integral formalism [D.J. Gross and R. Jackiw (1972); K. Fujikawa (1975); K. Fujikawa (1979); R.E. Gamboa Saravi et al. (1984)].

Very recently, an apparently new kind of anomaly, so called "multiplicative anomaly", has appeared in the physical literature [E. Elizalde et al. (1998e); A.A. Bytsenko and F.L. Williams (1998); A. Filippi (1997); J.J McKenzie-Smith and D.J. Toms (1998); J.S. Dowker (1998); T.S. Evans (1999); E. Elizalde et al. (1998b)]. The name, which is due to mathematicians [M. Wodzicki (1987); C. Kassel (1989); M. Kontsevich and S. Vishik (1993)] (they have been termed "defects" by some other outstanding mathematicians, as Isadore Singer), may create some confusion among physicists, since such an anomaly has little to do (at least in a direct manner) with the breaking of classical symmetries. It can be thought of as a "measure" of the lack of a factorization property for the determinant of the product of two (pseudo)differential operators, regularized by means of zeta function.

In many physical situations, one is directly confronted with the validity of the multiplicative property for the determinant of an operator

$$\det(AB) = \det A \, \det B \,, \tag{7.1}$$

which is of course known to hold for non-singular matrices in finite-dimensional vector spaces. We will assume all the time that the operators here considered admit a zeta function and shall adopt the definition of determinant obtained from it (as described in Chapters 1 and 2). It would seem clear that, if we have a product of two commuting operators,

$$\begin{aligned}
\det_\zeta(AB) &= \exp\left[\sum_{i\in I}\log(\lambda_i\mu_i)\right] = \exp\left[\sum_{i\in I}(\log\lambda_i + \log\mu_i)\right] \\
&= \exp\left[\sum_{i\in I}\log\lambda_i + \sum_{i\in I}\log\mu_i\right] \qquad (7.2) \\
&= \exp\left[\sum_{i\in I}\log\lambda_i\right]\exp\left[\sum_{i\in I}\log\mu_i\right] = \det_\zeta A\ \det_\zeta B.
\end{aligned}$$

But this is *not* true and the reader should understand why before continuing to read this Section. One (and *only one*) of these steps fails to be true.

Very much related with this is the fact that the zeta-function trace

$$\mathrm{Tr}_\zeta A = \sum_{i\in I}\lambda_i = \zeta_A(s=-1) \qquad (7.3)$$

is in fact not a trace, since in general it fails to satisfy the additive property

$$\mathrm{Tr}_\zeta(A+B) \neq \mathrm{Tr}_\zeta A + \mathrm{Tr}_\zeta B, \qquad (7.4)$$

for, again, this is a *regularized* trace (it involves analytical continuation) which is used, in general, with non trace-class operators (see also the recent work [V. Mathai et al. (2002)] for the general definition of the trace).

For an example, consider the following commuting linear operators in an infinite-dimensional space, given in diagonal form by

$$A_1 = \mathrm{diag}\ (1,2,3,4,\ldots), \qquad A_2 = \mathrm{diag}\ (1,1,1,1,\ldots), \qquad (7.5)$$

and their sum

$$A_1 + A_2 = \mathrm{diag}\ (2,3,4,5,\ldots). \qquad (7.6)$$

The corresponding zeta-traces are easily obtained and in fact

$$\begin{aligned}
&\mathrm{Tr}_\zeta A_1 = \zeta_R(-1) = -\frac{1}{12}, \qquad \mathrm{Tr}_\zeta A_2 = \zeta_R(0) = -\frac{1}{2}, \\
&\mathrm{Tr}_\zeta(A_1+A_2) = \zeta_R(-1) - 1 = -\frac{13}{12}, \qquad (7.7)
\end{aligned}$$

ζ_R being the Riemann zeta function. The last trace has been calculated according to the rules of infinite series summation (see e.g., Hardy [K. Knopp (1931); G.H. Hardy (1949)]). We observe that

$$\mathrm{Tr}_\zeta(A_1 + A_2) - \mathrm{Tr}_\zeta A_1 - \mathrm{Tr}_\zeta A_2 = -\frac{1}{2} \neq 0. \tag{7.8}$$

To summarize, unlike for ordinary, finite dimensional determinants, for which we have the property: $\det(AB) = \det(A)\det(B)$, for zeta determinants one rather has to consider, in general, an additional piece (called *anomaly* or *defect*). It is usually written as

$$a(A, B) = \log \frac{\det(AB)}{\det(A)\det(B)}, \tag{7.9}$$

or

$$a\,(A, B) = \zeta'_A\,(0) + \zeta'_B\,(0) - \zeta'_{AB}\,(0)\,. \tag{7.10}$$

Thus the anomaly $a(A, B)$ will vanish if the derivatives at $s = 0$ of the respective zeta function satisfy the additive property. There is an explicit expression, due to Wodzicki, for $a\,(A, B)$, that simplifies enormously the calculation of the multiplicative anomaly in many cases [M. Wodzicki (1987)]. It will be considered later.

An elementary example of this issue, in the physical context, is the following. Consider a free vector-valued scalar field ϕ_i in R^4, with a broken $O(N)$ symmetry —owing to the mass terms m_i^2. The Euclidean action is

$$S = \int dx^4 \phi_i \left[\left(-\Delta + m_i^2\right)\phi_i\right] \tag{7.11}$$

and the related Euclidean operator reads

$$L_{ij} = \left(-\Delta + m_i^2\right)\delta_{ij}\,, \tag{7.12}$$

in which Δ is the Laplace operator. Thus, one is actually dealing with a matrix-valued elliptic differential operator. In this case, the Euclidean partition function is given by

$$\log Z = -\frac{1}{2}\log\det\left|\ell^2 L_{ij}\right| = -\frac{1}{2}\log\det\left[\ell^2(-\Delta + m_1^2)\cdots\ell^2(-\Delta + m_N^2)\right]\,. \tag{7.13}$$

Another, not so trivial, example concerns the finite-temperature effects for a gas of free relativistic charged bosons. With regard to this case, it is possible to show that the logarithm of the grand canonical partition function, choosing as parametrization of the charged boson field the two real scalar fields ϕ_i, can be expressed as [K. Benson et al. (1991)]

$$\log Z_{\beta,\mu} = -\frac{1}{2}\log\det\left(\ell^4 L_+ L_-\right) , \tag{7.14}$$

with

$$L_\pm = -\partial_\tau^2 - \Delta + m^2 + \mu^2 \pm 2\mu\sqrt{-\Delta + m^2} , \tag{7.15}$$

τ being the imaginary time compactified with period β, the inverse equilibrium temperature and μ the chemical potential. Note however that, if one chooses as parametrization of the charged boson field the complex scalar fields ϕ and ϕ^*, one has

$$\log Z_{\beta,\mu} = -\frac{1}{2}\log\det\left(\ell^4 K_+ K_-\right) , \tag{7.16}$$

with

$$K_\pm = -\partial_\tau^2 - \Delta + m^2 + \mu^2 \pm 2\mu\partial_\tau . \tag{7.17}$$

S. Dowker [J.S. Dowker (1998)] shed doubt on the recipe used in computing the partition functions, Eqs. (7.13) and (7.14). In our opinion, the above prescription is correct and we refer to [E. Elizalde et al. (1998c)], where this issue was discussed in detail.

As further examples, we simply recall that in the evaluation of the one-loop Vilkovisky-DeWitt effective action (see, for example, [G.A. Vilkovisky (1984); S.D. Odintsov (1990); I.L. Buchbinder et al. (1992); J. Balakrishnan and D.J. Toms (1992)] for details) and in some GUT-like models (see, for example, [G. Cognola and I.L. Shapiro (1998)]), one deals with matrices of higher-order differential operators, which give rise to products of the same, when one needs to compute functional determinants.

As a quite general rule in the physical literature, starting with the examples we have recalled —as well as in many other which we do not mention here, for the sake of conciseness— the way one usually proceeds is by formally assuming the validity of the multiplicative relation (7.1), without much concerns. Needless to say, this may be dangerous. One has to use always some regularization procedure and it turns out, in fact, that the regularized determinants do *not* satisfy in general the multiplicative property, Eq. (7.1). Even when one is dealing with commuting operators the multiplicative anomaly may easily appear [C. Kassel (1989);

M. Kontsevich and S. Vishik (1993)]. In terms of

$$F(A,B) \equiv \det(AB)/(\det A \det B)$$

this anomaly has been defined as [M. Kontsevich and S. Vishik (1993)],

$$a(A,B) = \log F(A,B) = \log\det(AB) - \log\det(A) - \log\det(B)\,. \quad (7.18)$$

It should be noted, on passing, that the non-vanishing of the multiplicative anomaly seems to imply that the relation

$$\log\det A = \mathrm{Tr}\,\log A \quad (7.19)$$

does not hold, in general, for elliptic operators. In fact, if one assumes that $\mathrm{Tr}\,\log A$ is a linear functional, also in the simplest case of $[A,B]=0$, one should have

$$\mathrm{Tr}\,\log AB = \mathrm{Tr}(\log A + \log B) = \mathrm{Tr}\,\log A + \mathrm{Tr}\,\log B\,. \quad (7.20)$$

Thus, if Eq. (7.19) holds, one seems to arrive at a contradiction with Eq. (7.18), as soon as the multiplicative anomaly is non-vanishing. However, it should be noted (as already advanced before), that here we are dealing with a zeta-regularized trace, which in fact is not a trace in the usual sense (e.g., it is highly non-linear in general, and the last step in (7.20) does not hold). With regard to this issue, taking as a definition Eq. (7.19), one might start with the partition function

$$Z = \exp\left(-\frac{1}{2}\,\mathrm{Tr}\,\log L_D\right)\,, \quad (7.21)$$

and then make use of some regularization, while preserving the linear property of the trace. In this case, in [T.S. Evans (1999)] the absence of the multiplicative anomaly has been claimed and doubt has been shed on the use of zeta-function regularization. In our opinion, this is not justified (see [E. Elizalde et al. (1998c)]). What actually happens —as has been explained in [E. Elizalde (1999)]— is again that the trace here in question is *not* linear, owing to the fact that it is a *regularized* trace, the zeta trace

$$\mathrm{Tr}\,A = \mathrm{Tr}_\zeta\,A = \zeta_A(-1). \quad (7.22)$$

The point is that the operators involved are in general not of the trace class: the trace is divergent and it is given a meaning through the process of zeta regularization. This will be explained in more detail in the next section.

The presence of the multiplicative anomaly in the commuting case might turn out as a surprise. However, in [E. Elizalde et al. (1998d)] and in this and the next chapter, it will be shown that its existence is actually unavoidable in order to have independence of the partition function on the choice of bosonic degrees of freedom (see, also [E. Elizalde et al. (1998c)]). We will show some explicit, very basic examples with harmonic oscillators and Laplacian operators. It must be pointed out, however, that in some cases the multiplicative anomaly gives a contribution to the effective action which can be readily absorbed into the renormalization procedure, but in some other systems it certainly produces physical consequences, which have to be carefully analyzed. A non-trivial example has been discussed in [E. Elizalde et al. (1998d)], where after a renormalization of the charge operator, one sees that, starting from the natural definition, Eq. (1.7), the multiplicative anomaly gives rise to a new contribution —overlooked in previous treatments— to the high-temperature expansion of the free energy of the relativistic boson gas in the symmetric unbroken phase.

The present Chapter is devoted to the definition, discussion and applications of the multiplicative anomaly. The next one will deal with further examples and applications. In order to introduce the argument, we shall begin in Section 7.2 with a very down-to-earth presentation of the problem [E. Elizalde (1999)]. In Section 7.3 we treat a perturbative derivation of the multiplicative anomaly for a particular class of differential operators in arbitrary dimensions, while the general formula in the case of lower dimensions $D \leq 4$ will be given in 7.3.1. In Appendix B we shall recall the definition of the Wodzicki residue, together with some related results, which will be used in Section 7.4 in order to render explicit its relation with the multiplicative anomaly formula. Then, we show in 7.4.1 that such a general formula can be notably simplified in lower dimensions, confirming completely the result obtained by using perturbation theory. In Subsection 7.4.2, we show how the multiplicative anomaly might be used in order to compute the heat-kernel coefficients. Finally, we discuss some basic physical examples where the multiplicative anomaly plays an important role and we point out its relation with the symmetry-breaking anomalies, thus leading the way to Chapter 8, where a number of additional examples and physical applications will be presented.

7.2 Zeta Trace, Determinant and the Multiplicative Anomaly

The mathematical theory of divergent series has been very fruitful in taming the infinities that appeared in quantum field theory, from the very beginning of its conception. Its role is very essential, at least in the first stage of the regularization/renormalization procedure. Euler and Borel summation methods, and analytic continuation techniques are commonly used there. But some difficulties exist that are inherent to the theory of divergent series (see, for instance, [K. Knopp (1931); G.H. Hardy (1949)]). One of them is the well known fact that, sometimes, by using different schemes, different results are obtained. In a well-posed physical situation, the 'right' one can then only be chosen after experimental validation. Another problem is to understand, in physical terms, what you are doing, while performing say an analytic continuation from one region of the complex plane to another. This has prevented the zeta-function regularization procedure from getting general acceptance among physicists.

The situation concerning infinite determinants is even worse, in a sense. There is no book on the subject to be compared, for instance, with the above mentioned one by Hardy and we see every day that dubious manipulations are being performed at the level of the eigenvalues, that are then translated to the determinant itself and elevated sometimes to the category of standard results —when not of lore theorems. The first problem is the definition of the determinant itself. Let us quote in this respect from a paper by Edward Witten [E. Witten (1989)]: *The determinant of the Dirac operator is defined roughly as*

$$\det \mathcal{D} = \prod_i \lambda_i, \tag{7.23}$$

where the infinite product is regularized with (for example) zeta-function or Pauli-Villars regularization. The zeta-function definition of the determinant

$$\det_\zeta \mathcal{D} = \exp\left[-\zeta_{\mathcal{D}}'(0)\right], \tag{7.24}$$

maybe the one that has more firm mathematical grounds [D.B. Ray (1970); D.B. Ray and I.M. Singer (1971); D.B. Ray and I.M. Singer (1973)]. As already discussed in the Introduction, in spite of starting from the identity: $\log\det = \mathrm{Tr}\,\log$, it is known to develop the so-called *multiplicative anomaly*: the determinant of the product of two operators is not equal, in general, to the product of the determinants (even if the operators commute). This happens already with very simple operators (as two one-dimensional harmonic oscillators only differing in a constant term, Laplacians plus different mass terms, etc.). It may look incredible, at first sight, from the Tr log property and the (supposed) additivity of the trace. As advanced already, we must just take into account that the zeta trace is *no* ordinary trace (for it involves regularization), namely:

$$\mathrm{Tr}_\zeta \mathcal{D} = \zeta_{\mathcal{D}}(-1), \tag{7.25}$$

so that $\mathrm{Tr}_\zeta (A+B) \neq \mathrm{Tr}_\zeta A + \mathrm{Tr}_\zeta B$, in general. Not to understand this rather obvious fact has originated a considerable amount of errors in the specialized literature —falsely attributed to misfunctions of the rigorous and elegant zeta-function method! [E. Elizalde (1994d); E. Elizalde (1994a); E. Elizalde et al. (1995); E. Elizalde (1989a); E. Elizalde (2001a)]

7.2.1 *The zeta determinant*

Let A be a pseudo-differential operator with a spectral decomposition: $\{\varphi_i, \lambda_i\}_{i\in I}$, where I is some set of indices. The definition of determinant starts by trying to make sense of the product $\prod_{i\in I} \lambda_i$, which can be easily transformed into a "sum": $\log \prod_{i\in I} \lambda_i = \sum_{i\in I} \log\lambda_i$. From the definition of the zeta function of A: $\zeta_A(s) = \sum_{i\in I} \lambda_i^{-s}$, by taking the derivative at $s=0$: $\zeta_A'(0) = -\sum_{i\in I} \log\lambda_i$, we arrive at the above definition [D.B. Ray (1970); D.B. Ray and I.M. Singer (1971); D.B. Ray and I.M. Singer (1973)] of determinant, Eq. (7.24). An older definition (due to Weierstrass) is obtained by subtracting in the series above (when it is such) the leading behavior of λ_i as a function of i, as $i \to \infty$, until the series $\sum_{i\in I} \log\lambda_i$ is made to converge. The shortcoming is here —for physical applications— that these additional terms turn out to be *non-local* and, thus, are non-admissible in any renormalization procedure.

As a very simple example of the anomaly that may arise in the zeta-definition of the determinant, consider the following commuting linear operators in an infinite-dimensional space, given in diagonal form by:

$$O_1 = \mathrm{diag}\ (1,2,3,4,\ldots), \qquad O_2 = \mathrm{diag}\ (1,1,1,1,\ldots) \equiv I, \tag{7.26}$$

and their sum

$$O_1 + O_2 = \mathrm{diag}\ (2,3,4,5,\ldots). \tag{7.27}$$

The corresponding zeta-traces are easily obtained, namely

$$\begin{aligned} \mathrm{Tr}_\zeta O_1 &= \zeta_R(-1) = -\frac{1}{12}, \qquad \mathrm{Tr}_\zeta O_2 = \zeta_R(0) = -\frac{1}{2}, \\ \mathrm{Tr}_\zeta (O_1+O_2) &= \zeta_R(-1) - 1 = -\frac{13}{12}, \end{aligned} \tag{7.28}$$

the last trace having been calculated according to the rules of infinite series summation (see *e.g.* [K. Knopp (1931); G.H. Hardy (1949)]). We observe that

$$\mathrm{Tr}_\zeta (O_1+O_2) - \mathrm{Tr}_\zeta O_1 - \mathrm{Tr}_\zeta O_2 = -\frac{1}{2} \neq 0. \tag{7.29}$$

But, if this happens in such a simple situation, involving the identity operator, one can easily imagine that any precaution one can take in manipulating infinite sums might turn out to be insufficient. Moreover, since the multiplicative anomaly —as has been pointed out before— originates precisely in the failure of this addition property for the regularized trace, we can already guess that it can also show up in very simple situations, as will now be proven. The appearance of the multiplicative anomaly prevents, in particular, naive manipulations with the eigenvalues in the determinant, as reorderings and splittings, what a number of physicists seem not to be aware of.

7.2.2 *Simple examples of infinite determinants*

Let us calculate some simple determinants with the zeta-function method. To start with, take

$$\Delta_1 = \prod_{n=1}^{\infty} n, \tag{7.30}$$

which is called $\infty!$ sometimes and corresponds to the determinant of an operator having for eigenvalues: $\lambda_n = n, n = 1, 2, 3, \ldots$. We have in this case

$$\zeta_1(s) = \sum_{n=1}^{\infty} n^{-s} = \zeta_R(s), \qquad {\zeta_R}'(0) = -\frac{1}{2}\log(2\pi), \tag{7.31}$$

so that

$$\Delta_1 = \exp\left[\frac{1}{2}\log(2\pi)\right] = \sqrt{2\pi}, \tag{7.32}$$

a nice result and rather well known. In the same way, we obtain

$$\Delta_2 = \prod_{n=1}^{\infty} n^{-1} = \frac{1}{\sqrt{2\pi}}, \tag{7.33}$$

from $\zeta_2(s) = \zeta_R(-s)$, as should be expected.

Let us now consider the apparently more simple case:

$$\Delta_3 = \prod_{n=1}^{\infty} \lambda, \tag{7.34}$$

that is, the determinant of a multiple of the identity: λI. This poses a problem to the zeta-function method, which must be modified somehow to cope with such situation. In fact, the corresponding zeta function,

$$\zeta_3(s) = \sum_{n=1}^{\infty} \lambda^{-s}, \tag{7.35}$$

has no abscissa of convergence in the complex plane (since the sequence of eigenvalues is neither increasing nor decreasing). This can be, however, naturally solved as follows, namely by taking logarithms (what is in fact inherent with the definition of the zeta-function method). And using again the rules for infinite series, it is plain that the result is, in this case,

$$\log \Delta_3 = \log \lambda \sum_{n=1}^{\infty} 1 = -\frac{1}{2} \log \lambda, \tag{7.36}$$

where the factor in front of $\log \lambda$ may be interpreted as the 'zeta measure' of the set of positive natural numbers, thus

$$\Delta_3 = \prod_{n=1}^{\infty} \lambda = \frac{1}{\sqrt{\lambda}}. \tag{7.37}$$

This leads, in particular, to the following specific results:

$$\Delta_4 = \prod_{n=1}^{\infty} 1 = 1^{-\frac{1}{2}} = \pm 1, \tag{7.38}$$

for the determinant of the identity operator, I and

$$\Delta_5 = \prod_{n=1}^{\infty} (-1) = (-1)^{-\frac{1}{2}} = \mp i, \tag{7.39}$$

for the determinant of the operator $-I$. As it seems clear that, the determinant of the identity operator should be 1, this tells us (by choice everywhere of the *same* determination of the logarithm in the complex plane) that the determinant of $-I$ is $-i$ and, that of λI, the inverse of the corresponding square root of λ in (7.37). Notice that, in this way, we are starting to build up a set of consistency rules that are reminiscent, in some manner, of the corresponding rules for infinite series [K. Knopp (1931); G.H. Hardy (1949)]. More than this, by use of the logarithm, *all* the ordinary rules for infinite series are applicable to the series of logs of eigenvalues, in particular, the ones concerning multiplication by a common factor (used before already), or of splitting out a finite number of first terms from the series (that is to say, a finite number of first factors from the determinant). However, the splitting of an *infinite* number of terms —or of the whole series into two— is *not* allowed in general. That is, again, the lesson we have learned from the existence of the multiplicative anomaly of the determinant when evaluated by the zeta-function procedure. An additional comment is in order: in dealing with infinite series we always take logarithms and this introduces an ambiguity in the zeta-function definition of the determinant. This fact is well known [M. Kontsevich and S. Vishik (1993)] and is common to other regularization methods (as Pauli-Villars), under different names. It can duely be taken care of by sticking to one and the same determination of the logarithm during the whole calculation.

An apparent problem —or virtue perhaps?— of the zeta-function definition of infinite determinants is posed by the following example. It turns out that the determinants

$$\Delta_6 = \prod_{n=1}^{\infty} (-1)^{2n+1}, \qquad \Delta_7 = \prod_{n=1}^{\infty} (-1)^{4n+1}, \qquad \dots \tag{7.40}$$

and Δ_5 are all different. This originates from

$$\begin{aligned} \zeta_6(-1) &= 2\zeta_H(-1,1/2) - 1 = -B_2(1/2) - 1 = -\frac{11}{12}, \\ \zeta_7(-1) &= 4\zeta_H(-1,1/4) - 1 = -2B_2(1/4) - 1 = -\frac{23}{24}, \quad \dots \end{aligned} \tag{7.41}$$

and can be interpreted as due to the change of the 'zeta measure' of the number of factors in the product leading to the determinant. We could try to avoid this problem by sticking always to the most simple characterization of the eigenvalues series (in this case $-1, -1, -1, \dots$, any reference to n

being superfluous). Things are, regretfully, not that simple. Consider the determinants Δ_4, Δ_5 and

$$\Delta_8 = \prod_{n=1}^{\infty} (-1)^n, \tag{7.42}$$

and try to make compatible the apparently obvious fact that

$$\Delta_8 = \sqrt{\Delta_4 \Delta_5}. \tag{7.43}$$

The determinant Δ_8 can be obtained in three different ways, that yield the same result.

(i) We have, through the corresponding zeta function,

$$\begin{aligned}
\zeta_8(s) &= \sum_{n=1}^{\infty} (-1)^{-ns} = \frac{1}{(-1)^s - 1}, \\
\zeta_8{}'(s) &= \log(-1) \left[\frac{1}{\pi^2 s^2} + \frac{1}{12} + \mathcal{O}(s) \right], \\
\Delta_8 &= \exp[- \, \zeta_8{}'(0)\big|_{reg}] = (-1)^{-1/12}.
\end{aligned} \tag{7.44}$$

(ii) On the other hand, taking logs as before, from the zeta function measure of the set of exponents, we get

$$\begin{aligned}
\log \Delta_8 &= \log(-1) \sum_{n=1}^{\infty} n = \log(-1) \zeta_R(-1), \\
\Delta_8 &= (-1)^{\zeta_R(-1)} = (-1)^{-1/12}.
\end{aligned} \tag{7.45}$$

(iii) Also, we may instead choose to take derivatives, term by term, in the first expression for the zeta function

$$\begin{aligned}
&\zeta_8(s) = \sum_{n=1}^{\infty} (-1)^{-ns}, \\
&\zeta_8{}'(s) = -\sum_{n=1}^{\infty} n(-1)^{-ns} \log(-1), \quad \zeta_8{}'(0) = -\log(-1) \sum_{n=1}^{\infty} n = \frac{\log(-1)}{12}, \\
&\Delta_8 = \exp[-\zeta_8{}'(0)] = (-1)^{-1/12}.
\end{aligned} \tag{7.46}$$

Remarkably enough, in all three cases we obtain the *same* result for this determinant. Let us now try to fulfill the factorization condition (7.43).

We have

$$\Delta_4 = \prod_{n=1}^{\infty} 1 = 1^{-1/2} = \pm 1, \qquad \Delta_5 = \prod_{n=1}^{\infty} (-1) = (-1)^{-1/2} = \mp i,$$
$$\Delta_8 = \prod_{n=1}^{\infty} (-1)^n = (-1)^{-1/12} = \sqrt{\Delta_4 \Delta_5} = 1^{-1/4}\,(-1)^{-1/4}. \tag{7.47}$$

The only way to fulfill this property (7.43) is to accept that

$$\prod_{n=1}^{\infty} 1 = -1 \qquad !! \tag{7.48}$$

On the contrary, if we insists (as almost everybody would agree on) that the determinant of the identity is 1, then we must give up the property that the determinant of the alternating series of eigenvalues $1, -1, 1, -1, \ldots$ is equal to the subdeterminant product of the 1s, times the subdeterminant product of the -1's. This is perhaps the most simple reflection one could ever have expected to obtain of the *multiplicative anomaly* of the determinant!

It may seem a joke, but if we choose to preserve, at any price, the multiplication property of the determinant and give sense to the strange fact that det $I = -1$, then we do attain compatibility in Eqs. (7.47) by setting

$$\prod_{n=1}^{\infty} 1 = e^{i\pi}, \qquad \prod_{n=1}^{\infty} (-1) = e^{i\pi/2},$$
$$\prod_{n=1}^{\infty} (-1)^n = e^{3i\pi/4}, \qquad \prod_{n=1}^{\infty} (-1)^{4n} = e^{3i\pi}, \quad \ldots. \tag{7.49}$$

All these are compatible zeta-function definitions of the determinant (they can be fixed from acceptable roots of 1 or -1 as given by the zeta-function exponents) satisfying the multiplication rule. However, it is easy to see that this process cannot go forever (and thus eliminate the anomaly): the following dets cannot possibly fulfill the multiplicative property

$$\prod_{n=1}^{\infty} (-1)^{2n+1} = (-1)^{-11/12} \neq \prod_{n=1}^{\infty} (-1)^{2n} \prod_{n=1}^{\infty} (-1). \tag{7.50}$$

But maybe this is asking too much, as has been observed before, in which case we are still left with the compatible (albeit really weird) choice (7.49).

Anyhow, it is easy to check that we do not find problems in factorizations like

$$\prod_{n=1}^{\infty}(-n) = \prod_{n=1}^{\infty}(-1) \prod_{n=1}^{\infty} n, \tag{7.51}$$

since

$$\begin{aligned} &\zeta(s) = (-1)^{-s}\zeta_R(s), \qquad \zeta'(0) = -\frac{1}{2}\log(-1) - \frac{1}{2}\log(2\pi), \\ &\prod_{n=1}^{\infty}(-n) = e^{-\zeta'(0)} = (-1)^{-1/2}\sqrt{2\pi} = \prod_{n=1}^{\infty}(-1) \prod_{n=1}^{\infty} n, \end{aligned} \tag{7.52}$$

or in the more involved one

$$\prod_{n=1}^{\infty}(-1)^n n = \prod_{n=1}^{\infty}(-1)^n \prod_{n=1}^{\infty} n, \tag{7.53}$$

because

$$\begin{aligned} &\zeta(s) = \sum_{n=1}^{\infty}(-1)^{-ns}n^{-s} = \Phi((-1)^{-s}, s), \\ &\zeta'(0) = \frac{i\pi}{12} + \Phi'(0,1) = \frac{i\pi}{12} - \frac{1}{2}\log(2\pi), \\ &\prod_{n=1}^{\infty}(-1)^n n = e^{-\zeta'(0)} = (-1)^{-1/12}\sqrt{2\pi} = \prod_{n=1}^{\infty}(-1)^n \prod_{n=1}^{\infty} n, \end{aligned} \tag{7.54}$$

being Φ the polylogarithm function. The factorization of the determinants holds here again, in the zeta-function prescription and this fact does not seem to be that immediate, in view of the last calculation as compared with what we had before.

All the above considerations may sound rather trivial, but actually they are warnings, and should be carefully taken into account before proceeding with the sort of manipulations of the eigenvalues and splittings of determinants that pervade the specialized literature.

7.3 Perturbative Derivation of the Multiplicative Anomaly

Here we shall consider a particular case involving two self-adjoint elliptic invertible operators H, $H_V = H + V$ and the related product

$$A = H(H + V) \tag{7.55}$$

and we shall also assume H and V to commute, that is $[H, V] = 0$. By definition, the corresponding multiplicative anomaly reads

$$a(H, H_V) = \log\det A - \log\det H - \log\det H_V\,, \tag{7.56}$$

where the functional determinants are evaluated by using zeta-function regularization. It is convenient to introduce the quantity

$$\mathcal{A}(s) = \zeta(s|H) + \zeta(s|H_V) - \zeta(s|A)\,. \tag{7.57}$$

Thus

$$a(H, H_V) = \mathcal{A}'(0) = \lim_{s\to 0} \frac{d}{ds}\left[\zeta(s|H) + \zeta(s|H_V) - \zeta(s|A)\right]\,. \tag{7.58}$$

Now we also suppose V to be a small perturbation potential. We indicate by $G = H^{-1}$ the inverse operator of H and by G_V the inverse of H_V. One has the well known operatorial equation $G_V = G - GVG_V$, whose solution is formally given by

$$G_V = G\sum_{n=0}^{\infty}(-VG)^n\,. \tag{7.59}$$

As H and V commute, then H and H_V are also commuting operators and we may compute the complex power by means of the binomial expansion,

$$G_V^s = G^s\left[1 + \sum_{n=1}^{\infty} f_n(s) V^n G^n\right]\,, \qquad f_n(s) = \frac{(-1)^n\Gamma(s+n)}{n!\Gamma(s)}\,. \tag{7.60}$$

Using the definition of the zeta function and the properties of the trace, we obtain

$$\zeta(s|H_V) \equiv \operatorname{tr} G_V^s = \zeta(s|H) + \sum_{n=1}^{\infty} f_n(s)\operatorname{tr}(V^n H^{-s-n}) \tag{7.61}$$

and, in a similar way,

$$\zeta(s|A) = \zeta(2s|H) + \sum_{n=1}^{\infty} f_n(s)\,\mathrm{tr}(V^n H^{-2s-n})\,. \tag{7.62}$$

Finally

$$\mathcal{A}(s) = \sum_{n=1}^{\infty} f_n(s)\,[\zeta(s+n|H) - \zeta(2s+n|H)]\ \mathrm{tr}\,V^n\,, \tag{7.63}$$

where tr is the trace on internal indices. Under the assumptions above, all operators here involved are invertible. If H has a zero eigenvalue this must be excluded from the definition of the zeta functions for the operators H and A, but it gives a contribution to $\zeta(s|H_V)$. Such a contribution modifies the anomaly by an additive logarithmic term since, in this case,

$$\mathcal{A}(s) = g_0 V^{-s} + \sum_{n=1}^{\infty} f_n(s)\,[\zeta(s+n|H) - \zeta(2s+n|H)]\ \mathrm{tr}\,V^n\,, \tag{7.64}$$

g_0 being the number of zero modes.

As is well known [R.T. Seeley (1967)], for elliptic operators the zeta function has only simple poles (see Eq. (1.14)). Then the multiplicative anomaly assumes the form

$$\begin{aligned} a(H,H_V) &= \mathcal{A}'(0) = -g_0 \log V \\ &\quad + \sum_{n\geq 2} (-1)^n \frac{\gamma+\psi(n)}{2n}\ \mathrm{Res}\,\zeta(s|H)|_{s=n}\,\mathrm{tr}\,V^n\,, \end{aligned} \tag{7.65}$$

where

$$\gamma + \psi(n) = \sum_{k=1}^{n-1} \frac{1}{k}\,, \tag{7.66}$$

the sum in Eq. (7.65) being extended to all positive integers where the zeta function has a simple pole. Here γ is the Euler-Mascheroni constant, while $\psi(x)$ is the digamma function. This expression has limited validity, mainly because it assumes that the operators involved in the product commute, but nevertheless, one should note that it holds as soon as one can define the zeta functions involved. This includes such general situations as, for instance, the case when the manifold is non-compact, or when it is compact and with a boundary.

Furthermore, when one deals with the last case, i.e. a compact D-dimensional manifold with a boundary, one can certainly proceed further, since the meromorphic structure of the zeta function corresponding to an elliptic operator H, of finite order h, is well known. From Eq. (1.14) one immediately obtains

$$\text{Res}\,\zeta(s|H)|_{s=n} = \frac{K_{D-hn}(H)}{(n-1)!} \tag{7.67}$$

and thus,

$$a(H, H_V) = -g_0 \log V + \sum_{n\geq 2, hn\leq D} (-1)^n \frac{\gamma + \psi(n)}{2n!} K_{D-hn}(H)\ \text{tr}\, V^n\,. \tag{7.68}$$

We close this Section with some additional remarks. The multiplicative anomaly formula above is valid for commuting H and V only. For a manifold without a boundary one has $K_r = 0$, when r is odd. Thus, for a second-order differential operator ($h = 2$) in any odd-dimensional compact manifold without boundary, the multiplicative anomaly vanishes. It also vanishes in two dimensions, but it is certainly present for $D = 4$, being proportional to the Seeley-DeWitt coefficient K_0. In the case of first order differential operators ($h = 1$), the anomaly is non-vanishing for $D = 2$, while for the physically more interesting case $D = 4$ one has (assuming zero modes to be absent)

$$a(H, H_V) = \frac{1}{4} K_2(H)\,\text{tr}\, V^2 + \frac{11}{288} K_0(H)\,\text{tr}\, V^4\,. \tag{7.69}$$

Note the presence of the first non-trivial Seeley-DeWitt coefficient, $K_2(H)$, which for a Dirac-like operator depends on the scalar curvature of the manifold one is dealing with and V may just be a mass difference. This could be interpreted as a (potentially) interesting finite effect of induced gravity –according to Sakharov– caused by the quantum fluctuation of the matter spinor field.

It is also interesting to note that the linear term in V does not contribute to the multiplicative anomaly. If one invokes dimensional arguments, this fact is not really surprising in the case of constant V. However, as we shall see in the following, this result is also true for an *arbitrary* potential. To show this important fact we now make use of standard perturbation theory.

7.3.1 *Explicit expression for the multiplicative anomaly from perturbation theory*

As above, we assume H to be an elliptic differential operator of known, non-degenerate spectral decomposition $\{\lambda_i, \varphi_i\}_{i\in I}$ and V a small (non constant in general) perturbing potential. Let us denote by μ_i and a_i the eigenvalues of the operator H_V and A, respectively. By using perturbation theory, up to second order in V, we have

$$\mu_i = \lambda_i + V_{ii} + \sum_{j\neq i} \frac{V_{ij}V_{ji}}{\lambda_i - \lambda_j} + \mathcal{O}(V^3)\,, \tag{7.70}$$

$$a_i = \lambda_i^2 + \lambda_i V_{ii} + \sum_{j\neq i} \frac{\lambda_i\lambda_j V_{ij}V_{ji}}{\lambda_i^2 - \lambda_j^2} + \mathcal{O}(V^3)\,, \tag{7.71}$$

$$V_{ij} = (\varphi_i, V\varphi_j)\,.$$

Let us assume that $\operatorname{Re} s$ is sufficiently large. Then, by definition $\zeta(s|H) = \sum_i \lambda_i^{-s}$ while, using again the binomial expansion

$$\zeta(s|H_V) = \sum_i \mu_i^{-s} = g_0 V^{-s} + \zeta(s|H) - s\sum_i V_{ii}\lambda_i^{-s-1}$$
$$+\frac{s(s+1)}{2}\sum_i V_{ii}^2\lambda_i^{-s-2} - s\sum_{i,j,i\neq j} \frac{V_{ij}V_{ji}}{\lambda_i - \lambda_j}\lambda_i^{-s-1} + \mathcal{O}(V^2)\,, \tag{7.72}$$

$$\zeta(s|A) = \sum_i a_i^{-s} = \zeta(2s|H) - s\sum_i V_{ii}\lambda_i^{-2s-1}$$
$$+\frac{s(s+1)}{2}\sum_i V_{ii}^2\lambda_i^{-2s-2} - s\sum_{i,j,i\neq j} \frac{\lambda_j V_{ij}V_{ji}}{\lambda_i^2 - \lambda_j^2}\lambda_i^{-2s-1} + \mathcal{O}(V^2)\,. \tag{7.73}$$

As a result, the quantity $\mathcal{A}(s)$ can be written as

$$\mathcal{A}(s) = g_0 V^{-s} + 2\zeta(s|H) - \zeta(2s|H) + s[\Phi_1(2s) - \Phi_1(s)]$$
$$-\frac{s(s+1)}{2}[\Phi_2(2s) - \Phi_2(s)] + s[\Psi_1(2s) - \Psi_1(s) + \Psi_2(2s)]\,, \tag{7.74}$$

where we have introduced the functions

$$\Phi_k(s) = \sum_i V_{ii}^k \lambda_i^{-s-k}\,, \tag{7.75}$$

$$\Psi_1(s) = \sum_{i,j,i\neq j} \frac{V_{ij}V_{ji}}{\lambda_i - \lambda_j}\lambda_i^{-s-1}\,, \tag{7.76}$$

$$\Psi_2(s) \;=\; \sum_{i,j,i\neq j} \frac{V_{ij}V_{ji}}{\lambda_i^2-\lambda_j^2}\lambda_i^{-s}\,. \tag{7.77}$$

Since we are interested in the derivative at $s = 0$, let us assume that the behavior of the analytic continuation of these functions, in a neighborhood of the origin, has the form

$$\Phi_k(s) \;=\; \frac{B_k}{s} + A_k + \mathcal{O}(s)\,, \tag{7.78}$$

$$\Psi_k(s) \;=\; \frac{C_k}{s} + D_k + \mathcal{O}(s), \qquad k = 1,2\,, \tag{7.79}$$

in agreement with the fact that the zeta function has only simple poles. From Eqs. (7.74)-(7.79), we easily obtain

$$a(H,H_V) = \frac{B_2}{4} + D_2, \tag{7.80}$$

again up to second order of perturbation theory. As anticipated before, at first order in V there is no contribution to the multiplicative anomaly (observe that the potential here is *not* necessarily constant).

When V is constant, one has

$$\Phi_k(s) = V^k\zeta(s+k|H)\,, \qquad\qquad \Psi_1(s) = \Psi_2(s) = 0 \tag{7.81}$$

and then

$$a(H,H_V) = \frac{V^2}{4(4\pi)^{D/h}}\,K_{D-2h}(H)\,, \tag{7.82}$$

which is the leading term in V in Eq. (7.65). To summarize, up to second order of perturbation theory and for the special class of a product of two operators, the multiplicative anomaly depends only on V^2 and not on the derivatives of V (as could have been the case, in principle). Moreover, for dimensional reasons, in a 4-dimensional manifold and for a second-order differential operator, the second-order approximation gives the exact result and thus, in such a case one may argue that the form of the anomaly reads

$$a(H,H_V) = \frac{1}{4(4\pi)^2}\int d^4x\, k_0(x|H)\;\mathrm{tr}\,V^2\,, \tag{7.83}$$

whatever be the potential (e.g. not necessarily constant). We will confirm this result by making use of a more powerful technique which we are going to introduce in the next Section.

7.4 The Multiplicative Anomaly Formula

In this Section we will present a quick proof of the multiplicative anomaly formula, following somehow the steps of the derivation in [M. Kontsevich and S. Vishik (1993)]. We consider two invertible, elliptic, self-adjoint operators A and B on M_D of orders a and b respectively and the quantity (anomaly or defect)

$$F(A,B) = \frac{\det(AB)}{(\det A)(\det B)} = e^{a(A,B)} . \tag{7.84}$$

Moreover, we introduce the family of pseudo-differential operators

$$\begin{array}{llll} A(t) = \eta^t B^{\frac{a}{b}} , & A(0) = B^{\frac{a}{b}} , & A(1) = A , & \eta = AB^{-\frac{a}{b}} , \\ \operatorname{ord} \eta = 0, & \operatorname{ord} A(t) = a, & \operatorname{ord} (A(t)B) = a+b , & \end{array}$$

and the function

$$F(A(t),B) = \frac{\det(A(t)B)}{(\det A(t))(\det B)} .$$

One trivially gets

$$F(A(0),B) = 1 , \qquad F(A(1)),B) = \frac{\det(AB)}{(\det A)(\det B)} = F(A,B) . \tag{7.85}$$

As a consequence, one is led to deal with the following expression for the anomaly

$$\begin{aligned} a(A(t),B) &= \log F(A(t),B) \\ &= -\lim_{s\to 0} \partial_s \left[\operatorname{Tr}(A(t)B)^{-s} - \operatorname{Tr} A(t)^{-s} - \operatorname{Tr} B^{-s} \right] . \end{aligned} \tag{7.86}$$

This quantity has the following properties: $a(A(0),B) = 0$ and $a(A(1),B) = a(A,B)$. The next step is to compute the first derivative of $a(A(t),B)$ with respect to t, the result being

$$\partial_t a(A(t),B) = \lim_{s\to 0} \partial_s \left\{ s \operatorname{Tr} \log\eta \left[(A(t)B)^{-s} - \operatorname{Tr} A(t)^{-s} \right] \right\} . \tag{7.87}$$

Furthermore, making use of Lemma B.24 in Appendix B, with $Q = \log\eta$, one obtains

$$\partial_t a(A(t),B) = \operatorname{res} \left[\log\eta \left(\frac{\log A(t)}{a} - \frac{\log[A(t)B]}{a+b} \right) \right] . \tag{7.88}$$

Finally, performing the integration with respect to t from 0 to 1, one gets the Kontsevich-Vishik multiplicative anomaly formula [M. Kontsevich and S. Vishik (1993)], namely

$$a(A,B) = \int_0^1 dt \;\text{res}\left[\log\eta\left(\frac{\log A(t)}{a} - \frac{\log[A(t)B]}{a+b}\right)\right] . \tag{7.89}$$

The multiplicative anomaly formula, Eq. (7.89), notably simplifies in the special case of commuting operators. In fact one has

$$\frac{\log[A(t)B]}{a+b} = \frac{t\log\eta}{a+b} + \frac{\log B}{b} , \qquad \frac{\log A(t)}{a} = \frac{t\log\eta}{a} + \frac{\log B}{b} \tag{7.90}$$

and

$$\frac{\log[A(t)B]}{a+b} - \frac{\log A(t)}{a} = t\frac{bt\log\eta}{a(a+b)} . \tag{7.91}$$

As a result,

$$a(A,B) = \frac{b}{2a(a+b)}\text{res}\left[(\log(AB^{-\frac{a}{b}}))^2\right] , \tag{7.92}$$

which can be rewritten as the celebrated Wodzicki multiplicative formula [C. Kassel (1989)]

$$a(A,B) = \frac{\text{res}\left[(\log(A^bB^{-a}))^2\right]}{2ab(a+b)} = a(B,A) , \tag{7.93}$$

where the symmetry property in A and B is manifest.

We would like to end this Section with the following simple but interesting remarks.

(1) The notion of non-commutative residue can be introduced only for classical pseudo-differential operators, namely the ones whose symbol admits the ordinary asymptotics Eq. (B.3). The quantity $a^{-1}\log A - b^{-1}\log B$ is a classical pseudo-differential operator, as well as $\log\eta$, since $\text{ord}(\eta) = 0$. Thus, all the formulae in which the non-commutative residue appears are well defined.

(2) For more than two operators, the recurrence equation

$$a(A,B,C) = a(AB,C) + a(A,B) \tag{7.94}$$

holds and, thus, the knowledge of the multiplicative formula for the product of two operators is sufficient for computing the multiplicative anomaly associated with the product of an arbitrary number of operators.

(3) The alternative computation of

$$a(A, B, C) = a(A, BC) + a(B, C) \tag{7.95}$$

must lead to the same final result, although the partial anomalies are, in general, clearly different.

7.4.1 *The multiplicative anomaly formula in lower dimensions*

In the commutative case, examples of the evaluation of the multiplicative anomaly have already been presented (see [E. Elizalde et al. (1998e); E. Elizalde et al. (1998d)]). For the non-commutative one, we have obtained some insights by using perturbative arguments. Here, we would like to compute the multiplicative anomaly in the lower dimensional cases $D \leq 4$, by making use of the Kontsevich-Vishik formula, Eq. (7.89). We shall consider a compact manifold, in order to make use of the Wodzicki local formula for the evaluation of the non-commutative residue. With this aim, we start from the following property for the product of the two non-commuting, self-adjoint, elliptic pseudo-differential operators, that we shall consider, Q and P (the Baker-Campbell-Hausdorff formula)

$$\begin{aligned} \log(PQ) &= \log P + \log Q + [\log P, G] \\ &\quad + \frac{1}{12}[\log Q, [\log Q, \log P]] + \mathcal{O}(\log^3 Q \log P)\,, \end{aligned} \tag{7.96}$$

where G is a suitable nested commutator. Making use of this operator identity, we have that the zero order operator Q_t, defined by

$$Q_t = \frac{\log(\eta^t B^{\frac{a}{b}})}{a} - \frac{\log(\eta^t B^{\frac{a+b}{b}})}{a+b}\,, \tag{7.97}$$

may be rewritten as

$$Q_t = -\frac{bt\log\eta}{a(a+b)} + [\log\eta, G] - \frac{t}{12b}[\log B, [\log B, \log\eta]] + \mathcal{O}(\log^3 B \log\eta)\,. \tag{7.98}$$

Since res(logηQ_t) is the integrand in the multiplicative anomaly formula, Eq. (7.89) and the non-commutative residue is a trace in the algebra of classical pseudo-differential operators, one has

$$\text{res}\,(\log\eta[\log\eta, G]) = 0\,. \tag{7.99}$$

As a consequence, the Kontsevich-Vishik formula, Eq. (7.89), yields

$$\begin{aligned} a(A,B) &= \frac{\text{res}\,(\log^2\eta)}{2ab(a+b)} - \frac{1}{12b}\text{res}\,(\log\eta[\log B, [\log B, \log\eta]]) \\ &\qquad + \mathcal{O}\left(\text{res}\,(\log^2\eta\log^3 B)\right)\,, \end{aligned} \tag{7.100}$$

whereby the term which is non-vanishing in the commutative case has been conveniently isolated.

Now, we recall that for a zeroth order pseudo-differential operator, η, one has $\sigma(F(\eta)) = F(\sigma(\eta))$ (being F analytic) and, moreover, that the symbol σ of the commutator of two operators reads

$$\begin{aligned} \sigma([Q,P]) &= 2\left[\frac{\partial}{\partial x^\mu}\sigma(Q)\frac{i\partial}{\partial k_\mu}\sigma(P) - \frac{i\partial}{\partial k_\mu}\sigma(Q)\frac{\partial}{\partial x^\mu}\sigma(P)\right] \\ &\qquad + \text{higher order derivatives}\,. \end{aligned} \tag{7.101}$$

If we assume, as in Section 7.3, that A and B have the same principal and sub-leading symbols except for the homogeneous one of degree zero, namely $a = b$ and $\sigma(A) - \sigma(B) = V(x)$, then

$$\begin{aligned} &\sigma(\log\eta) = \mathcal{O}(1/|k|^a)\,, &&\frac{\partial}{\partial x^\mu}\sigma(\log\eta) = \mathcal{O}(1/|k|^a)\,, \\ &\frac{\partial}{\partial x^\mu}\sigma(\log B) = \mathcal{O}(|k|^0)\,, &&\frac{\partial}{\partial k^\mu}\sigma(\log B) = \mathcal{O}(1/|k|), \end{aligned} \tag{7.102}$$

and it follows that

$$\sigma\left(\log\eta[\log B, [\log B, \log\eta]]\right) = \mathcal{O}(1/|k|^{2a+2})\,. \tag{7.103}$$

This means that, if the dimension of the compact manifold D satisfies $D < 2a + 2$, only the first term on the right hand side of Eq. (7.100) gives a non-vanishing contribution to the anomaly. Such a condition is satisfied in the particular but important case of second-order differential operators in four dimensions. On the other hand, for first-order differential operators, $a = b = 1$ and $D = 4$, also the second term of Eq. (7.100) gives a contribution. This exact analysis fully confirms the perturbative results of Section 7.3.

7.4.2 *Heat-kernel coefficients from the multiplicative anomaly*

As we we have shown in Section 7.3, Eq. (7.68), in the case of constant V the multiplicative anomaly $a(H, H_V)$ can be directly related to the heat-kernel coefficients $K_n(H)$, but it can also be computed by means of Wodzicki's formula, Eq. (7.93). This means that one could use the Wodzicki residue in order to compute heat-kernel coefficients (see [C. Kassel (1989); E. Elizalde (1997)]). Now we illustrate the method for an invertible differential operator H, of order h, on a manifold without boundary.

For this case, Eq. (7.93) reduces to

$$a_D(H_V, H) = \frac{1}{4h}\,\mathrm{res}\left[\log(H_V H^{-1})\right]^2 = \frac{1}{4h}\,\mathrm{res}\left[\log(I + VH^{-1})\right]^2 .$$

If V is a potential commuting with H, from Eq. (7.68) one also obtains

$$a_D(H_V, H) = \frac{1}{4} K_{D-2h}(H)\, V^2 + \mathcal{O}(V^3)\,. \tag{7.104}$$

Now, taking the limit $V \to 0$, one shows that

$$K_{D-2h}(H) = \lim_{V\to 0} \frac{a_D(H_V, H)}{4V^2} = \frac{1}{4h}\,\mathrm{res}\, H^{-2}\,, \tag{7.105}$$

in agreement with [E. Elizalde (1997)].

7.5 Multiplicative Anomaly for Locally Symmetric Spaces

In this Section, we compute the multiplicative anomaly related to the product of Laplace-like operators $\mathcal{L}_j$, defined on generic locally-symmetric spaces. The spectral zeta function associated with the product of two operators, $\mathcal{L}_1 \otimes \mathcal{L}_2$, has the form

$$\zeta(s|\mathcal{L}_1 \otimes \mathcal{L}_2) = \sum_{\ell\geq 0} n_\ell \prod_{j=1}^{2} (\lambda_\ell + b_j)^{-s}. \tag{7.106}$$

We will always assume that $b_1 \neq b_2$, say $b_1 > b_2$; otherwise, if $b_1 = b_2$ then $\zeta(s|\mathcal{L}_1 \otimes \mathcal{L}_2) = \zeta(2s|\mathcal{L})$ is a well-known function. For $b_1, b_2 \in \mathbb{R}$, set $b_+ \stackrel{def}{=} (b_1 + b_2)/2$, $b_- \equiv (b_1 - b_2)/2$, thus $b_1 = b_+ + b_-$ and $b_2 = b_+ - b_-$.

Lemma 7.1 ([A.A. Bytsenko and F.L. Williams (1998)]) *The spectral* *zeta function can be written as follows*

$$\zeta(s|\mathcal{L}_1 \otimes \mathcal{L}_2) = (2b_-)^{1/2-s} \frac{\pi^{1/2}}{\Gamma(s)} \int_0^\infty dt\, K_\Gamma(t|\mathcal{L}_+) K_{s-1/2}(b_- t), \qquad (7.107)$$

where $\mathcal{L}_+ = L + b_+$ *and the integral converges absolutely for* $\operatorname{Re} s > d/4$ ($D = \dim(G/K)$).

This formula is a main starting point to study the zeta function. It expresses $\zeta(s|\mathcal{L}_1 \otimes \mathcal{L}_2)$ in terms of the Bessel function $K_{s-1/2}(b_- t)$ and $K_\Gamma(t|\mathcal{L}_+)$, where the trace formula applies to $K_\Gamma(t|\mathcal{L}_+)$. Thus we can study the zeta function by the trace formula. One can use the trace formula for $K_\Gamma(t|\mathcal{L}_+)$, Eq. (4.37); as a result

$$\zeta(s|\mathcal{L}_1 \otimes \mathcal{L}_2) = (2b_-)^{1/2-s} \frac{\pi^{1/2}}{\Gamma(s)} \int_0^\infty dt\, t^{s-1/2} \left[\frac{\chi(1)\mathrm{Vol}(\Gamma\backslash G)}{4\pi} \right.$$

$$\left. \times \int_{\mathbb{R}} e^{-(r^2+b_+ +\rho_0^2)t} |C(r)|^{-2} dr + \theta_\Gamma(t|\mathcal{L}) \right] K_{s-1/2}. \qquad (7.108)$$

Then, for $\operatorname{Re} s > 0$, Fubini's theorem gives

$$\int_0^\infty dt\, \frac{\chi(1)\mathrm{Vol}(\Gamma\backslash G)}{4\pi} \int_{\mathbb{R}} dr\, e^{-(r^2+b_+ +\rho_0^2)t} |C(r)|^{-2} K_{s-1/2} t^{s-1/2} =$$

$$(2b_-)^{\frac{1}{2}-s} \frac{\chi(1)\mathrm{Vol}(\Gamma\backslash G)\Gamma(s)}{4\pi^{3/2}} \int_{\mathbb{R}} dr\, |C(r)|^{-2} \prod_j (r^2 + v_j)^{-s}. \qquad (7.109)$$

In order to analyze the last integral in Eq. (7.109) (for the possibility of a meromorphic continuation), it is useful to rewrite the function $|C(r)|^{-2}$ (see Eq. (4.7)), using the identity $\tanh(ar) \equiv 1 - 2(1+e^{2ar})^{-1}$. Then, one can calculate a suitable integral in terms of the hypergeometric function $F(\alpha, \beta; \gamma; z)$, namely

$$\int_0^\infty dr\, r^{2\ell+1} \prod_{j=1}^2 (r^2 + v_j)^{-s} = \frac{\pi^{1/2}\Gamma(2s-\ell-1)\ell!}{2^{2s}\Gamma(s)\Gamma(s+\frac{1}{2})} (v_1 v_2)^{-s} \left(\frac{2v_1 v_2}{v_1 + v_2} \right)^{\ell+1}$$

$$\times F\left(\frac{\ell+1}{2}, \frac{\ell+2}{2}; s + \frac{1}{2}; \left(\frac{v_1 - v_2}{v_1 + v_2} \right)^2 \right), \qquad (7.110)$$

which is holomorphic function on $\mathrm{Re}\, s > (\ell+1)/2$ and admits a meromorphic continuation to $\mathbb{C}$ with only simple poles at points $s = (\ell+1-n)/2,\ n \in \mathbb{N}$. Let us introduce the function

$$E_\ell(s) \equiv 2\int_0^\infty \frac{dr\, r^{2\ell+1}}{1+e^{2a(G)r}} \prod_j (r^2+v_j)^{-s}, \tag{7.111}$$

which is an entire function of s. Then a consequence of Eqs. (7.107)-(7.111) is

Theorem 7.1 ([A.A. Bytsenko and F.L. Williams (1998)]) *For* $Res > D/4$ *the explicit meromorphic continuation holds:*

$$\zeta(s|\mathcal{L}_1 \otimes \mathcal{L}_2) = \mathcal{F}(s) - 2V^{(G)} \sum_{\ell=0}^{D/2-1} a_{2\ell} E_\ell(s) + \mathcal{I}(s), \tag{7.112}$$

where $V^{(G)} \equiv \pi V C_G$,

$$\mathcal{F}(s) \equiv \frac{V^{(G)}}{(v_1v_2)^s} \sum_{\ell=0}^{D/2-1} \frac{a_{2\ell}\ell! \left(\frac{2v_1v_2}{v_1+v_2}\right)^{\ell+1} F\left(\frac{\ell+1}{2}, \frac{\ell+2}{2}; s+\frac{1}{2}; \left(\frac{v_1-v_2}{v_1+v_2}\right)^2\right)}{(2s-1)(2s-2)...(2s-(\ell+1))}, \tag{7.113}$$

$$\mathcal{I}(s) \equiv (2b_-)^{1/2-s} \frac{\pi^{1/2}}{\Gamma(s)} \int_0^\infty dt\, t^{s-1/2}\, \theta_\Gamma(t|\mathcal{L}_+) K_{s-1/2}(b_-t). \tag{7.114}$$

The goal now is to compute the derivative of the zeta function at $s=0$ [E. Elizalde (1986); E. Elizalde (1985); E. Elizalde (1993)]. We have

$$\frac{d}{ds}\zeta(s|\mathcal{L}_1 \otimes \mathcal{L}_2)\Big|_{s=0} = V^{(G)} \sum_{\ell=0}^{D/2-1} a_{2\ell} \sum_{q=1}^{4} \mathcal{E}_q, \tag{7.115}$$

where

$$\mathcal{E}_1 = \ell!(v_1^{\ell+1} + v_2^{\ell+1}) \sum_{k=0}^{\ell} \frac{(-1)^{k+1}}{k!(\ell-k)!(\ell+1-k)!}, \tag{7.116}$$

$$\mathcal{E}_2 = v_2^{\ell+1} \left(\frac{v_1-v_2}{2v_1}\right) \frac{(-1)^\ell}{(\ell+1)!} \sum_{k=1}^{\infty} \frac{(\ell+k+1)!}{(k+1)!} \sigma_k \left(\frac{v_1-v_2}{v_1}\right)^k, \tag{7.117}$$

$$\mathcal{E}_3 = \log(v_1 v_2)\frac{(-1)^\ell}{2(\ell+1)}(v_1^{\ell+1}+v_2^{\ell+1}) - 4\int_0^\infty \frac{dr\, r^{2\ell+1}\log\left(\frac{r^2+v_1}{r^2+v_2}\right)}{1+e^{2a(G)r}}, \quad (7.118)$$

$$\mathcal{E}_4 \equiv \frac{d}{ds}\mathcal{I}(s)|_{s=0} = T_\Gamma(0,b_1,\chi_1) + T_\Gamma(0,b_2,\chi_2), \quad (7.119)$$

and

$$T_\Gamma(0,b_j,\chi_j) \equiv \int_0^\infty dt\, \theta_\Gamma(t|\mathcal{L}_j)t^{-1}, \qquad \sigma_n \equiv \sum_{k=1}^{n} k^{-1}. \quad (7.120)$$

Similarly, one can show that

$$\begin{aligned} &\frac{d}{ds}\zeta(s|\mathcal{L}_j)|_{s=0} \\ &= V^{(G)}\sum_{\ell=0}^{D/2} a_{2\ell}\left\{\ell! v_j^{\ell+1}\left[\sum_{k=0}^{\ell}\frac{(-1)^{k+1}}{k!(\ell-k)!(\ell+1-k)^2} + \frac{(-1)^\ell}{(\ell+1)!}\log v_j\right]\right. \\ &\qquad \left. +4\int_0^\infty \frac{dr\, r^{2\ell+1}\log(r^2+v_j)}{1+e^{2a(G)r}}\right\} + T_\Gamma(0,b_j,\chi_j). \quad (7.121) \end{aligned}$$

We can put all these results together, in order to compute the multiplicative anomaly. Recall that

$$a(\mathcal{L}_1,\mathcal{L}_2) \equiv \frac{d}{ds}\left(\zeta(s|\mathcal{L}_1\otimes\mathcal{L}_2) - \zeta(s|\mathcal{L}_1) - \zeta(s|\mathcal{L}_2)\right)|_{s=0}. \quad (7.122)$$

Thus, we have

Theorem 7.2 ([A.A. Bytsenko and F.L. Williams (1998)]) *A preliminary form of the multiplicative anomaly is*

$$\begin{aligned} a(\mathcal{L}_1,\mathcal{L}_2) &= V^{(G)}\sum_{\ell=0}^{D/2-1} a_{2\ell}v_2^{\ell+1}\left(\frac{v_1-v_2}{2v_1}\right)\frac{(-1)^\ell}{(\ell+1)!} \\ &\quad\times\sum_{k=1}^{\infty}\sigma_k\left(\frac{v_1-v_2}{v_1}\right)^k\frac{(\ell+k+1)!}{(k+1)!} \\ &\quad+V^{(G)}\sum_j\log(v_j)\sum_{\ell=0}^{D/2-1} a_{2\ell}\frac{(-1)^{\ell+j}}{2(\ell+1)}\left(v_1^{\ell+1}-v_2^{\ell+1}\right). \quad (7.123) \end{aligned}$$

Now, in order to obtain the final explicit form of the multiplicative anomaly (see also the monograph [F.L. Williams (2003)]), we must compute the infinite series appearing in the statement of Theorem 7.2. For $0 < x < 1$, $\ell = 0, 1, 2, \ldots$, let

$$S_\ell(x) = \sum_{k=1}^{\infty} \frac{(\ell+k+1)!}{(k+1)!} \sigma_k x^k. \tag{7.124}$$

Thus, in Theorem 7.2 we choose $x = (v_1 - v_2)/v_1$ and, after a tedious but straightforward calculation, we obtain the following result.

Theorem 7.3 *The function $S_\ell(x)$ admits the representation*

$$(1-x)^{\ell+1} S_\ell(x) = P_\ell(x) + Q_\ell(x) \log(1-x), \tag{7.125}$$

where $P_\ell(x)$ and $Q_\ell(x)$ are polynomials of degree ℓ, given by

$$\frac{P_\ell(x)}{(\ell+1)!} = \frac{\ell x}{2}(1-x)^{\ell-1} + \frac{\ell(\ell-1)}{4} x^2 (1-x)^{\ell-2} + \sum_{q=3}^{\ell} \frac{\ell!}{(q+1)!(\ell-q)!}$$

$$\times \left[\frac{1}{q} + \frac{1}{q-1} + \sum_{k=1}^{q-2} \frac{1}{q-k-1} \right] x^q (1-x)^{\ell-q}, \tag{7.126}$$

$$\frac{x Q_\ell(x)}{(\ell+1)!} = -\frac{1}{\ell+1} \left[1 - (1-x)^{\ell+1} \right]. \tag{7.127}$$

Chapter 8

Applications of the Multiplicative Anomaly

Here we are going to consider some examples in which the multiplicative anomaly may actually have physical consequences. Our aim has been, in fact, to look for the most simple situations (from the point of view of the number of space dimensions and of the order and nature of the differential operators involved) that might already exhibit a non-zero anomaly. It should not come as a surprise that, among them, harmonic oscillators play an important role. They yield non-trivial results, easy to calculate analytically, in any number of dimensions.

8.1 Anomalies for Dirac-like Operators

8.1.1 *The one-dimensional case*

Consider the square root of the harmonic oscillator obtained by Delbourgo in [R. Delbourgo (1995)]. This example has potentially some interesting physical applications, for it is well known that a fermion in an external constant electromagnetic field has a similar spectrum (Landau spectrum). Exactly in the same way as when going from the Klein-Gordon to the Dirac equation and at the same price of doubling the number of components (e.g., introducing spin), Delbourgo has constructed a model for which there exists a square root of its Hamiltonian, which is very close to the one for the harmonic oscillator. It is in fact different from the Dirac oscillator introduced by several other authors, corresponding to the minimal substitution $\mathbf{p} \to \mathbf{p} - i\alpha\mathbf{r}$. The main difference lies in the introduction now of the parity operator, Q. Whereas creation and destruction operators for the harmonic oscillator, $a^{\pm} = P \pm iX$, are non-Hermitian, the combinations $\mathcal{D}^{\pm} = P \pm iQX$ are Hermitian and

$$H^{\pm} \equiv (\mathcal{D}^{\pm})^2 = P^2 + X^2 \mp Q = 2H_{\text{OSC}} \mp Q. \tag{8.1}$$

Notice that the parity term commutes with H_{OSC}. Doubling the components (σ_i are the Pauli matrices)

$$P \to -i\sigma_1 \frac{\partial}{\partial x}, \qquad X \to \sigma_1 x, \qquad Q \to \sigma_2, \tag{8.2}$$

the operators $\mathcal{D}^{\pm}$ are represented by

$$\mathcal{D}^{\pm} \to -i\sigma_1 \frac{\partial}{\partial x} \pm \sigma_3 x. \tag{8.3}$$

In the sequel, we will only consider the operator $\mathcal{D} \equiv \mathcal{D}^{+}$. It has for eigenfunctions and eigenvalues, respectively,

$$\begin{aligned}
\psi_n^{\pm}(x) &= \frac{-ie^{-x^2/2}}{\sqrt{2^{n+1}\,(n-1)!\sqrt{\pi}}} \begin{pmatrix} -i\left[H_{n-1}(x) \pm H_n(x)/\sqrt{2n}\right] \\ \left[H_{n-1}(x) \mp H_n(x)/\sqrt{2n}\right] \end{pmatrix}, \\
\lambda_n &= \pm\sqrt{2n}, \ n \geq 1, \\
\psi_0(x) &= \frac{e^{-x^2/2}}{\sqrt{2\sqrt{\pi}}} \begin{pmatrix} 1 \\ i \end{pmatrix}, \qquad \lambda_0 = 0,
\end{aligned} \tag{8.4}$$

where the $H_n(x)$ are Hermite polynomials.

The two operators we shall consider for the calculation of the anomaly are $\mathcal{D}$ and $\mathcal{D}_V = \mathcal{D} + V$, V being a real, constant potential with $|V| < \sqrt{2}$, that goes multiplied with the identity matrix in the two (spinorial) dimensions (omitted here). Notice that $\mathcal{D}$ and $\mathcal{D}_V$ are Hermitian, commuting operators. The zeta function for the operator $\mathcal{D}$ reads

$$\begin{aligned}
\zeta(s|\mathcal{D}) = \sum_i \lambda_i^{-s} &= \sum_{n=1}^{\infty} [1 + (-1)^{-s}] \left(\sqrt{2n}\right)^{-s} \\
&= [1 + (-1)^{-s}] 2^{-s/2} \zeta_R(s/2),
\end{aligned} \tag{8.5}$$

$\zeta_R(s)$ being the Riemann zeta function, which has a simple pole at $s = 1$. Furthermore, the manifold is not compact; thus, by direct use of Eq. (7.65) of Chapter 7, we readily get

$$a(\mathcal{D}, \mathcal{D}_V) = \frac{V^2}{2} - \log V. \tag{8.6}$$

The logarithmic term is due to the presence of a zero mode.

How does this match with the Wodzicki formula? First of all, we point out that one has to deal with zero modes. Thus some care must be taken in order to treat them properly. Secondly, since we are working in a non-compact manifold, we cannot use the local formula to evaluated the non-commutative residue. A direct check, modulo the zero mode problem, that makes use of Eq. (B.12), yields the multiplicative anomaly above. It thus seems that Wodzicki's expression requires only a small modification in order to deal with spinorial operators, as the ones we have here.

We conclude by pointing out that we have here, before us, the first and most simple example of the presence of a non-trivial anomaly for operators of degree one in a space of dimension one (spinorial, however).

8.1.2 *Generalization to arbitrary dimensions*

Referring again to the work by Delbourgo in [R. Delbourgo (1995)], the above operators $\mathcal{D}^+$ and $\mathcal{D}^-$ admit possible extensions to D dimensions, which have for eigenvalues, respectively,

$$\lambda_+ = \pm\sqrt{2(2n_r + l)}, \quad \lambda_- = \pm\sqrt{2(2n_r + l + D)}, \quad n_r, l = 0, 1, 2, \ldots \tag{8.7}$$

Notice that $\mathcal{D}^+$ exhibits a zero mode (for $n_r = l = 0$), what is not the case with $\mathcal{D}^-$. Each of these two operators provides a different example for the calculation of the anomaly and we will denote the corresponding partners by $\mathcal{D}_V^{\pm}$ (see the preceding subsection).

The basic zeta functions are now

$$\begin{aligned}
\zeta_+(s|H) &= \left[1 + (-1)^{-s}\right] 2^{-s/2} \sum_{n_r,l=0}^{\infty}{}' (2n_r + l)^{-s/2} \\
&= \left[1 + (-1)^{-s}\right] 2^{-s/2} \zeta_2\left(\frac{s}{2}, 0|\,(2,1)\right),
\end{aligned} \tag{8.8}$$

$$\begin{aligned}
\zeta_-(s|H) &= \left[1 + (-1)^{-s}\right] 2^{-s/2} \sum_{n_r,l=0}^{\infty} (2n_r + l + D)^{-s/2} \\
&= \left[1 + (-1)^{-s}\right] 2^{-s/2} \zeta_2\left(\frac{s}{2}, D|\,(2,1)\right).
\end{aligned} \tag{8.9}$$

As usual, the prime means in the first expression that the zero mode has to be excluded from the sum (i.e., the term with $n_r = l = 0$). Here, $\zeta_2(s, b|(a_1, a_2))$ is the Barnes zeta function in 2 dimensions. It has simple poles at the points $s = 1, 2$ (in general at $s = N, N-1, \ldots$ for ζ_N) and the residues are well known, as given by generalized Bernoulli polynomials at those points (see [E.W. Barnes (1903); J.S. Dowker (1994); K. Kirsten and D.J. Toms (1996d)]). Again using Eq. (7.65) of Chapter 7, one gets

$$\begin{aligned} a_+ &\equiv a(\mathcal{D}^+, \mathcal{D}_V^+) \\ &= \sum_{j=1}^{2} \frac{V^{2j}}{2^j} \frac{\psi(2j)+\gamma}{j} \operatorname{Res} \zeta_2\left(j, 0|\,(2,1)\right) - \log V, \end{aligned} \tag{8.10}$$

$$\begin{aligned} a_- &\equiv a(\mathcal{D}^-, \mathcal{D}_V^-) \\ &= \sum_{j=1}^{2} \frac{V^{2j}}{2^j} \frac{\psi(2j)+\gamma}{j} \operatorname{Res} \zeta_2\left(j, D|\,(2,1)\right), \end{aligned} \tag{8.11}$$

where, as before, the logarithmic term is due to the presence of the zero mode and ψ is the digamma function. For $D = 2$, the result is

$$a_+ = \frac{V^2}{4}\left(\frac{11V^2}{12} - 1\right) - \log V, \tag{8.12}$$

$$a_- = \frac{V^2}{4}\left(\frac{11V^2}{12} - 1\right). \tag{8.13}$$

Let us now check with the result obtained from Wodzicki's formula

$$\begin{aligned} a_W(A_1, A_2) &= \frac{\operatorname{res}\left[\left(\log(A_1^{a_2} A_2^{-a_1})\right)^2\right]}{2a_1a_2(a_1+a_2)} \\ &= \frac{1}{4} \operatorname{res}\left\{\left[\log\left(I + \frac{V}{\mathcal{D}^\pm}\right)\right]^2\right\}. \end{aligned} \tag{8.14}$$

Here $a_1 = a_2 = 1$. Looking for the non-commutative residue, making use again of Eq. (B.12), of course one obtains Eqs. (8.11) and (8.11) (modulo the logarithmic term corresponding to the zero mode, which is immediate to supply). From the physical point of view, it is not completely clear if these operators make sense for any value of D. In fact, in [R. Delbourgo (1995)] some doubts were arisen concerning their precise physical meaning in four and higher dimensions.

8.1.3 *Harmonic oscillator in D-dimensions*

Let us recall the case when H corresponds to a harmonic oscillators in D dimensions, with angular frequencies $(\omega_1, \ldots, \omega_D)$. The eigenvalues read

$$\lambda_{\bar{n}} = \bar{n} \cdot \bar{\omega} + b\,, \quad \bar{n} \equiv (n_1, ..., n_D)\,, \bar{\omega} \equiv (\omega_1, ..., \omega_D)\,, \quad b = \frac{1}{2}\sum_{k=1}^{D} \omega_k \quad (8.15)$$

and the related zeta function is the Barnes one $\zeta_D(s, b|\bar{\omega})$, whose poles are to be found at the points $s = k$ $(k = D, D-1, \ldots, 1)$. Their corresponding residua can be expressed in terms of generalized Bernoulli polynomials $B^{(D)}_{D-k}(b|\bar{\omega})$. They are defined by [N.E. Nörlund (1922)]

$$\frac{t^D e^{-at}}{\prod_{i=1}^{D}(1-e^{-b_i t})} = \frac{1}{\prod_{i=1}^{D} b_i} \sum_{n=0}^{\infty} B_n^{(D)}(a|b_i) \frac{(-t)^n}{n!}. \qquad (8.16)$$

The residua of the Barnes zeta function at $k = D, D-1, ...$ are then

$$\text{Res}\, \zeta_D(k, b|\bar{\omega}) = \frac{(-1)^{D+k}}{(k-1)!(D-k)! \prod_{j=1}^{D} \omega_j} B^{(D)}_{D-k}(b|\bar{\omega}). \qquad (8.17)$$

Now, if V is a constant potential, from Eq. (7.65) we easily obtain the anomaly associated with the pair of operators H and $H_V = H + V$:

$$a(H, H_V) = \frac{(-1)^D}{2\prod_{j=1}^{D}\omega_j} \sum_{k=1}^{[D/2]} \frac{[\gamma + \psi(D-2k)]\, B^{(D)}_{2k}(b|\bar{\omega})}{(2k)!\,(D-2k)!} V^{2k}. \qquad (8.18)$$

Notice that in our case the generalized Bernoulli polynomials of odd order vanish $B^{(D)}_1(b|\bar{\omega}) = B^{(D)}_3(b|\bar{\omega}) = \cdots = 0$, for any D. On the other hand, the remaining generalized Bernoulli polynomials are *never* zero, in fact

$$\begin{aligned}
B_0^{(D)}(b|\bar{\omega}) &= 1, \qquad\qquad B_2^{(D)}(b|\bar{\omega}) = -\frac{1}{12}\sum_{i=1}^{D}\omega_i^2, \\
B_4^{(D)}(b|\bar{\omega}) &= \frac{1}{24}\left[\frac{7}{10}\sum_{i=1}^{D}\omega_i^4 + \sum_{i<j}\omega_i^2\omega_j^2\right], \qquad (8.19) \\
B_6^{(D)}(b|\bar{\omega}) &= -\frac{5}{96}\left[\frac{31}{70}\sum_{i=1}^{D}\omega_i^6 + \frac{7}{10}\sum_{i\neq j}\omega_i^4\omega_j^2 + \sum_{i<j<k}\omega_i^2\omega_j^2\omega_k^2\right], \quad \ldots
\end{aligned}$$

As a consequence, the anomaly does not vanish in any case, not for D odd or $D = 2$, whatever the frequencies ω_i be. Moreover, only even powers of the potential V appear. Again, the manifold is non-compact, but a direct use of the Wodzicki formula (see Eq. (B.12) of Appendix B) confirms its validity also in the present case.

8.2 The Massive Dirac Operator

Here we use the notations of Chapter 5. If four dimensions, it is usually assumed that $\det K = \det K^\dagger = \sqrt{\det K^\dagger K}$ (see, for example, the recent papers [R. Soldati and L. Sorbo (1998); D.G.C. McKeon and C. Schubert (1998); G. Dunne and T.L. Hall (1999)]). However, a multiplicative anomaly may be present sometimes, what will expoil the second equality, while the first one could be spoiled by a phase of the determinant [A. Asada (1999)].

To begin with, it should be noted here that, strictly speaking, the Wodzicki multiplicative anomaly formula we have discussed so far is *not* valid for Dirac-like operators, since they are (first order) non-positive operators. However, we may use the definition of the multiplicative anomaly, i.e.

$$a(K^\dagger, K) = \log\det K^\dagger K - \log\det K^\dagger - \log\det K\,. \tag{8.20}$$

The determinant of the differential operator $K^\dagger K = \mathfrak{D}^2 + m^2$ has been calculated in the preceding Chapter 5, namely it is defined through the zeta-function regularization by means of

$$\log\det K^\dagger K = \log\det(\mathfrak{D}^2 + m^2) = -\zeta'(0|\mathfrak{D}^2 + m^2)\,, \tag{8.21}$$

while, for K and $K^\dagger$ themselves, we have to set

$$\log\det K = -\zeta'_+(0|K)\,, \qquad \log\det K^\dagger = -\zeta'_-(0|K^\dagger)\,. \tag{8.22}$$

The choice ζ_+ and ζ_- for K and $K^\dagger$, respectively, is a consequence of the discussion we have carried out in Chapter 5. First, we observe that, for this regularized determinant, one gets the nice property

$$\log\det K^\dagger = (\log\det K)^*, \tag{8.23}$$

which is valid for the determinant of non-Hermitian matrices. Coming back to the multiplicative anomaly, making use of the result obtained in Chapter 5, we obtain a non-vanishing result, indeed the multiplicative anomaly for the massive Dirac operator and its adjoint in a space of even dimension, $D = 2p$, reads

$$a_{2p}(K^\dagger, K) = 2\sum_{j=1}^{p} \frac{(-1)^j m^{2j} c_j}{j!} A_{p-j}(\mathfrak{D}^2) , \tag{8.24}$$

while in a space of odd dimension, $D = 2p + 1$, it is given by

$$a_{2p+1}(K^\dagger, K) = \pi \sum_{j=0}^{p} (-1)^{j+1} \frac{m^{2j+1}}{\Gamma(j+\frac{3}{2})} A_{p-j}(\mathfrak{D}^2) , \tag{8.25}$$

where $A_{p-j}(\mathfrak{D}^2)$ are the Seeley-DeWitt coefficients corresponding to the operator $\mathfrak{D}^2$. Of course, when $m \to 0$ the multiplicative anomaly vanishes. Directly in terms of the determinants of the Dirac operators, we can express our results as follows. In the even dimensional case

$$\left.\frac{\det K}{\det K^\dagger}\right|_{even} = \exp[i\pi\zeta(0|\mathfrak{D}^2 + m^2)],$$

$$\left.\frac{\det K}{\sqrt{\det(\mathfrak{D}^2 + m^2)}}\right|_{even} = \exp\left[-i\frac{\pi}{2}\zeta(0|\mathfrak{D}^2 + m^2)\right] \cdot \exp\left[-\frac{1}{2}a_{2p}(K^\dagger, K)\right] .$$

In the odd dimensional case, the eta function of the Laplacian undertakes the role of the zeta function. In fact, we have

$$\left.\frac{\det K}{\det K^\dagger}\right|_{odd} = \exp[-i\pi\eta(0|A)] \cdot \exp[- \left(\eta'(0|K) - \eta'(0|K^\dagger)\right)],$$

$$\begin{aligned}\left.\frac{\det K}{\sqrt{\det(\mathfrak{D}^2 + m^2)}}\right|_{odd} &= \exp\left[-i\frac{\pi}{2}\eta(0|A)\right] \cdot \exp\left[-\frac{1}{2}a_{2p+1}(K^\dagger, K)\right] \\ &\quad \times \exp\left[-\frac{1}{2}\left(\eta'(0|K) - \eta'(0|K^\dagger)\right)\right] .\end{aligned}$$

8.3 Consistent, Covariant and Multiplicative Anomalies

In order to show the physical relevance of the multiplicative anomaly, we consider as an example the vector-axial-vector model, and we show that the multiplicative anomaly is strictly related to the well known consistent and covariant anomalies [G. Cognola and S. Zerbini (1999)]. For the two dimensional case we are able to perform an exact computation.

In [E. Langmann and J. Mickelsson (1996)] (see also [J. Mickelsson (1996)], in connection with lattice anomalies), it has been shown that the non-Abelian chiral anomaly can be evaluated as a regularized trace of the commutator of pseudo-differential operators. The connection of this elegant and general result with the multiplicative anomaly is that the latter is strictly related to the lack of cyclicity of the regularized trace. In particular, one can see more directly why the multiplicative anomaly leads to the chiral anomaly. The argument goes as follows. The effective action may be written as

$$W(A) \simeq \log\det(\mathfrak{D}_0^\dagger \mathfrak{D}(A))\,, \tag{8.26}$$

where $\mathfrak{D}_0$ is a suitable, free Dirac operator and $\mathfrak{D}(A)$ is the chiral Dirac operator mapping left spinors into right spinors. Performing a gauge transformation, one has that

$$W(A_g) \simeq \log\det(\mathfrak{D}_0^\dagger g^{-1}\mathfrak{D}(A)g)\,. \tag{8.27}$$

Thus, the non-vanishing of the multiplicative anomaly leads to the loss of gauge invariance of the effective action and this implies the existence of the chiral anomaly. The relevance of the multiplicative anomaly in situations of this kind has also been noticed in a completely different context, namely in non-relativistic quantum statistical mechanics [J.J McKenzie-Smith and D.J. Toms (1998)].

Here we shall not be able to show (at least unequivocally) that the multiplicative anomaly is related to directly observable effects. Nevertheless, we will show, indeed, that it does have indirect physical importance, since it relates consistent and covariant current anomalies in a vector-axial-vector model. In this respect, qualifying it as an 'anomaly' is not totally inappropriate.

We recall that the current anomalies of the vector-axial-vector model have been derived for the first time by Bardeen in [W.A. Bardeen (1969)]. Later, all these results have been confirmed and extended to curved manifolds by several other authors [M. Reuter (1985); A.P. Balachandran et al. (1982); A. Andrianov and L. Bonora (1984); M.B. Einhorn and D.R.T. Jones (1984); S.H. Hu and B.L. Young (1984); G. Cognola and P. Giacconi (1989); G. Cognola et al. (1990a)]. In particular, the local polynomials which one has to add to the currents in order to pass from the consistent anomaly to the covariant one, have been explicitly obtained in [G. Cognola et al. (1990a)].

Now, we briefly summarize the results concerning the vector-axial-vector model. For details and notations see [G. Cognola and P. Giacconi (1989); G. Cognola et al. (1990a)]. One considers a massless Dirac particle with arbitrary internal degrees of freedom in interaction with gauge and axial gauge potentials.

The model is described by the classical action

$$S = \int d^4x \left(\frac{1}{2} i\bar{\psi}\gamma^k \nabla_k \psi - \frac{1}{2} i \nabla_k \bar{\psi} \gamma^k \psi + i\bar{\psi}\gamma^k\gamma^5 A_k \psi \right) , \tag{8.28}$$

where

$$\nabla_k \psi = (\partial_k + V_k)\psi , \qquad V_k = -iV_k^a \tau_a , \qquad A_k = -iA_k^a \tau_a , \tag{8.29}$$

are the covariant derivative, the gauge and the axial-gauge potentials respectively. By τ_a we indicate the generators of the gauge group and by γ^k the (Euclidean) Dirac matrices.

The one-loop effective action, W, is given by

$$W = -\log \det \mathfrak{D} = \zeta'(0|\mathfrak{D}) , \tag{8.30}$$

with

$$\mathfrak{D} = i\gamma^k \nabla_k + i\gamma^k\gamma^5 A_k , \qquad \mathfrak{D}^\dagger = i\gamma^k \nabla_k - i\gamma^k\gamma^5 A_k . \tag{8.31}$$

Note that we are considering the Euclidean section and thus $\gamma^k = \gamma^{k\dagger}$ and $\mathfrak{D} = \mathfrak{D}^\dagger + 2i\gamma^k\gamma^5 A_k$.

From a classical viewpoint, both the vector and axial-vector currents are conserved, but quantum corrections break the symmetry and the currents acquire anomalous terms. If δ_V and δ_A are the infinitesimal vector and axial-vector transformations, which classically give the conserved currents, that is

$$\delta_V \psi = -\varepsilon\psi , \qquad \delta_V V_k = \partial_k \varepsilon + [V_k, \varepsilon] , \qquad \delta_V A_k = [A_k, \varepsilon] ,$$

$$\delta_A \psi = -\gamma^5 \varepsilon \tau_a \psi , \qquad \delta_A A_k = \partial_k \varepsilon + [V_k, \varepsilon] , \qquad \delta_A V_k = [A_k, \varepsilon] , \tag{8.32}$$

$$\delta_V \mathfrak{D} = -(\mathfrak{D}\varepsilon - \varepsilon\mathfrak{D}) , \qquad \delta_A \mathfrak{D} = -(\mathfrak{D}\varepsilon + \varepsilon\mathfrak{D}) ,$$

where $\varepsilon = \varepsilon^a(x)\tau_a$ is an infinitesimal matrix (depending on x), then the related currents satisfy the (modified) continuity equations

$$\partial_k J^k + [V_k, J^k] + [A_k, {}^5J^k] = \mathcal{A}_a \tau^a \,, \tag{8.33}$$

$$\partial_k {}^5J^k + [V_k, {}^5J^k] + [A_k, J^k] = {}^5\mathcal{A}_a \tau^a \,. \tag{8.34}$$

In the literature, two kinds of anomalies concerning this currents have been well studied. The first one is the so-called consistent anomaly, which has been first derived by Bardeen from perturbation theory and can be easily obtained by a path integral, upon regularizing the one-loop effective action according to the rule

$$W_1 = \frac{1}{2} \lim_{s\to 0} \frac{d}{ds}\, \zeta(s|\mathfrak{D}^2) \,, \tag{8.35}$$

with the implicit continuation $A_k \to iA_k$, in order to have a Hermitian operator $\mathfrak{D}$. Gauge invariance is lost, but the Wess-Zumino consistency conditions are satisfied (see, for example, [M. Reuter (1985); K. Fujikawa (1979)]). In this case only the axial symmetry is broken.

Using Eqs. (8.32), one easily obtains

$$\delta_V \operatorname{Tr} \mathfrak{D}^{-2s} = 0 \,, \qquad\qquad \delta_A \operatorname{Tr} \mathfrak{D}^{-2s} = -2s \operatorname{Tr}(\mathfrak{D}^{-2s}\gamma^5\varepsilon) \,. \tag{8.36}$$

Since the analytic continuation of $\operatorname{Tr}(\mathfrak{D}^{-2s}\gamma^5\varepsilon)$ is regular at $s = 0$, for the consistent anomalies we have

$$\mathcal{A}_a^{cons} = \frac{\delta_V}{\delta\varepsilon^a}\, W_1 = 0 \,, \tag{8.37}$$

$${}^5\mathcal{A}_a^{cons} = \frac{\delta_A}{\delta\varepsilon^a}\, W_1 = \frac{1}{8\pi^2} \operatorname{Tr}[\tau_a \gamma^5 a_2(x|\mathfrak{D}^2)] \,, \tag{8.38}$$

where $a_2(x|H)$ is the local spectral coefficient relative to the second-order operator H. It is understood that in arbitrary even dimensions $2N$, in the latter equation one has to replace $a_2/(4\pi)^2$ with $a_N/(4\pi)^N$.

The second one, which is called "covariant anomaly", can be formally obtained from the "covariant functional"

$$W_2 = \frac{1}{2} \lim_{s\to 0} \frac{d}{ds}\, \zeta(s|\mathfrak{D}^\dagger\mathfrak{D}) \,. \tag{8.39}$$

In this last case, both currents, J^k and ${}^5J^k$, acquire an anomalous "covariant" term, but the Wess-Zumino consistency conditions are not satisfied. This is due to the fact that, in zeta-function regularization one has to work with an arbitrary argument, s and only at the end of the calculation is one allowed to take the limit $s \to 0$. Such a "breaking of the consistency" is strictly related to the operator one has to deal with and to the necessity of working with regularized quantities. With regard to this, the form of the variation of the operator is crucial.

Now, again using Eqs. (8.32) and the "intertwining identity",

$$\mathfrak{D}\, f(\mathfrak{D}^\dagger \mathfrak{D})\, \mathfrak{D}^\dagger = \mathfrak{D}\mathfrak{D}^\dagger\, f(\mathfrak{D}\mathfrak{D}^\dagger)\,, \tag{8.40}$$

one gets

$$\begin{aligned} \delta_V \operatorname{Tr}(\mathfrak{D}^\dagger \mathfrak{D})^{-s} &= -2s \operatorname{Tr}\left\{\left[(\mathfrak{D}^\dagger \mathfrak{D})^{-s} - (\mathfrak{D}\mathfrak{D}^\dagger)^{-s}\right] \varepsilon\right\}\,, \\ \delta_A \operatorname{Tr}(\mathfrak{D}^\dagger \mathfrak{D})^{-s} &= -2s \operatorname{Tr}\left\{\left[(\mathfrak{D}^\dagger \mathfrak{D})^{-s} + (\mathfrak{D}\mathfrak{D}^\dagger)^{-s}\right] \gamma^5 \varepsilon\right\}\,. \end{aligned} \tag{8.41}$$

The traces in the formulae above have regular analytic continuations to $s = 0$ and the first derivative with respect to s, at $s = 0$, gives the following formulae for the covariant anomaly:

$$\mathcal{A}_a^{cov} = \frac{1}{16\pi^2} \operatorname{Tr}\{\tau_a [a_2(x|\mathfrak{D}^\dagger \mathfrak{D}) - a_2(x|\mathfrak{D}\mathfrak{D}^\dagger)]\}\,, \tag{8.42}$$

$$^5\mathcal{A}_a^{cov} = \frac{1}{16\pi^2} \operatorname{Tr}\{\tau_a \gamma^5 [a_2(x|\mathfrak{D}^\dagger \mathfrak{D}) + a_2(x|\mathfrak{D}\mathfrak{D}^\dagger)]\}\,. \tag{8.43}$$

Explicit expressions for the anomalies have been computed using different methods also in curved spacetime with torsion [G. Cognola and P. Giacconi (1989); G. Cognola et al. (1990a)]). For a non-Abelian gauge group, the difference reads

$$\mathcal{A}_a^{cov} - \mathcal{A}_a^{cons} = \mathcal{A}_a^{cov} = \frac{1}{8\pi^2} \operatorname{Tr}[F^*G + G^*F]\,, \tag{8.44}$$

$$\begin{aligned} {}^5\mathcal{A}_a^{cov} - {}^5\mathcal{A}_a^{cons} &= \frac{1}{12\pi^2} \operatorname{Tr}\{G^*G - 8e_{ijrs}A^i A^j A^r A^s + \\ &\quad 4[{}^*F_{ij} A^i A^j + A^i A^j\, {}^*F_{ij} + A^i\, {}^*F_{ij} A^j]\}\, \tau_a\,, \end{aligned} \tag{8.45}$$

where

$$F_{ij} = \partial_i V_j - \partial_j V_i + [V_i, V_j] + [A_i, A_j]\,, \tag{8.46}$$

$$G_{ij} = \partial_i A_j - \partial_j A_i + [V_i, A_j] - [V_j, A_i]\,, \tag{8.47}$$

and

$$^*F^{ij} = e^{ijrs}\, F_{rs}\,, \qquad G^*G = e^{ijrs} G_{ij} G_{rs}\,. \tag{8.48}$$

It is possible to pass from the consistent anomaly to the covariant one by adding local polynomials to the currents. Such polynomials $\chi^k = \chi^k_a \tau^a$ and ${}^5\chi^k = {}^5\chi^k_a \tau^a$ are chosen to satisfy [G. Cognola et al. (1990a)] (there is a misprint in that paper, χ^k and ${}^5\chi^k$ are exchanged.)

$$\partial_k \chi^k + [V_k, \chi^k] + [A_k, {}^5\chi^k] = (\mathcal{A}^{cov}_a - \mathcal{A}^{cons}_a)\tau^a \,, \tag{8.49}$$

$$\partial_k {}^5\chi^k + [V_k, {}^5\chi^k] + [A_k, \chi^k] = ({}^5\mathcal{A}^{cov}_a - {}^5\mathcal{A}^{cons}_a)\tau^a \,. \tag{8.50}$$

In this way, if J^k and ${}^5J^k$ satisfy continuity equations with the consistent anomalous term, then $J^k + \chi^k$ and ${}^5J^k + {}^5\chi^k$ satisfy continuity equations with the covariant anomalous term. The polynomials one has to add to the currents are chosen "ad hoc" in order to satisfy the previous conditions. Here we show that they are related to the corresponding multiplicative anomaly (connections between current terms and the Wodzicki residue have been pointed out also in [J. Mickelsson (1993)]).

To this aim, we first observe that

$$\zeta'(0|\mathfrak{D}^\dagger\mathfrak{D}) = \zeta'(0|\mathfrak{D}^\dagger) + \zeta'(0|\mathfrak{D}) + \, a(\mathfrak{D}^\dagger, \mathfrak{D}) \,. \tag{8.51}$$

Moreover, since $\mathfrak{D}$ and $\mathfrak{D}^\dagger$ only differ in the sign of A_k, we also have

$$\delta_V \zeta'(0|\mathfrak{D}^\dagger) \;=\; -\delta_V \zeta'(0|\mathfrak{D}) = 0 \,, \tag{8.52}$$

$$\delta_A \zeta'(0|\mathfrak{D}^\dagger) \;=\; \delta_A \zeta'(0|\mathfrak{D}) = \frac{1}{2}\delta_A \zeta'(0|\mathfrak{D}^2), \tag{8.53}$$

and, finally, using Eqs. (8.51)-(8.53) and recalling the definitions of the anomalies, we obtain the result

$$(\mathcal{A}^{cov}_a - \mathcal{A}^{cons}_a) = \frac{\delta_V}{2\delta\varepsilon^a}\, a(\mathfrak{D}^\dagger, \mathfrak{D}) \,, \tag{8.54}$$

$$({}^5\mathcal{A}^{cov}_a - {}^5\mathcal{A}^{cons}_a) = \frac{\delta_A}{2\delta\varepsilon^a}\, a(\mathfrak{D}^\dagger, \mathfrak{D}) \,. \tag{8.55}$$

The multiplicative anomaly is a functional of V_k and A_k, then by performing the variations and comparing the result with Eqs. (8.49) and (8.50), we obtain

$$\chi^k_a = \frac{1}{2}\,\frac{\delta a(\mathfrak{D}^\dagger, \mathfrak{D})}{\delta V^a_k} \,, \qquad {}^5\chi^k_a = \frac{1}{2}\,\frac{\delta \mathcal{A}(\mathfrak{D}^\dagger, \mathfrak{D})}{\delta A^a_k} \,. \tag{8.56}$$

In order to check the above relations, which we have rigorously proved using zeta-function techniques, one has to evaluate the multiplicative anomaly by using the general formula, Eq. (7.93). This can be done in principle, but, since one is dealing with non-commuting operators, it is not an easy task, in general. However, it is so in two dimensions, where in general complicated expression reduces to the simple formula

$$a(\mathfrak{D}, \mathfrak{D} + M) = \frac{1}{8\pi} \int d^2x \, \mathrm{Tr}\, M^2 \,, \tag{8.57}$$

where Tr is the trace over internal or Dirac matrices.

Thus, let us consider the simple case of an Abelian vector-axial vector field in two dimensions. Such a model is described by a classical action similar to the one in Eq. (8.28), with $\gamma^0 = \sigma_1$, $\gamma^1 = \sigma_2$ and $\gamma^5 = \sigma_3$. Since we are working in two dimensions, now we have

$$\begin{aligned} {}^5\mathcal{A}_a^{cov} - {}^5\mathcal{A}_a^{cons} &= \frac{1}{4\pi} \mathrm{Tr} \left\{ \gamma^5 \left[a_1(x|\mathfrak{D}^\dagger \mathfrak{D}) + a_1(x|\mathfrak{D}\mathfrak{D}^\dagger) - 2a_1(x|\mathfrak{D}^2) \right] \right\} \\ &= -\frac{1}{\pi} \, \partial_k A^k \,. \end{aligned} \tag{8.58}$$

With regard to the multiplicative anomaly, in two dimensions we can use Eq. (8.57) with $M = 2i\gamma^k \gamma^5 A_k$, and the final result reads

$$a(\mathfrak{D}^\dagger, \mathfrak{D}) = \frac{1}{\pi} \int d^2x \, A_k A^k \,, \tag{8.59}$$

from which, the relation

$${}^5\mathcal{A}_a^{cov} - {}^5\mathcal{A}_a^{cons} = \frac{\delta_A}{2\delta\varepsilon} a(\mathfrak{D}^\dagger, \mathfrak{D}) = -\frac{1}{\pi} \, \partial_k A^k \tag{8.60}$$

easily follows.

8.4 Interacting Charged Scalar Model

We include in this Section an application of the multiplicative anomaly while revisiting the charged relativistic Bose gas.

The one-loop finite-temperature effective potential with a non-vanishing chemical potential has been considered quite often during the last years. The case of a free relativistic charged bosonic field was investigated in [J.I. Kapusta (1981); J.I. Kapusta (1989); H.E. Haber and H.A. Weldon (1981); H.E. Haber and H.A. Weldon (1982a)], while the self-interacting charged scalar field has been studied in [K. Benson et al. (1991); K. Kirsten and D.J. Toms (1995)]. Recently, the issue has been reconsidered and critically analyzed in [A. Filippi (1997)], where use has been made of zeta-function regularization in the calculations and a different approach for including the chemical potential has been proposed.

Here, we shall take full advantage of a rigorous use of zeta-function regularization in investigating the finite-temperature effects in the presence of a non-vanishing chemical potential. In order to illustrate the new issue we shall be dealing with, we start with the partition function for a free charged field in $\mathbb{R}^D$, described by two real components ϕ_i. The Euclidean action is

$$S = \int dx^D \left[\phi_i \left(-\Delta_D + m^2\right) \phi_i\right] , \tag{8.61}$$

where $\phi^2 = \phi_k \phi_k$ is $O(2)$ invariant.

As it is well known, in order to study finite-temperature effects, one observes that the grand canonical partition function may be written as

$$Z_\beta(\mu) = \operatorname{Tr} e^{-\beta(H-\mu Q)} , \tag{8.62}$$

in which H is the Hamiltonian, Q the conserved charge operator and β the inverse of the equilibrium temperature. Making use of a path integral representation and integrating over the momenta, one arrives at the following recipe [J.I. Kapusta (1981); J.I. Kapusta (1989); K. Benson et al. (1991); A. Filippi (1997); H.E. Haber and H.A. Weldon (1982b)]: (i) compactify the imaginary time τ in the interval $[0, \beta]$, (ii) assume a periodic boundary condition in β and (iii) include the chemical potential μ by adding to the action the term

$$i\mu e \epsilon_{ik} \phi_i \partial_\tau \phi_k - \frac{1}{2} e^2 \mu^2 \phi^2 , \tag{8.63}$$

where e is the elementary charge. As a result, the partition function reads

$$Z_\beta(\mu) = \int_{\phi(\tau)=\phi(\tau+\beta)} [d\phi_i] e^{-\frac{1}{2}\int_0^\beta d\tau \int d^N x \phi_i A_{ij} \phi_j} , \tag{8.64}$$

where $D = N + 1$ and A is given by

$$A_{ij} = L_{ij} + 2e\mu\epsilon_{ij}\sqrt{L_\tau} , \tag{8.65}$$

with

$$L_{ij} = \left(L_\tau + L_N - e^2\mu^2\right) \delta_{ij}, \quad L_N = -\Delta_N + m^2 , \tag{8.66}$$

in which Δ_N is the Laplace operator on $\mathbb{R}^N$ (continuous spectrum $\mathbf{k}^2$) and $L_\tau = -\partial_\tau^2$ (discrete spectrum $\omega_n^2 = \frac{4\pi^2 n^2}{\beta^2}$), the Laplace operator on $\mathbb{S}^1$. Thus, one is actually dealing with a matrix-valued elliptic differential non self-adjoint operator acting on scalar fields in $\mathbb{S}^1 \times \mathbb{R}^N$. In this case, the partition function may be written as [K. Benson et al. (1991)]

$$\log Z_\beta(\mu) = -\frac{1}{2} \log\det \left\| A_{ik}\ell^2 \right\| = -\frac{1}{2} \log\det \left[L_+\ell^2 L_-\ell^2 \right] , \qquad (8.67)$$

where

$$L_\pm = L_\tau + L_N + e^2\mu^2 \pm 2e\mu(L_N)^{\frac{1}{2}} . \qquad (8.68)$$

Another possible factorization (see, for example, [A.A. Bytsenko et al. (1996)]) is

$$K_\pm = L_N + L_\tau - e^2\mu^2 \pm 2ie\mu(L_\tau)^{\frac{1}{2}} . \qquad (8.69)$$

Of course we have $L_+L_- = K_+K_-$ and, in both cases, one is dealing with a couple of pseudo-differential operators, L_+ and L_- being also formally self-adjoint.

As it is clear, the product of two elliptic ΨDOs, say A_+ and A_-, appears in any case and, consequently one has the multiplicative anomaly. As a result, the partition function, chosen, for example, the factorization A_+A_-, is given by

$$\log Z_\beta(A_+, A_-) = -\frac{1}{2} \log\det \left[A_+\ell^2 A_-\ell^2 \right] = \qquad (8.70)$$

$$\frac{1}{2}\zeta'(0|A_+) + \frac{1}{2}\zeta'(0|A_-) - \frac{\log \ell^2}{2} \left[\zeta(0|A_+) + \zeta(0|A_-) \right] - \frac{1}{2} a_D(A_+, A_-) .$$

Here one can see the crucial role of the multiplicative anomaly: as, in the factorization, different operators may enter, the multiplicative anomaly is necessary in order to have the same regularized partition function in both cases, namely $\log Z_\beta(A_+, A_-) = \log Z_\beta(B_+, B_-)$, which is an obvious physical requirement.

One could argue that the presence of a multiplicative anomaly is strictly linked with the zeta-function regularization employed and, thus, that it might be an artifact of it. However, it is possible to prove that the multiplicative anomaly is present indeed in a large class of regularizations of functional determinants appearing in the one-loop effective action.

8.4.1 *Chemical potential in the non-interacting case revised*

As a first step, we shall treat here the non-interacting bosonic model at finite temperature with a non-vanishing chemical potential. This situation has been extensively investigated in the past. We will reconsider it here making rigorous use of the zeta-function regularization procedure, which will allow us to clarify some subtle points overlooked in previous studies.

Again, it is convenient to work in $\mathbb{S}^1 \times \mathbb{R}^N$, with $D = N + 1$ in order to exploit the role of the dimension of the spacetime. As discussed before, here one has to deal with the functional determinant of the product of the two operators

$$L_\pm = L_\tau + \left(\sqrt{L_N} \pm e\mu\right)^2 , \tag{8.71}$$

while a second factorization is obtained in terms of the operators

$$K_\pm = L_N + \left(\sqrt{L_\tau} \pm ie\mu\right)^2 , \tag{8.72}$$

with $L_N = -\Delta_N + m^2$.

Let us start with the latter case, since it renders calculations easier. For the real part of s sufficiently large, one has

$$\zeta(s|K_\pm) = \frac{1}{\Gamma(s)} \int_0^\infty dt\; t^{s-1} \mathrm{Tr}\, \exp\left(-tK_\pm\right) , \tag{8.73}$$

where $\mathrm{Tr}\,\exp(-tK_\pm)$ is given by

$$\mathrm{Tr}\, e^{-tK_\pm} = \mathrm{Tr}\, e^{-tL_N} \sum_n e^{-t(\omega_n \pm ie\mu)^2} . \tag{8.74}$$

For notational simplicity, we write

$$\mathrm{Tr}\, e^{-tL_N} = \sum_i e^{-t\lambda_i} = \frac{V_N}{(4\pi t)^{\frac{N}{2}}} e^{-tm^2} , \tag{8.75}$$

with λ_i the eigenvalues of the operator L_N, namely $\mathbf{k}^2 + m^2$. Thus, via Mellin transform, one has the standard result

$$\zeta(z|L_N) = \frac{V_N}{(4\pi)^{\frac{N}{2}}} \frac{\Gamma(z - \frac{N}{2})}{\Gamma(z)} m^{N-2z} . \tag{8.76}$$

The Poisson-Jacobi resummation formula gives

$$\sum_n e^{-t(\omega_n \pm i\mu)^2} = \frac{\beta}{(4\pi t)^{\frac{1}{2}}} \left(1 + 2\sum_{n=1}^{\infty} \cosh(ne\mu\beta) e^{-\frac{n^2\beta^2}{4t}}\right) . \quad (8.77)$$

As a result, even though $K_+ \neq K_-$, one gets

$$\begin{aligned} \mathrm{Tr}\, e^{-tK_+} &= \mathrm{Tr}\, e^{-tK_-} & (8.78) \\ &= \mathrm{Tr}\, e^{-tL_N} \frac{\beta}{(4\pi t)^{1/2}} \left[1 + 2\sum_{n=1}^{\infty} \cosh(ne\mu\beta) e^{-\frac{n^2\beta^2}{4t}}\right] , \end{aligned}$$

and $\zeta(s|K_+) = \zeta(s|K_-)$. Therefore, making use of Eqs. (8.73), (8.74), (8.77) and of the well known integral representation of the MacDonald function $K_s(x)$,

$$K_s(\sqrt{b}x) = \frac{1}{2}\left(\frac{x}{2\sqrt{b}}\right)^s \int_0^{\infty} dt\, e^{-bt - \frac{x^2}{4t}} t^{-s-1} , \quad (8.79)$$

one has

$$\frac{1}{2}\left[\zeta(s|K_+) + \zeta(s|K_-)\right] = \frac{\beta V_N}{(4\pi)^{\frac{D}{2}}} \frac{\Gamma(s - \frac{D}{2})}{\Gamma(s)} m^{D-2s}$$

$$+ \frac{2^{3/2-s}\beta}{\sqrt{\pi}\Gamma(s)} \sum_{n=1}^{\infty} \cosh(ne\mu\beta)(n\beta)^{s-\frac{1}{2}} \sum_i \lambda_i^{\frac{1-2s}{4}} K_{s-\frac{1}{2}}(n\beta\sqrt{\lambda_i}). \quad (8.80)$$

The second term in this equation represents the statistical sum contribution, which depends non-trivially on the temperature and chemical potential and can be calculated in arbitrary dimension. Near $s = 0$, it can be written as $\mathcal{S}(\beta, \mu)\, s + \mathcal{O}(s^2)$, with

$$\mathcal{S}(\beta, \mu) = -\sum_i \left[\log\left(1 - e^{-\beta(\sqrt{\lambda_i} + e\mu)}\right) + \log\left(1 - e^{-\beta(\sqrt{\lambda_i} - e\mu)}\right)\right] . \quad (8.81)$$

Taking Eq. (8.75) into account, this contribution can also be rewritten as follows

$$\mathcal{S}(\beta, \mu) = \frac{4V_N\beta}{(4\pi)^{\frac{D}{2}}} \sum_{n=1}^{\infty} \cosh(ne\mu\beta)(n\beta)^{-\frac{D}{2}} (2m)^{\frac{D}{2}} K_{\frac{D}{2}}(n\beta m) . \quad (8.82)$$

It should be noted that this statistical sum contribution is a series involving the very well behaved MacDonald's functions (or special Bessel functions of the third kind). This series is always convergent in any $D > 1$ spacetime dimension and for any β and $e|\mu| \leq m$, even in the critical limit $e|\mu_c| = m$, m being the lowest eigenvalue in the spectrum of L_N. This follows from the asymptotics of $K_\nu(z)$ for large z:

$$K_\nu(z) \simeq \left(\frac{\pi}{2z}\right)^{\frac{1}{2}} e^{-z} \left[1 + \mathcal{O}(\frac{1}{z})\right] . \tag{8.83}$$

Coming back to Eq. (8.80), the first term (the vacuum contribution) has to be considered, as usual, for D odd and D even separately. For D odd, we have

$$\zeta(0|K_+) + \zeta(0|K_-) = 0 , \tag{8.84}$$

and

$$\frac{1}{2}\left(\zeta'(0|K_+) + \zeta'(0|K_-)\right) = \frac{\beta V_N}{(4\pi)^{\frac{D}{2}}}\Gamma(-\frac{D}{2})m^D + \mathcal{S}(\beta,\mu) . \tag{8.85}$$

For D even, we shall restrict ourselves to the two cases $D = 2$ and $D = 4$. We obtain respectively

$$\begin{aligned}
\zeta(0|K_+)\log M^2 + \zeta'(0|K_+) &= -\frac{\beta V_1 m^2}{4\pi}\left(\log\frac{m^2}{M^2} - 1\right) + \mathcal{S}(\beta,\mu) , \\
\zeta(0|K_+)\log M^2 + \zeta'(0|K_+) &= -\frac{\beta V_3 m^4}{32\pi^2}\left(\log\frac{m^2}{M^2} - \frac{3}{2}\right) + \mathcal{S}(\beta,\mu) .
\end{aligned}$$

Let us now consider the first factorization.

$$\mathrm{Tr}\, e^{-tL_\pm} = \mathrm{Tr}\, e^{-t(\sqrt{L_N}\pm e\mu)^2} \sum_n e^{-t\omega_n^2} , \tag{8.86}$$

with

$$\mathrm{Tr}\, e^{-t(\sqrt{L_N}\pm e\mu)^2} = \sum_i e^{-t(\sqrt{\lambda_i}\pm e\mu)^2} . \tag{8.87}$$

Poisson resummation gives

$$\mathrm{Tr}\, e^{-tL_\pm} = \mathrm{Tr}\, e^{-t(\sqrt{L_N}\pm e\mu)^2} \frac{\beta}{(4\pi t)^{1/2}} \left(1 + 2\sum_{n=1}^\infty e^{-\frac{n^2\beta^2}{4t}}\right)$$

and, as a consequence,

$$\zeta(s|L_\pm) = \frac{\beta}{2\sqrt{\pi}}\frac{\Gamma(s-\frac{1}{2})}{\Gamma(s)}\zeta(s-\frac{1}{2}|(\sqrt{L_N}\pm e\mu)^2)$$

$$- \quad 2s \sum_i \log\left(1 - e^{-\beta(\sqrt{\lambda_i} \pm e\mu)}\right) + \mathcal{O}(s^2) . \qquad (8.88)$$

Now, for $e|\mu| < m$, the binomial theorem yields

$$\zeta(z|(\sqrt{L_N} + e\mu)^2) + \zeta(z|(\sqrt{L_N} - e\mu)^2) = 2\zeta(z|L_N)$$
$$+2\sum_{r=1}^{\infty} \frac{\Gamma(2z + 2r)}{(2r)!\Gamma(2z)}(e\mu)^{2r}\zeta(z + r|L_N) , \qquad (8.89)$$

and this leads to

$$\frac{1}{2}\left(\zeta(s|L_+) + \zeta(s|L_-)\right) = \frac{\beta V_N m^{D-2s}}{(4\pi)^{\frac{D}{2}}} \left\{ \frac{\Gamma(s - \frac{D}{2})}{\Gamma(s)} \right.$$
$$\left. + \sum_{r=1}^{\infty} \frac{\Gamma(2s + 2r - 1)\Gamma(s - \frac{1}{2})}{(2r)!\Gamma(s)\Gamma(s - \frac{1}{2} + r)\Gamma(2s - 1)} \left(\frac{e\mu}{m}\right)^{2r} \Gamma(s + r - \frac{D}{2}) \right\}$$
$$-\mathcal{S}(\beta, \mu)\, s + \mathcal{O}(s^2) . \qquad (8.90)$$

As a result, for D odd, taking the derivative with respect to s and the limit $s \to 0$, the two factorizations give

$$\log Z_{\beta,\mu}(L_+, L_-) = \frac{\beta V_N m^D}{(4\pi)^{\frac{D}{2}}}\Gamma(-\frac{D}{2}) + \mathcal{S}(\beta, \mu) - \frac{1}{2}a_D(L_+, L_-) \qquad (8.91)$$

and

$$\log Z_{\beta,\mu}(K_+, K_-) = \frac{\beta V_N m^D}{(4\pi)^{\frac{D}{2}}}\Gamma(-\frac{D}{2}) + \mathcal{S}(\beta, \mu) - \frac{1}{2}a_D(K_+, K_-) , \qquad (8.92)$$

respectively. It should be noted that one gets

$$\log Z_{\beta,\mu}(L_+, L_-) = \log Z_{\beta,\mu}(K_+, K_-)$$

and thus the standard textbook result, as soon as one is able to prove that the two multiplicative anomalies vanish, for D odd. We can anticipate that this indeed happens.

For D even however, the situation is different because, within the first factorization, the vacuum sector depends explicitly on the chemical potential μ. In fact, for example, for $D = 2$ one has

$$\frac{1}{2}\left\{\zeta'(0|L_+) + \zeta'(0|L_-) + [\zeta(0|L_+) + \zeta(0|L_+)] \log M^2\right\}$$
$$= \frac{\beta V_1}{\pi}\left[m^2(\log\frac{m^2}{M^2} - 1)\right] + \frac{\beta V_1}{2\pi}e^2\mu^2 + \mathcal{S}(\beta, \mu) \qquad (8.93)$$

and for $D = 4$

$$\frac{1}{2}\left\{\zeta'(0|L_+) + \zeta'(0|L_-) + [\zeta(0|L_+) + \zeta(0|L_+)]\log M^2\right\}$$
$$= \frac{\beta V_3}{32\pi^2}\left[m^4(\log\frac{m^2}{M^2} - 3/2)\right] + \frac{\beta V_3}{8\pi^2}2\left[\frac{e^4\mu^4}{3} - e^2\mu^2 m^2\right] + \mathcal{S}(\beta,\mu)\,. \quad (8.94)$$

As a result, for $D = 2$ the two factorizations give

$$\log Z_{\beta,\mu}(L_+, L_-) = \frac{\beta V_1 m^2}{4\pi}(\log\frac{m^2}{M^2} - 1) + \mathcal{S}(\beta,\mu) + \frac{V_1\beta}{2\pi}e^2\mu^2 - \frac{1}{2}a_2(L_+, L_-) \quad (8.95)$$

and

$$\log Z_{\beta,\mu}(K_+, K_-) = \frac{\beta V_1 m^2}{4\pi}(\log\frac{m^2}{M^2} - 1) + \mathcal{S}(\beta,\mu) - \frac{1}{2}a_2(K_+, K_-)\,, \quad (8.96)$$

respectively, while for $D = 4$ one has

$$\log Z_{\beta,\mu}(L_+, L_-) = \frac{\beta V_3}{32\pi^2}\left[m^4(\log\frac{m^2}{M^2} - 3/2)\right] + \frac{\beta V_3}{8\pi^2}\left(\frac{e^4\mu^4}{3} - e^2\mu^2 m^2\right) + \mathcal{S}(\beta,\mu) - \frac{1}{2}a_4(L_+, L_-) \quad (8.97)$$

and

$$\log Z_{\beta,\mu}(K_+, K_-) = \frac{\beta V_3}{32\pi^2}\left[m^4(\log\frac{m^2}{M^2} - 3/2)\right] + \mathcal{S}(\beta,\mu) - \frac{1}{2}a_4(K_+, K_-)\,. \quad (8.98)$$

In general for D even, one gets

$$\log Z_{\beta,\mu}(L_+, L_-) = -\beta V_N \mathcal{E}_V(m, M) + \beta V_N \mathcal{E}_D(m,\mu) + \mathcal{S}(\beta,\mu) - \frac{1}{2}a_D(L_+, L_-) \quad (8.99)$$

and

$$\log Z_{\beta,\mu}(K_+, K_-) = -\beta V_N \mathcal{E}_V(m, M) + \mathcal{S}(\beta,\mu) - \frac{1}{2}a_D(K_+, K_-)\,, \quad (8.100)$$

where $\mathcal{E}_V(m, M)$ is the naive vacuum energy density, $D = 2Q$ and

$$\mathcal{E}_D(m, \mu) = \sum_{r=1}^{Q} c_{Q,r} (e\mu)^{2r} m^{D-2r} \,. \tag{8.101}$$

Here the $c_{Q,r}$ are computable coefficients, given by

$$c_{Q,r} = \frac{\Gamma(2r-1)}{\Gamma(2r+1)\, \pi^{\frac{N}{2}}} \frac{2^{1-N}}{\Gamma(r-\frac{1}{2})} \frac{(-1)^{Q-r}}{(Q-r)!} \,. \tag{8.102}$$

In the cases $D = 2$, $D = 4$ and $D = 6$ one obtains, respectively,

$$\begin{aligned} \mathcal{E}_2(m, \mu) &= \frac{1}{2\pi} e^2 \mu^2 \,, \\ \mathcal{E}_4(m, \mu) &= -\frac{1}{8\pi^2} \left[e^2 \mu^2 \left(m^2 - \frac{e^2 \mu^2}{3} \right) \right] \\ \mathcal{E}_6(m, \mu) &= -\frac{1}{16\pi^3} \left[e^2 \mu^2 \left(-\frac{1}{4} m^4 + \frac{e^2 \mu^2 m^2}{6} - \frac{2 e^4 \mu^4}{45} \right) \right] . \end{aligned} \tag{8.103}$$

8.4.2 *The interacting case in the one-loop approximation*

In this subsection, we will review the interacting charged boson model and we will compute, within the one-loop approximation, the corresponding operators appearing in the two possible factorizations of the partition function.

The Euclidean action for a self-interacting charged field, again described by two real components ϕ_i, is

$$S = \int dx^D \left[\phi_i \left(-\Delta_D + m^2 \right) \phi_i + \lambda_D (\phi^2)^{\frac{D}{D-2}} \right] , \tag{8.104}$$

where $\phi^2 = \phi_k \phi_k$ is $O(2)$ invariant and $\lambda_D = \frac{\lambda}{D!}$ is a dimensionless coupling constant. In the important case $D = 4$ one has

$$S = \int dx^4 \left[\phi_i \left(-\Delta + m^2 \right) \phi_i + \frac{\lambda}{4!} (\phi^2)^2 \right] . \tag{8.105}$$

Finite-temperature effects with non-vanishing chemical potential are accounted for in the following one-loop partition function [J.I. Kapusta (1981); J.I. Kapusta (1989); H.E. Haber and H.A. Weldon (1982b); K. Benson et al. (1991)]

$$Z_\beta(\mu) = e^{-S_0} \int_{\phi(\tau)=\phi(\tau+\beta)} [d\phi_i] e^{-\frac{1}{2}\int_0^\beta d\tau \int d^N x \phi_i A_{ij} \phi_j} \,, \tag{8.106}$$

where

$$S_0 = \beta V_N \left[\frac{1}{2}(m^2 - e^2\mu^2)\Phi^2 + \lambda_D (\Phi^2)^{\frac{D}{D-2}} \right] \,, \tag{8.107}$$

Φ being the background field, assumed to be constant and A the Euclidean small disturbances operator, given by

$$A_{ij} = L_{ij} + 2e\mu\epsilon_{ij}\sqrt{L_\tau} + \frac{8D\lambda_D}{(D-2)^2}(\Phi^2)^{\frac{4-D}{D-2}} \Phi_i \Phi_j \,, \tag{8.108}$$

with

$$L_{ij} = \left(L_\tau + L_N - e^2\mu^2 + \frac{2D\lambda_D}{D-2}(\Phi^2)^{\frac{2}{D-2}} \right) \delta_{ij} \,. \tag{8.109}$$

Again, we are dealing with a matrix-valued, elliptic, differential, non self-adjoint operator acting on scalar fields in $\mathbb{S}^1 \times \mathbb{R}^N$. As a consequence, the partition function reads [K. Benson et al. (1991)]

$$\log Z_\beta(\mu) = -\frac{1}{2} \log \det \left\| A_{ik} \ell^2 \right\| = -\frac{1}{2} \log \det \left[L_+ \ell^2 L_- \ell^2 \right] \,, \tag{8.110}$$

where

$$L_\pm = L_\tau + L_N + e^2\mu^2 + h_D \pm \left[\frac{4}{D^2} h_D^2 + 4e^2\mu^2 (L_N + h_D) \right]^{\frac{1}{2}} \,, \tag{8.111}$$

with

$$h_D = \frac{2D^2\lambda_D}{(D-2)^2}(\Phi^2)^{\frac{2}{D-2}} \,, \qquad h_4 = \frac{\lambda}{3}\Phi^2 \,. \tag{8.112}$$

The other factorization is

$$K_\pm = L_N + L_\tau - e^2\mu^2 + h_D \pm \left(\frac{4}{D^2} h_D^2 - 4e^2\mu^2 L_\tau \right)^{\frac{1}{2}} \,, \tag{8.113}$$

and, again, we have $L_+ L_- = K_+ K_-$, which are the couples of ΨDOs which enter the multiplicative anomaly calculation in this case.

8.4.3 *The interacting O(2) model*

Here we come back to the problem of the computation of the multiplicative anomaly in this model. Putting $h_D = 0$, which is proportional to the coupling constant, we also get the results valid for the free case. Strictly speaking, the results of the last section are valid for a compact manifold, but in the case of $\mathbb{R}^D$, or in the finite-temperature case $\mathbb{S}^1 \times \mathbb{R}^{D-1}$, the divergence is trivial, being contained in the spatial volume factor.

Since the multiplicative anomaly depends on the regularization of the ultraviolet divergences (in our approach this is equivalent to the asymptotic behavior of the eigenvalues at infinity and to the related divergence of the functional determinant) it follows that, due to the fact that the Minkowski spacetime is ultrastatic ($g_{00} = -1$), its dependence on the temperature is simply proportional to β, as we will see.

First, let us consider the self-adjoint factorization. Now, the order of $L_\pm$ is 2. The Wodzicki formula gives

$$a_D(L_+, L_-) = \frac{1}{8} \operatorname{res} \left[(\log(L_+ L_-^{-1}))^2 \right] . \tag{8.114}$$

There is no ordering problem because we are dealing with commuting operators. We have to construct the complete symbol $A(x,k)$ of the classical ΨDO of zero order $[\log(L_+ L_-^{-1})]^2$. The complete symbol reads

$$A(x,k) =$$

$$\left[\log \left(k^2 + m^2 + e^2\mu^2 + h_D + \left(\frac{4}{D^2} h_D^2 + 4e^2\mu^2 (\mathbf{k}^2 + m^2 + h_D) \right)^{\frac{1}{2}} \right) \right.$$

$$\left. - \log \left(k^2 + m^2 + e^2\mu^2 + h_D - \left(\frac{4}{D^2} h_D^2 + 4e^2\mu^2 (\mathbf{k}^2 + m^2 + h_D) \right)^{\frac{1}{2}} \right) \right]^2 ,$$

where $k^2 = \mathbf{k}^2 + k_\tau^2$. Let us denote its asymptotics for large $|k|$ as

$$A(x,k) \equiv \sum_j A_{-j}(k) . \tag{8.115}$$

According to Wodzicki, we have to select the component A_{-D} and use the multiplicative anomaly formula, namely

$$a_D(L_+, L_-) = \frac{1}{8} \int_{M_D} dk \, \frac{dx}{(2\pi)^D} \int_{|k|=1} A_{-D}(k) = \frac{1}{8} \frac{\beta V_N}{(2\pi)^D} \int_{|k|=1} dk \, A_{-D}(k) .$$

For large k^2 and k_τ^2, due to the k^2 dependence, it follows that the homogeneous components with odd indices vanish. As a consequence, from the latter equation one immediately gets that, for D odd, the multiplicative anomaly vanishes. This result is consistent with the general theorem contained in [M. Kontsevich and S. Vishik (1993)].

For even D, however, the anomaly is present and the asymptotic expansion, from which one can easily read off the even homogeneous components, is

$$\begin{aligned} A(x,k) &= \frac{1}{(k^2)^2}\left[16e^2\mu^2\mathbf{k}^2 - 32(m^2+e^2\mu^2+h_D)\frac{e^2\mu^2\mathbf{k}^2}{k^2} + h_D^2 \right. \\ &\left. +16e^2\mu^2(m^2+h_D) + \frac{128}{3}\frac{e^4\mu^4(\mathbf{k}^2)^2}{(k^2)^2} + \mathcal{O}(\frac{1}{k^2})\right] . \end{aligned} \quad (8.116)$$

Thus, the first non-vanishing even components are

$$A_{-2}(x,k) = \frac{16e^2\mu^2\mathbf{k}^2}{(k^2)^2}, \quad (8.117)$$

$$\begin{aligned} A_{-4}(x,k) &= \frac{1}{(k^2)^2}\left[-32(m^2+e^2\mu^2+h_D)\frac{e^2\mu^2\mathbf{k}^2}{k^2} + h_D^2 \right. \\ &\left. +16e^2\mu^2(m^2+h_D) + \frac{128}{3}\frac{e^4\mu^4(\mathbf{k}^2)^2}{(k^2)^2}\right] . \end{aligned} \quad (8.118)$$

It follows that

$$a_2(L_+,L_-) = \frac{\beta V_1}{2\pi}e^2\mu^2 , \quad (8.119)$$

and

$$a_4(L_+,L_-) = \frac{\beta V_3}{8\pi^2}\left[e^2\mu^2(\frac{e^2\mu^2}{3} - m^2 - h_D) + \frac{h_D^2}{8}\right] . \quad (8.120)$$

If $e=0$, we recover the results of [E. Elizalde et al. (1998e)], in particular the absence of the multiplicative anomaly for $D=2$. In the free case $h_D=0$ and we have

$$a_4(L_+,L_-) = \frac{\beta V_3}{8\pi^2}\left[e^2\mu^2(\frac{e^2\mu^2}{3} - m^2)\right] . \quad (8.121)$$

Let us consider the other factorization. The complete symbol reads now

$$A(x,k) = \left[\log\left(k^2+m^2-e^2\mu^2+h_D+\left(\frac{4}{D^2}h_D^2 - 4e^2\mu^2k_\tau^2\right)^{\frac{1}{2}}\right)\right.$$

$$- \log \left(k^2 + m^2 - e^2\mu^2 + h_D - \left(\frac{4}{D^2} h_D^2 - 4e^2\mu^2 k_\tau^2 \right)^{\frac{1}{2}} \right) \Bigg]^2 . \quad (8.122)$$

For large k^2 and k_τ^2, due to the k^2 dependence, it follows again that the homogeneous components with odd indices are vanishing and, for D odd, the multiplicative anomaly is absent, as in the other case.

For D even, the first non-vanishing even components are

$$A_{-2}(x,k) = -\frac{16e^2\mu^2 k_\tau^2}{(k^2)^2}, \quad (8.123)$$

$$\begin{aligned} A_{-4}(x,k) &= \frac{1}{(k^2)^2} \left[32(m^2 + h_D - e^2\mu^2) \frac{e^2\mu^2 k_\tau^2}{k^2} \right. \\ &\left. + h_D^2 + \frac{128}{3} \frac{e^4\mu^4 (k_\tau^2)^2}{(k^2)^2} \right] . \end{aligned} \quad (8.124)$$

A straightforward calculation gives

$$a_2(K_+, K_-) = -\frac{\beta V_1}{2\pi} e^2\mu^2 \quad (8.125)$$

and

$$a_4(K_+, K_-) = \frac{\beta V_3}{8\pi^2} \left[e^2\mu^2 (m^2 + h_D - \frac{e^2\mu^2}{3}) + \frac{h_D^2}{8} \right] . \quad (8.126)$$

Again, for $e \to 0$ we get the result of [E. Elizalde et al. (1998e)].

Some remarks are in order. According to the results obtained so far, it seems quite natural to make the conjecture that, in the free case and for D even, one has

$$a_D(K_+, K_-) = -a_D(L_+, L_-) . \quad (8.127)$$

We have proved this conjecture for $D = 2$ and $D = 4$, with an explicit computation, namely

$$\begin{aligned} a_2(K_+, K_-) &= -\frac{\beta V_1}{2\pi} e^2\mu^2 , \\ a_4(K_+, K_-) &= \frac{\beta V_3}{8\pi^2} \left[e^2\mu^2 (m^2 - \frac{e^2\mu^2}{3}) \right] . \end{aligned} \quad (8.128)$$

Thus, in the free case, we also have

$$\log Z_\beta(L_+, L_-) = \log Z_\beta(K_+, K_-) . \quad (8.129)$$

On the other hand, assuming Eq. (8.129) and Eq. (8.127) one obtains a general expression for the multiplicative anomaly, namely

$$a_D(K_+, K_-) = -a_D(L_+, L_-) = -\beta V_N \mathcal{E}_D(m, \mu) , \tag{8.130}$$

where $\mathcal{E}_D(m, \mu)$ is given by Eq. (8.101). For instance, for $D = 6$, we have

$$a_6(K_+, K_-) = \frac{\beta V_5}{16\pi^3} \left[e^2\mu^2(-\frac{1}{4}m^4 + \frac{e^2\mu^2 m^2}{6} - \frac{2e^4\mu^4}{45}) \right] . \tag{8.131}$$

As a consequence, in the free case, one gets

$$\log Z_{\beta,\mu}(K_+, K_-) = -\beta V_N \mathcal{E}_V(m, M) + V_N \mathcal{S}(\beta, \mu) + \frac{1}{2}\beta V_N \mathcal{E}_D(m, \mu) ,$$

where

$$\mathcal{E}_V = \frac{1}{V_N} E_V , \qquad \mathcal{S}(\beta, \mu) = \frac{1}{V_N} S(\beta, \mu) .$$

8.4.4 *The free charged bosonic model at finite temperature*

The presence of this new anomaly term in the grand canonical partition requires a re-analysis of the statistical mechanical behavior. This is better accomplished with the use of the effective potential formalism, at finite temperature and charge. We shall keep working in arbitrary spacetime dimensions, returning to the physically important $4D$ case at the end of this subsection.

In order to study the effective action in the presence of a non-vanishing mean field, it is necessary to introduce the one-loop grand partition function in the presence of external (constant) sources J_i and a chemical potential μ. This is done with the addition of the coupling term $-\int dx^D (J_i\phi_i)$ in the action. As the action is quadratic in the field degrees of freedom, a trivial Gaussian integration gives

$$Z_\beta(\mu, J_i) = \exp\left[\beta V_N \frac{J_i J_i}{2(m^2 - e^2\mu^2)}\right] Z_\beta(\mu) , \tag{8.132}$$

where V_N is the spatial volume. In terms of generating functionals per unit volume

$$V(\beta, \mu, J_i) = -\frac{1}{\beta V_N} \log Z_\beta(\mu, J_i) ,$$

$$V_0(\beta, \mu) = -\frac{1}{\beta V_N} \log Z_\beta(\mu) ,$$

we have

$$V(\beta,\mu,J_i) = V_0(\beta,\mu) - \frac{J_iJ_i}{2(m^2-e^2\mu^2)}\,, \tag{8.133}$$

which is nothing but the thermodynamical potential of the non-interacting boson gas. The effective potential in the presence of external sources is the Legendre transform of the generating functional per unit volume (that is to say, the thermodynamical free-energy density) and reads

$$F(\beta,\rho,\Phi_i) = V(\beta,\mu,J_i) + \mu\rho + J_i\Phi_i, \tag{8.134}$$

where μ and J_i have to be expressed as functions of the charge density, ρ and the mean field Φ_i, by solving the defining equations

$$\rho = -\frac{\partial V_\beta}{\partial \mu} = \frac{< Q >}{V_N}\,, \qquad \Phi_i = -\frac{\partial V_\beta}{\partial J_i} = \frac{1}{V_N}\int dx^N < \phi_i >\,, \tag{8.135}$$

where V_β is a shortened notation for $V(\beta,\mu,J_i)$. As a consequence, with $x = |\mathbf{\Phi}|$, eliminating the external sources, one has

$$F(\beta,\rho,x) = V_0(\beta,\mu) - \mu\frac{\partial V_0(\beta,\mu)}{\partial \mu} + \frac{1}{2}(m^2+e^2\mu^2)x^2\,, \tag{8.136}$$

$$\rho = -\frac{\partial V_0(\beta,\mu)}{\partial \mu} + e^2\mu x^2\,, \tag{8.137}$$

where μ is assumed to be a function, $\mu = \mu(\beta,\rho,x)$, obtained by solving Eq. (8.137) in the unknown μ.

The possible equilibrium states correspond to minima of the effective potential with respect to x, at fixed temperature and charge density. Making use of Eq. (8.136) and Eq. (8.137), one arrives at

$$\frac{\partial F}{\partial x} = x(m^2-e^2\mu^2)\,, \qquad \frac{\partial^2 F}{\partial x^2} = (m^2-e^2\mu^2) - 2xe^2\mu\frac{\partial \mu}{\partial x}. \tag{8.138}$$

Requiring the first derivative to vanish gives either:

(i) The unbroken phase, with solution $x = 0$. From the expression of the second derivative, this will be a minimum, provided $e\mu < m$ and the related free energy density will be

$$\begin{aligned}\mathcal{F}_\beta &= \min_x F(\beta,\rho,x) = V_0(\beta,\mu) - \mu\frac{\partial V_0(\beta,\mu)}{\partial \mu} = V_0(\beta,\mu) + \mu\rho\,,\\ \rho &= -\frac{\partial V_0(\beta,\mu)}{\partial \mu}\,.\end{aligned}$$

(ii) The symmetry breaking solution

$$e\mu = \pm m\,,$$

here derived as an extremal property of the effective potential. In this case $|\Phi| > 0$ and assuming $e\mu = m$, one has

$$\begin{aligned} \mathcal{F}_\beta &= \min_x F(\beta,\rho,x) = V_0(\beta, e\mu = m) + \frac{m}{e}\rho\,, \\ emx^2 &= \rho + \left.\frac{\partial V_0(\beta,\mu)}{\partial\mu}\right|_{e\mu=m}. \end{aligned}$$

In this symmetry breaking phase, one has the phenomenon of Bose-Einstein condensation (BEC). From the last equation and the fact that $V_0(\beta,\mu)$ is a monotonically decreasing function of β, it follows that $x^2 = \Phi_i\Phi_i$ is non-vanishing when $\beta > \beta_c$, the inverse of the critical temperature of the BEC transition, as given by the condition

$$\rho = -\left.\frac{\partial V_0(\beta_c,\mu)}{\partial\mu}\right|_{e\mu=m}, \tag{8.139}$$

in agreement with the criterion for relativistic Bose-Einstein condensation proposed in [K. Kirsten and D.J. Toms (1996c); K. Kirsten and D.J. Toms (1997)]. For $\beta \le \beta_c$, on the other hand, $|\Phi| = 0$ and the $U(1)$-symmetry is restored.

Now, in the free case we know V_β. In the unbroken phase, $x = 0$, $e|\mu| < m$, $\beta < \beta_c$ and including the anomaly term, Eq. (8.101), with $2Q = D$, one has

$$\begin{aligned} \mathcal{F}_\beta &= \mathcal{E}_V - \frac{4m^{\frac{D}{2}}}{(2\pi)^{\frac{D}{2}}}\sum_{r=1}^{\infty}\left(\frac{1}{\beta r}\right)^{\frac{D}{2}-1}\cosh(re\mu\beta)K_{\frac{D}{2}}(rm\beta) \\ &\quad +\mu\rho + \frac{1}{2}\sum_{r=1}^{Q} c_{Q,r}(e\mu)^{2r}m^{2Q-2r}\,, \end{aligned} \tag{8.140}$$

$$\begin{aligned} \rho &= \frac{4em^{\frac{D}{2}}}{(2\pi)^{\frac{D}{2}}}\sum_{r=1}^{\infty}\left(\frac{1}{\beta r}\right)^{\frac{D}{2}-1}\sinh(re\mu\beta)K_{\frac{D}{2}}(rm\beta) \\ &\quad -e\sum_{r=1}^{Q} c_{Q,r}r(e\mu)^{2r-1}m^{2Q-2r}\,. \end{aligned} \tag{8.141}$$

The last term in both equations accounts for the anomaly.

In the broken phase, $e|\mu| = m$, $\beta > \beta_c$, one obtains

$$\begin{aligned} \mathcal{F}_\beta &= \mathcal{E}_V - \frac{4m^{\frac{D}{2}}}{(2\pi)^{\frac{D}{2}}} \sum_{r=1}^{\infty} \left(\frac{1}{\beta r}\right)^{\frac{D}{2}-1} \cosh(rm\beta) K_{\frac{D}{2}}(rm\beta) \\ &\quad + \frac{1}{2}\sum_{r=1}^{Q} c_{Q,r} m^{2Q} + \frac{m}{e}\rho, \end{aligned} \tag{8.142}$$

$$\begin{aligned} \rho &= \frac{4em^{\frac{D}{2}}}{(2\pi)^{\frac{D}{2}}} \sum_{r=1}^{\infty} \left(\frac{1}{\beta r}\right)^{\frac{D}{2}-1} \sinh(rm\beta) K_{\frac{D}{2}}(rm\beta) \\ &\quad - e\sum_{r=1}^{Q} r c_{Q,r} m^{2Q-1} + emx^2 . \end{aligned} \tag{8.143}$$

In this case, if we formally take the zero-temperature limit $\beta \to \infty$, we get

$$\rho \to -e\sum_{r=1}^{Q} r c_{Q,r} m^{2Q-1} + emx^2 . \tag{8.144}$$

Now, observe that the anomaly contribution leads to some problems, since it does not have a definite sign. For example, for any fixed $\rho = 0$, x^2 in (8.144) could be made negative when the anomaly contribution is negative. We thus arrive at a contradiction, as happens, for example, for $D = 6$.

At this point, however, we should notice that our discussion has involved "unrenormalized" quantities only, just regularized by using the zeta function procedure. In particular, the charge operator has not been properly defined and the above result indicates that this operator was not normally ordered relative to the Minkowski vacuum. Now, the charge operator appears in the Hamiltonian multiplied by μ. Therefore, any normal ordering ambiguity in the charge operator gives rise to an ambiguity in the effective action, which must be a linear homogeneous function of μ, namely $K\mu$, with a freely disposable constant K. We have therefore the freedom to redefine the generating functional $V(\beta, \mu, J_i)$ up to the linear term $K\mu$. This will not change the effective action, the Legendre transform being unaffected by linear terms. We then introduce the renormalized generating functional

$$V^R = V + \mu K = \mathcal{E}_V + \frac{1}{2}\mathcal{E}_D(m,\mu) - \frac{\mathcal{S}(\beta,\mu)}{\beta} - \frac{J_i J_i}{2(m^2 - e^2\mu^2)} + \mu K,$$

where $\mathcal{E}_D(m,\mu)$ is the anomaly, given by Eq. (8.101) and $\mathcal{S}(\beta,\mu)$ was defined in Eq. (8.82). So, we have also a finite renormalization of the charge density

$$\rho^R = -\frac{\partial V_0^R}{\partial \mu} + e\mu x^2 = \rho - K. \tag{8.145}$$

Since the effective potential is the same, the extremals are unchanged and we may rewrite the free energy in the symmetry breaking phase as

$$\begin{aligned}
\mathcal{F}_\beta &= \mathcal{E}_V - \frac{1}{\beta}\mathcal{S}(\beta, e\mu = m) + \frac{m}{e}\rho^R + \frac{m}{e}K + \frac{1}{2}\sum_{r=1}^{Q} c_{Q,r}m^{2Q}, \\
\rho^R &= -\frac{1}{\beta}\frac{\partial \mathcal{S}(\beta,\mu)}{\partial \mu}|_{e\mu=m} - K - \frac{e}{m}\sum_{r=1}^{Q} r c_{Q,r}m^{2Q} + emx^2 \,. \qquad (8.146)
\end{aligned}$$

On the other hand, in the unbroken, symmetric phase we have,

$$\begin{aligned}
\mathcal{F}_\beta &= \mathcal{E}_V - \frac{1}{\beta}\mathcal{S}(\beta, \mu) + \mu\rho^R + \mu K + \frac{1}{2}\sum_{r=1}^{Q} c_{Q,r}(e\mu)^{2r}m^{2Q-2r}, \\
\rho^R &= -\frac{1}{\beta}\frac{\partial \mathcal{S}(\beta,\mu)}{\partial \mu} - K - e\sum_{r=1}^{Q} r c_{Q,r}(e\mu)^{2r-1}m^{2Q-2r} \,. \qquad (8.147)
\end{aligned}$$

The constant K can now be fixed by imposing normalization conditions. By doing this, we see how the spacetime dimension affects renormalization and how special the case $D = 4$ is. We do so just by demanding that, at zero temperature and charge density, the symmetry be unbroken, which sounds as a very natural normalization condition. Using (8.146), this fixes K to be

$$K = -e\sum_{r=1}^{Q} r c_{Q,r}m^{2Q-1} \,. \tag{8.148}$$

As a result, in the broken phase ($|e\mu| = m$) one obtains

$$\begin{aligned}
\mathcal{F}_\beta &= \mathcal{E}_V - \frac{1}{\beta}\mathcal{S}(\beta, e\mu = m) + \frac{m}{e}\rho^R + \frac{1}{2}\sum_{r=1}^{Q} c_{Q,r}(1-2r)m^{2Q}, \\
\rho^R &= -\frac{1}{\beta}\frac{\partial \mathcal{S}(\beta,\mu)}{\partial \mu}|_{e\mu=m} + emx^2 \,, \qquad (8.149)
\end{aligned}$$

while, in the symmetric phase, one gets

$$\begin{aligned}\mathcal{F}_\beta &= \mathcal{E}_V - \frac{1}{\beta}\mathcal{S}(\beta,\mu) + \mu\rho^R - e\mu\sum_{r=1}^{Q} r c_{Q,r} m^{D-1} \\ &\quad + \frac{1}{2}\sum_{r=1}^{Q} c_{Q,r}(e\mu)^{2r} m^{2Q-2r}, \\ \rho^R &= -\frac{1}{\beta}\frac{\partial \mathcal{S}(\beta,\mu)}{\partial\mu} + e\sum_{r=1}^{Q} r c_{Q,r} m^{2Q-1} \\ &\quad - e\sum_{r=1}^{Q} r c_{Q,r}(e\mu)^{2r-1} m^{2Q-2r}. \end{aligned} \tag{8.150}$$

Note that in this symmetric phase, one can now take the limit $e \to 0$ to reach the usual expression of the free-energy density for the uncharged boson gas. For general $D = 2Q$, however, the free energy receives contributions from the anomaly term in both phases. On the other hand, with the above charge renormalization, the critical temperature is given implicitly by (choosing $\mu > 0$)

$$\rho^R + \frac{1}{\beta}\frac{\partial \mathcal{S}(\beta_c,\mu)}{\partial\mu}\bigg|_{e\mu=m} = 0\,, \tag{8.151}$$

which is the usual condition for the critical temperature [K. Kirsten and D.J. Toms (1996c); K. Kirsten and D.J. Toms (1997)] and it remains unaffected by the anomaly. It is important to recognize that the normalization condition just discussed is not the only one that is possible. Another physically natural condition would be to demand that, at zero coupling, the free energy reduces to the free energy of uncharged bosons. This then fixes K to be

$$K = -\frac{e}{2}\sum_{r=1}^{Q} c_{Q,r} m^{2Q-1}\,. \tag{8.152}$$

Now it is the free energy in the broken phase that remains unaffected, while β_c changes.

Let us consider now the physically important case $D = 4$. Then, from the expression for the anomaly, we have in addition the remarkable identity

$$\sum_{r=1}^{2} r c_{2,r} = \frac{1}{2}\sum_{r=1}^{2} c_{2,r} = \frac{1}{24\pi^2}\,. \tag{8.153}$$

This shows that the two normalization conditions just discussed are actually equivalent and fix a *unique* value $K = -\frac{em^3}{24\pi^2}$. As a result, in the symmetry breaking phase and for $D = 4$, we obtain

$$\begin{aligned} \mathcal{F}_\beta &= \mathcal{E}_V - \frac{1}{\beta}\mathcal{S}(\beta, m) + \frac{m}{e}\rho^R, \\ ex^2 &= \frac{1}{m}\left(\rho^R + \frac{1}{\beta}\frac{\partial \mathcal{S}(\beta,\mu)}{\partial\mu}\bigg|_{e\mu=m}\right), \end{aligned} \tag{8.154}$$

which means that the broken phase is totally unaffected by the anomaly. On the other hand, in the symmetric phase $\beta < \beta_C$, $x = 0$ and $e\mu < m$, the result is

$$\begin{aligned} \mathcal{F}_\beta &= \mathcal{E}_V - \frac{1}{\beta}\mathcal{S}(\beta,\mu) + \mu\rho^R - \frac{\mu e m^3}{12\pi^2} + \frac{1}{8\pi^2}\left(e^2\mu^2 m^2 - \frac{1}{3}e^4\mu^4\right), \\ \rho^R &= -\frac{1}{\beta}\frac{\partial \mathcal{S}(\beta,\mu)}{\partial\mu} - \frac{1}{8\pi^2}\left(2e^2\mu m^2 - \frac{4}{3}e^4\mu^3\right) + \frac{em^3}{12\pi^2}\,. \end{aligned} \tag{8.155}$$

We observe that the anomaly term (the term in square brackets in (8.155)) is present in the free energy above T_C. Moreover, it is again vanishing as $e \to 0$ and the correct expression of the free energy density for the uncharged boson gas is recovered. Although the multiplicative anomaly term itself may be non-negligible (as a numerical study of it shows), the renormalization used in order to preserve the vacuum structure has rendered the anomaly term quite harmless near T_C and below. On the other hand, at ultra-relativistic temperatures, $T \gg m$, it is indeed negligible in comparison with the Planckian T^4-term, since the anomaly term increases with mass precisely as m^4 (however this holds whenever the charge density is much less than T^3e^{-6}, in order to neglect the Coulomb interaction). Our conclusion is that it gives a relevant correction at intermediate temperatures, say of the order of $T \simeq m$, which are relativistic for any known massive elementary particle.

We conclude this subsection with the following remarks. The presence of a multiplicative anomaly term, whose existence had been completely overlooked in the literature, has been found in the partition function associated with a relativistic charged Bose gas. Using known mathematical techniques, this anomaly term can be computed exactly. Its explicit form has been derived here for $D = 2$ and $D = 4$ (while a simple program yields it for *any* desired value of D). We have shown it to be vanishing for D odd and also its fundamental importance in getting factorization invariance (in the operational sense) of the regularized one-loop effective action. On the other hand, assuming factorization invariance, we have obtained, in the free case, a general expression for the multiplicative anomaly, valid for any even D.

The existence of this new contribution has led us to revisit and discuss the non-interacting case in detail. In particular, we have re-examined the spontaneous symmetry breaking issue and the related relativistic Bose-Einstein condensation phenomenon. A renormalization of the charge density has been introduced in arbitrary even dimension and has led to the usual expression for the critical temperature. In particular, in the broken phase, we have seen that only in the physically important case $D = 4$ may the new contribution due to the multiplicative anomaly be absorbed in the charge renormalization process. However, in the symmetric unbroken phase, the multiplicative anomaly term gives a non-vanishing contribution, which has been overlooked in previous investigations. Although it is a non-leading term in the ultra-high temperature regime, nevertheless it can give relevant corrections at intermediate temperatures of order $T \simeq m$, m being the mass of the charged boson.

8.5 Concluding Remarks

In this Chapter we have considered the issue of the evaluation of functional determinants that typically are present in the one-loop approximation of the Euclidean partition function. We have stressed that the (zeta) regularized determinant of a product of two elliptic operators has, generically, a multiplicative anomaly factor. Usually, such factor is non-vanishing and it should be taken into account in the associated physical applications, since it can play an important role in obtaining the correct physical results. The non-commutative case of the multiplicative anomaly formula has been discussed in some detail, making use of different techniques. We have started with an elementary application of standard perturbation theory of linear operators, in the particular case involving a product of two of them.

Later, more powerful techniques comprising ΨDOs and the Wodizcki theory of the non-commutative residue have been employed in a useful re-derivation of the general multiplicative-anomaly formula.

One of our most interesting results deals with the possibility of getting a remarkable simplification of the general formula for the non-commuting case in the lower-dimensional situations, $D = 2$, $D = 4$, both of which turn out to be most relevant for physical applications.

Chapter 9

The Casimir Effect

9.1 Introduction

The Casimir effect has acquired considerable importance during the past five years or so. It was proposed in 1948 by the late H.B.G. Casimir, at the instance of Niels Bohr, as a fundamental description —in terms of measurable fluctuations of the quantum vacuum— of the well known Van der Waals interactions [H.B.G. Casimir (1948)]. The same year, a first derivation of an analogous result had been published by Casimir and Polder, but interpreting it in more classical terms, as a retarded Van der Waals force, the retardation being a consequence of the finite velocity of light (the residual electromagnetic dipole interactions [H.B.G. Casimir and D. Polder (1948)]). In fact, it was shortly after this first paper, by Casimir and Polder, that Bohr suggested to Casimir that the zero-point energy was responsible for the intermolecular force, and that there indeed might be an equivalent viewpoint stemming from the more fundamental quantum theory [K.A. Milton (2001)]. The Casimir and Polder paper gained more acceptance in the years to come, but the one by Casimir was rather forgotten for some decades, in spite of the early attempts to measure the Casimir force experimentally in physical conditions that pretended to approach sufficiently well the ideal, mathematically clean quantum formulation (infinite conductance of the plates, no roughness, zero-temperature or no low-T dependence, and so on). In fact, from the fundamental point of view as a quantum vacuum effect, there is no force between the plates which could come from classical electrodynamics. In this ideal situation, at zero temperature, no real photons between the plates are present, which can cause their attraction. Thus, it is only the variation of the vacuum state of quantum electrodynamics (QED) —e.g., the one with the plates present as compared with the one corresponding to the absence of the plates— which causes the plates to attract each other. It is remarkable that a macroscopic quantum force shows up in this way [M. Bordag et al. (2001)]. The deep roots for the explanation of such behavior go back to the introduction by Planck, in 1911, of the so-called half quanta [M. Planck and D. Verh (1911)] (the ubiquitous additional 1/2 in the occupation numbers of the spectrum for the quantum harmonic oscillator), whose precise physical meaning remained a mystery and was the subject of strong debate for several decades, until it gained the well-established status that it has nowadays.

In the late eighties there was an important revival of the Casimir effect, but rather of theoretical nature, more or less coinciding with the 40-year anniversary, in which some of the authors of this book took a very active part [G. Plunien et al. (1986); E. Elizalde and A. Romeo (1991)]. In parallel, concerning the other mentioned approach, an important thrust was the description in terms of the so-called Lifschitz theory [I.E. Dzyaloshinskii et al. (1961)], which takes into consideration, in an effective way, the main physical characteristics of the media, as mentioned above [P. Yam (1997); V.M. Mostepanenko and N.N. Trunov (1997); L. Spruch (1996); P.W. Milonni (1994)]. Taking into account the corrections imposed by the experimental setup, it was clearly realized in the eighties that the pretended experimental verifications of the Casimir effect carried out to those dates were not such. Nowadays, the 1997 experiment by Lamoreaux [S.K. Lamoreaux (1997)] is considered, by many, to be actually the first reliable measurement of the Casimir force. It has also been marked as the origin of the recent, explosive revival of the issue in several different domains of experimental and theoretical physics, technology (MEM devices), and cosmology.

There is here no place to described in detail all those important developments, which we will now, however, summarize briefly. In a nutshell, what has occurred in such a short time interval can be viewed as the usual transition, common to many new discoveries, which start, at the beginning, as very tiny effects almost impossible to measure and, the moment technology advances and they begin to be clearly measurable in many laboratories worldwide is the same in which they evolve into a nuisance, an undesired background that, on the one hand, prevents the detection of even finer effects and, what is even worse, may lead to malfunctioning of the nanotechnological devices which permit their detection. This has indeed been the fate of the Casimir force*, when the relevant distances for electromechanical devices have reached the nanometer region where, in the absence of electrostatic interactions and other residual forces (which are well under control), it is by far the dominant effect that remains. In particular, it prevents checking Newton's law at small distances and leads to *stiction* and correspondingly to a malfunctioning of such mechanical devices at the nanoscale. Fortunately this is, in our case, only part of the story. Being a bit more specific (and positive), the recent developments that have taken place in the Casimir issue can be summarized as follows.

(1) From the experimental viewpoint, there have been many different determinations of the Casimir force during the last five years [T. Ederth (2000); B.W. Harris et al. (1998); U. Mohideen and A. Roy (1998)], albeit only a most recent one [G. Bressi et al. (2002)] uses two flat plates (the original setting), due to the extreme difficulty of actually measuring the distance between two parallel plates, and of keeping them parallel all the time (the error involved in this measurement is no less than 15 %). Recent experiments which carry out a rigorous analysis of the influence of roughness, finite conductivity, and finite-temperature corrections, have been able to bring this figure down to a mere 1 %, and the possibility of .1 % error measurements is already envisaged for the near future.†

(2) New fundamental issues relating the Casimir effect with the cosmological constant have arisen recently. This connection was well known since many years ago, in the sense that, in spite of being the cosmological constant a constant indeed and, as such, independent of any typical dimension of the universe (as an average intergalactic distance at some given time, or the like), it turns out that in a quantum field theory calculation the contribution of the vacuum fluctuation and that of the cosmological constant add together and give

$$\frac{\Lambda\, c^2}{8\pi G} + \frac{1}{\mathrm{Vol}} \frac{\hbar c}{2} \sum_i \omega_i \,. \tag{9.1}$$

*Emilio Elizalde thanks Joseph Sucher for such a vivid description of this quite common feature (private communication).

†What would be really fantastic, see the lectures presented at the recent Workshop organized by Babb and Mohideen [J. Babb and U. Mohideen (2002)].

This means that at a given time (say now) an observational measure of this contribution cannot distinguish between the separate, relative value of the two parts. In an extreme case, it could be that the whole contribution (loosely called 'cosmological constant') were actually coming from fluctuations of the quantum vacuum (a Casimir energy density or pressure) [J. Silk (1994); J.D. Bjorken (2003)]. This issue has acquired an impressive boost from the recent determination of the acceleration of the observable universe, using standard candle supernovae. A large number of different computations of the vacuum energy density in various fundamental or effective models: SUSY strings, branes, topological defects, braneworld universes, etc., have been carried out recently, with the aim of evaluating the importance of the contribution, in each case, of a Casimir like energy to the value of the dark energy, as obtained from cosmological and astrophysical observations. One such calculation, using zeta-function regularization techniques, will be discussed in this Chapter.

(3) The first practical applications of the Casimir effect in nanotechnology (a nonlinear micromechanical Casimir oscillator) have been proposed in impacting papers from a Bell-Labs group [H.B. Chan et al. (2001a)]. This practical application has followed a very precise measurement of the Casimir force effects in a MEMS (micro electro mechanical) oscillatory system, by the same group [H.B. Chan et al. (2001b)], which led to the first example of a quantum mechanical *actuator* of a MEMS by the Casimir force. But this is by no means the whole story, since it turns out that, at this scale (the present or near-future scale of small devices) the Casimir force is a problem, rather than a solution, since it causes *stiction*, the malfunctioning of these small system due to the tendency of the small pieces of the appliance to stick together in an uncontrolled way. Thus, the Casimir effect has reached the stage succinctly described before, of an omnipresent, undesired background at this level. As a consequence, considerable effort is being invested nowadays in solid state physics in the search for specific materials that would develop a *repulsive* Casimir force, for usual simple configurations (two plates, a plate and a sphere, or a plate and a cylinder). It is well known, on the other hand, that closed topologies, as in the case of a spher-

ical surface, may lead to a repulsive force for ordinary materials, but in the opinion of the experts these configurations are neither experimentally feasible, nor useful, at least at present [J. Babb and U. Mohideen (2002)].

(4) Further issues, as the so-called lateral Casimir force, stemming from the roughness of the plates, the temperature dependence of the effect [K. Kirsten and E. Elizalde (1996a); E. Elizalde et al. (2002a); E. Elizalde and A.C. Tort (2002b); E. Elizalde et al. (2003c); E. Elizalde and A.C. Tort (2003d); E. Elizalde and A.C. Tort (2003e)], the finite conductivity of the setting and the introduction of different media between the plates, are being systematically investigated, showing again the experimental maturity of this subject [V.B. Bezerra et al. (2000); T. Emig et al. (2001); M. Schaden and L. Spruch (2002); C. Genet et al. (2002)]. Those are very important lines of research but will not be dealt with here any further. We shall rather concentrate on some precise theoretical issues, as the ones already considered before and those to come in the next point.

(5) New problems with the mathematical computation of the Casimir energy density (QFT infinities) have appeared [N. Graham et al. (2002)], giving birth to a renovated and very strong controversy in the community of Casimir specialists. The prompt answer given by Milton [K.A. Milton (2002)] involves some of our own work dealing with the issue of a physically meaningful and rather down-to-earth renormalization of the infinites [E. Elizalde et al. (1998a); M. Bordag et al. (1997)]. The main question to be answered concerns, in our opinion, the possible physical sense of the new infinities that appear in a rigorous QFT renormalization treatment: if they are indeed physically plausible, they should indefectibly give rise to a rather intense, measurable effect. If they are not, on the contrary, they should be eventually removed by a convenient, new kind of regularization/renormalization procedure. This situation is actually not very special. Indeed, there have been all sort of longstanding problems involving the vacuum energy for different fields with topologies as simple as a sphere or even a closed box (instabilities of the finite, repulsion result in the first case against even the least deformation of the perfectly spherical shape or its acquiring of the tiniest finite thickness, or the difficult issue of the computation of the domain exterior to the closed box, in the sec-

ond case). In the second situation envisaged before (e.g., if the new infinities are in fact irrelevant), the problem could possibly be solved by using a more sophisticated regularization method. In particular, Hadamard regularization, which is very widely employed nowadays both in physics [L. Blanchet and G. Faye (2000); H. Sahlmann and R. Verch (2001); R. Verch (1995); R. Brunetti and K. Fredenhagen (2002)] and mathematics [R. Estrada and R. Kanwal (2002); P.L. Butzer et al. (2002); N.A. Guz and V.V. Zozulya (2001); G. Criscuolo (1997); J.D. Elliott (1995); N. Mastronardi and D. Occorsio (1996)], has recently been advocated ([R. Verch (1995); L. Blanchet and G. Faye (2000); H. Sahlmann and R. Verch (2001); J. Babb and U. Mohideen (2002); R. Brunetti and K. Fredenhagen (2002)]) as being able to resolve the existing controversy —provided that a proof is found that the new infinities are not physically relevant (what recent experimental results do seem to suggest [K.A. Milton (2002)]).

Owing to the importance that the Casimir effect is still going to acquire from the experimental point of view, as already described, here we would like to recall the essential features of the phenomena and study some of the techniques leading to its theoretical computation, in particular those using zeta-function regularization [E. Elizalde (1989b); E. Elizalde and A. Romeo (1989a); E. Elizalde and A. Romeo (1989b); I. Brevik and E. Elizalde (1994)]. Some applications of the regularization to particle physics were pioneered by one of the authors with his students [E. Elizalde (1984); E. Elizalde and J. Soto (1985b); E. Elizalde and J. Soto (1985a)]. For some geometries, in particular for the case of quantum fields in an N-dimensional ball, we shall explain in some detail a very powerful set of mathematical techniques, which have been developed in the seminal papers [M. Bordag et al. (1996a); M. Bordag et al. (1997); M. Bordag et al. (1996c)] (for a recent general review on this subject see [K. Kirsten (2002)]). They permit the computation of the Casimir energy to be easily performed.

The Casimir effect arises as vacuum fluctuations of quantum fields in spaces with physical boundaries, where the fields have to satisfy suitable boundary conditions. In such situations, the difference between the vacuum energy computed for the two configurations —with and without boundaries— is non-vanishing and gives rise to a net force which can be measured. In the original Casimir configuration [H.B.G. Casimir (1948)], one measures an attractive force between two (really close) parallel conducting planes in Minkowski spacetime.

Several methods have been used for the explicit computation of the Casimir effect. Whereas in the earlier times the Green function formalism was preferred, in recent years different approaches —which make use of contour integral representations of the involved spectral sums— are commonplace. Although the idea for this method, in the specific context of Casimir energies, goes back to the early days of the subject [B. Davies (1972)], a systematical, effective and simple application of this approach in various contexts has only recently been achieved [M. Bordag et al. (1997); S. Leseduarte and A. Romeo (1996); V.V. Nesterenko and I.G. Pirozhenko (1997); A.O. Barvinsky et al. (1991); M. Bordag et al. (1996a); M. Bordag et al. (1996c); E. Elizalde et al. (1996); E. Elizalde et al. (1997); E. Elizalde and D.V. Vassilevich (1999); M.E. Bowers and C.R. Hagen (1999)].

The spectral sum which actually appears in the calculation depends on the regularization used and may include a cutoff function, to dampen high frequency contributions [M.E. Bowers and C.R. Hagen (1999)] or, as in the zeta-regularization technique [E. Elizalde et al. (1994); E. Elizalde (1995)], complex powers of the eigenvalues [M. Bordag et al. (1997); S. Leseduarte and A. Romeo (1996); M. Bordag et al. (1996a); M. Bordag et al. (1996b)]. As a result, the details of the computation may differ slightly, e.g., in the specific integration contour chosen, but all of them share the elegance and power of this method.

In recent contributions, zeta-function regularization sometimes combined with several contour integral representations, has been very much developed. Moreover, given the deep connections among zeta functions, heat kernels and functional determinants [P.B. Gilkey (1995); D.B. Ray (1970); D.B. Ray and I.M. Singer (1971); D.B. Ray and I.M. Singer (1973); J.S. Dowker and R. Critchley (1976); S.W. Hawking (1975)], one advantage of the method is that it can be applied, alternatively, to the calculation of heat-kernel coefficients [M. Bordag et al. (1996a)], functional determinants [M. Bordag et al. (1996c); M. Bordag et al. (1996b)] (see also [J.S. Dowker (1996a); J.S. Dowker (1996b)]), as well as Casimir energies [E. Elizalde et al. (1998a); M. Bordag et al. (1997); S. Leseduarte and A. Romeo (1996)]. This clearly shows that zeta functions serve as a unified framework in different areas of interest.

9.2 The Casimir Energy

The Casimir energy of a quantum scalar field $\Phi(t, \mathbf{x})$ satisfying field equations of the kind

$$(\partial_t^2 + A)\Phi(t, \mathbf{x}) = 0\,, \tag{9.2}$$

and fulfilling suitable boundary conditions, in an ultrastatic spacetime, is formally given by

$$E_{Cas} = \frac{1}{2}\sum_k \omega_k, \tag{9.3}$$

with the one-particle energies $\omega_k = \sqrt{\lambda_k}$ being obtained from

$$A\phi_k(\mathbf{x}) = \lambda_k \phi_k(\mathbf{x}), \tag{9.4}$$

where $\Phi(t, \mathbf{x}) = e^{-i\omega t}\phi(\mathbf{x})$ has been set.

The Casimir energy, as given by the formal expression (9.3), is ill defined and has to be regularized. In the zeta-function regularization procedure, one has

$$E_{Cas} = \frac{1}{2}\zeta(s|A)\Big|_{s=-1/2}. \tag{9.5}$$

The latter expression could be divergent and, in such case, one has to use the more general expression (1.67), which reduces to Eq. (9.5) when also the latter is mathematically well defined. It has to be noted that in the typical physical configurations, Eq. (9.5) can be seen as the sum of two contributions due to the internal and external domains, where the field satisfy different boundary conditions. This means that

$$E_{Cas} = \frac{1}{2}\left[\zeta^{int}(s|A) + \zeta^{ext}(s|A)\right]\Big|_{s=-1/2}. \tag{9.6}$$

This is what is actually meant with Eq. (9.3) (the ω's being dependent on the boundary conditions). In order to calculate E_{Cas} according to the previous definition, one needs information on the zeta function $\zeta(s|A)$ in a neighborhood of $s = -1/2$. This is a technical difficulty, which strictly depends on the geometry of the problem one is dealing with.

With just a few modifications, which involve the phase of the zeta function (see [E. Elizalde (1999)] for precise details), all the considerations above can be extended to the Dirac operator.

9.3 The Casimir Energy in the Ball

Calculations of Casimir energies in spherically symmetric situations have attracted the interest of physicists for well over thirty years now. Since the calculation of Boyer [T.H. Boyer (1968)], who computed the Casimir energy for a conducting spherical shell and found, to his surprise, a repulsive force, many different situations in the spherically symmetric context have been considered. For example, dielectrics were included [K.A. Milton (1980a)] (for the case of plane, parallel surfaces see [J. Schwinger et al. (1978)]) and used later for possible explanations of sonoluminescence [K.A. Milton and Y.J. Ng (1997); K.A. Milton and Y.J. Ng (1998); S. Liberati et al. (2000); I. Brevik et al. (1999); G. Barton (1999)]. Moreover, enormous interest has been attracted by the MIT bag model in QCD [A. Chodos et al. (1974b); A. Chodos et al. (1974a); G.E. Brown and M. Rho (1979); G.E. Brown et al. (1984); K.A. Milton (1980b); K.A. Milton (1983); J. Baacke and Y. Igarashi (1983); E. Elizalde et al. (1998a)] and, also, the influence of different boundary conditions has been considered in detail [M. Bordag et al. (1997); S. Leseduarte and A. Romeo (1996); V.V. Nesterenko and I.G. Pirozhenko (1997)].

Finally we would like to mention again the importance that the Casimir energy assumes in cosmological models [S.M. Carroll (2001); J.D. Bjorken (2003)]. Any energy density is a pressure. In the Casimir case it is the forced absence of certain modes that causes the negative Casimir pressure. The Casimir energy density is negative in all cases actually measured but —as already discussed— it is predicted to be positive in the closed spherical case (electromagnetic field) and also, in the case of a scalar field, for some other closed, no-boundary topologies of possibly great cosmological importance [E. Elizalde (1994a); E. Elizalde (1995); E. Elizalde (2001b)]. From this point of view, and taking into account the expression (9.1) above, we might perfectly interpret the preliminary claims of an accelerating universe as a first proof for the existence (and actual measurement) of a positive Casimir energy density pervading the universe. The problem is that it is rather difficult to interpret how this can have anything to do with QFT as we understand it today, since it is many orders of magnitude too small to possibly match. In other words, it is difficult to explain such a small but non-zero value of the cosmological constant. This has been termed by Weinberg the *second* cosmological constant problem [S. Weinberg (2000); J. Silk (1994); J.D. Bjorken (2003)].

9.3.1 *The method*

Now, we shall describe the method, first appeared in [M. Bordag et al. (1996a)], which permits to compute the zeta function when the eigenvalues are implicitly known as the zeros of an algebraic equation. To be more explicit, we focus our attention on the important physical case of a scalar field in an N-dimensional ball, embedded in the $D = (N+1)$-dimensional Minkowski spacetime. This means that the field has to satisfy suitable boundary conditions on the $(N-1)$-dimensional sphere. The method has been used for spinor [K. Kirsten and G. Cognola (1996b)] as well as for electromagnetic fields [G. Cognola et al. (2001)].

The operator for the scalar case is the Laplacian in the N-ball, and the eigenvalues are given as the zeros of a polynomial $P(\tilde{Z}_\nu, \tilde{Z}'_\nu)$ involving Bessel (inside) or Hankel (outside) functions and their derivatives [G. Cognola et al. (2001)]. That is

$$\tilde{P}(\tilde{Z}_{\nu_l}(\omega_{nl}R), \tilde{Z}'_{\nu_l}(\omega_{nl}R)) = 0\,, \qquad \lambda_{nl} = \omega_{nl}^2\,, \tag{9.7}$$

where $n, l \geq 0$ are the principal and azimuthal quantum numbers respectively. The degeneracy $d(l)$ of the eigenvalues and the index ν_l of the Bessel functions depend on l and on the dimension. Their explicit forms are strictly related to the fields and to the boundary conditions.

The zeta function can be expressed as an integral in the complex plane, that is

$$\zeta(s|A) = \sum_l \frac{d(l)}{2\pi i} \int_\gamma dk\, k^{-2s} \frac{\partial}{\partial k} \log\left[k^{-b\nu_l} \tilde{P}(\tilde{Z}_{\nu_l}(k), \tilde{Z}'_{\nu_l}(k))\right]\,, \tag{9.8}$$

where the open contour γ has to be chosen to run counterclockwise and to enclose all strictly positive solutions of Eq. (9.7). The latter integral is convergent for $\quad \mathrm{Re}\, s > \frac{D}{2}$. The additional factor $k^{-b\nu_l}$ has been inserted in order to cancel the pole at the origin, which is important when deforming the contour in the next step. In this way, γ can also include the origin. Here, b is a number which depends on the asymptotic behavior of $\tilde{P}$ at the origin: in our cases it will be ± 1 for scalars, but for spin $1/2$ with mixed boundary conditions it turns out to be ± 2.

For explicit calculation, it is convenient to write Eq. (9.8) as an integral on the real axis. This can be done by deforming the contour γ to the imaginary axis and by making the substitution $k \to iy$. In general, one has to be careful when deforming the contour, in that no poles in the plane

$\mathrm{Re}\, k \geq 0$ are hit. In fact, for Robin boundary conditions, one has

$$\tilde{P}(\tilde{Z}_\nu, \tilde{Z}'_\nu) = \alpha \tilde{Z}_\nu(k) + k\tilde{Z}'_\nu(k) = (\alpha - \nu)\tilde{Z}_\nu(k) + k\tilde{Z}_{\nu-1}(k) = 0\,, \qquad (9.9)$$

which may have solutions for $k \notin \mathbb{R}$ too, if $\alpha > \nu$. To avoid these cases, in the following we will consider $\alpha \leq \nu_0$ only, ν_0 corresponding to the smallest eigenvalue.

With this assumption, we can write the zeta function in the more simple form

$$\zeta(s|A) = \frac{\sin \pi s}{\pi} \sum_{l=0}^{\infty} d(l) \int_0^\infty dy\, y^{-2s} \frac{\partial}{\partial y} \log[y^{-b\nu_l} P(\nu_l, y)]\,, \qquad (9.10)$$

which is valid for $1/2 < \mathrm{Re}\, s < 1$ (for details see [M. Bordag et al. (1996a)]). Here, $P(\nu, y) = P(Z_\nu(y), Z'_\nu(y))$ is a polynomial like $\tilde{P}$ (aside from, possibly, an irrelevant sign) and the $Z_\nu(y) = \tilde{Z}_\nu(iy)$ are the modified Bessel functions corresponding to $\tilde{Z}$. In order to compute the Casimir energy, we need the ζ-function at $s = -1/2$ and so we have to do an analytic continuation of Eq. (9.10).

With this aim, let us now employ the asymptotic expansion of the modified Bessel functions. For large values of ν, we have [M. Abramowitz and I.A. Stegun (1970)]

$$I_\nu(\nu z) \sim \frac{1}{\sqrt{2\pi\nu}} \frac{e^{\nu\eta}}{(1+z^2)^{\frac{1}{4}}}\, \Sigma_1\,, \qquad \Sigma_1 = \sum_{k=0}^{\infty} \frac{u_k}{\nu^k}\,, \qquad (9.11)$$

$$I'_\nu(\nu z) \sim \frac{1}{\sqrt{2\pi\nu}} \frac{e^{\nu\eta}(1+z^2)^{\frac{1}{4}}}{z}\, \Sigma_2\,, \qquad \Sigma_2 = \sum_{k=0}^{\infty} \frac{v_k}{\nu^k}\,, \qquad (9.12)$$

$$K_\nu(\nu z) \sim \sqrt{\frac{\pi}{2\nu}} \frac{e^{-\nu\eta}}{(1+z^2)^{\frac{1}{4}}}\, \Sigma_3\,, \qquad \Sigma_3 = \sum_{k=0}^{\infty} (-1)^k \frac{u_k}{\nu^k}\,, \qquad (9.13)$$

$$K'_\nu(\nu z) \sim -\sqrt{\frac{\pi}{2\nu}} \frac{e^{-\nu\eta}(1+z^2)^{\frac{1}{4}}}{z}\, \Sigma_4\,, \qquad \Sigma_4 = \sum_{k=0}^{\infty} (-1)^k \frac{v_k}{\nu^k}\,, \qquad (9.14)$$

where $\eta = \sqrt{1+z^2} + \log[z/(1+\sqrt{1+z^2})]$. The first few coefficients u_k and v_k are listed in [M. Abramowitz and I.A. Stegun (1970)], while higher order coefficients are immediate to obtain by using the recursion relations

$$u_{k+1}(t) = \frac{1}{2} t^2 (1-t^2) u'_k(t) + \frac{1}{8} \int_0^t d\tau\, (1 - 5\tau^2) u_k(\tau)\,,$$

$$\begin{aligned} v_{k+1}(t) &= u_{k+1}(t) - \frac{1}{2}t(1-t^2)u_k(t) - t^2(1-t^2)u'_k(t)\,, \\ t &= \frac{1}{\sqrt{1+z^2}}\,, \qquad z = \frac{\sqrt{1-t^2}}{t}\,. \end{aligned} \tag{9.15}$$

As we shall see explicitly in the following, the above behavior of Bessel functions permits us to write

$$\log P(\nu, z\nu) \sim \log F(\nu, z) + \sum_{n=1}^{N} \frac{D_n(t)}{\nu^n}\,, \tag{9.16}$$

an expression which is valid for large values of ν. The function F is related to the exponential factors in Eqs. (9.11)-(9.14), while the coefficients $D_n(t)$ are related to Σ_k and are polynomials in t. More precisely,

$$D_n(t) = \sum_{k=0}^{2n} x_{nk} t^{n+k}\,. \tag{9.17}$$

Note that, when $b = \pm 1$, all x_{nk} with odd k vanish. Of course, F, D_n and x_{nk} depend on the specific problem under consideration. In [G. Cognola et al. (2001)] one can find all the cases, treated in detail.

The trick consists now in subtracting the asymptotic behavior from the integrand function and in integrating the asymptotic part, with arbitrary s, exactly. We thus get

$$\zeta(s|A) = Z_0(s) + Z(s) + \sum_{n=-1}^{N} A_n(s). \tag{9.18}$$

Here,

$$Z_0(s) = \delta(D-2)\, d[0]\, \frac{\sin(\pi s)}{\pi} \int_0^\infty dz\ z^{-2s} \frac{\partial}{\partial z} \log P(0, z) \tag{9.19}$$

is the contribution due to $\nu_l = 0$, which is present only in two dimensions and has to be treated specifically for any case. $Z(s)$ represents all the other terms, with the asymptotic contributions subtracted, that is

$$\begin{aligned} Z(s) &= \frac{\sin(\pi s)}{\pi} \sum_{\nu_l > 0} d(l) \int_0^\infty dz\ (z\nu)^{-2s} \\ &\quad \times \frac{\partial}{\partial z} \left\{ \log P(\nu_l, z\nu_l) - \log F(\nu_l, z) - \sum_{n=1}^{N} \frac{D_n(t)}{\nu_l^n} \right\}\,, \end{aligned} \tag{9.20}$$

and A_n are the integrals of the asymptotic part. They read [M. Bordag et al. (1996c)]

$$A_n(s) = -\frac{1}{\Gamma(s)}\zeta_{\mathcal{N}}(s+n/2)\sum_{k=0}^{2n} x_{nk}\frac{\Gamma(s+\frac{n+k}{2})}{\Gamma(\frac{n+k}{2})}\,, \qquad n \geq 1\,, \quad (9.21)$$

$$\begin{aligned} A_{-1}+A_0 &= \frac{\sin(\pi s)}{\pi}\sum_{\nu_l>0} d(l)\int_0^\infty dz\ (z\nu_l)^{-2s}\frac{\partial}{\partial z}\log[(z\nu_l)^{-b\nu_l}F(\nu_l,z)] \\ &= c_{-1}(s)\zeta_{\mathcal{N}}(s-1/2)+c_0(s)\zeta_{\mathcal{N}}(s)\,, \qquad (9.22) \end{aligned}$$

$$\zeta_{\mathcal{N}}(s) = \sum_{\nu_l>0} d(l)\nu_l^{-2s}\,.$$

Equation (9.20) is convergent for $(D-2-N)/2 < \operatorname{Re} s < 1$, thus for our aims it is sufficient to subtract $N = D$ asymptotic terms. This means that, with $N = D$, we can directly put $s = -1/2$ in Eq. (9.20) and perform the integral numerically.

As one can see in the explicit examples [G. Cognola et al. (2001)], the base zeta function, $\zeta_{\mathcal{N}}$, can be conveniently expressed in terms of the Barnes zeta function [J.S. Dowker (1994)], defined as [E.W. Barnes (1903)]

$$\begin{aligned} \zeta_B(s,a;d) &= \sum_{\mathbf{m}=0}^\infty \frac{1}{(a+m_1+\ldots+m_d)^s} = \sum_{n=0}^\infty e_n(d)(a+n)^{-s}\,, \\ e_n(d) &= \frac{(d+n-1)!}{n!(d-1)!}\,, \end{aligned}$$

for $\operatorname{Re} s > d$. Obviously, there is an expansion of the kind

$$e_n(d) = \sum_{\alpha=0}^{d-1} g_\alpha(d)(a+n)^\alpha$$

and this yields the expansion of the Barnes zeta function in terms of Hurwitz zeta functions [E.W. Barnes (1903); F.W. Olver (1954)]

$$\zeta_B(s,a;d) = \sum_{\alpha=0}^{d-1} g_\alpha(d)\zeta_H(s-\alpha,a). \qquad (9.23)$$

One can show that the $g_\alpha(d)$ are connected with the generalized Bernoulli polynomials [N.E. Nörlund (1922)]. This allows to determine, in a direct way, the residues and finite parts of the Barnes zeta function of the problem at hand. As a result, the asymptotic contributions in (9.18) are readily computed.

9.3.2 *An explicit example: the scalar field with Dirichlet boundary conditions*

In polar coordinates, the solutions read in this case

$$\phi_{l,m,n}(r,\Omega) = r^{1-D/2} f_{\nu_l}(\omega_{l,n} r) Y_{l+D/2}(\Omega) \tag{9.24}$$

with namely $\nu_l = l + (D-2)/2$, the $f_\nu(r)$ being Bessel functions of the first kind, $J_{\nu_l}(r)$, for the internal solution, and Hankel functions, $H_{\nu_l}(r)$, for the external one, while the $Y_{l+D/2}(\Omega)$ are hyperspherical harmonics [A. Erderlyi (1955)].

This means that, inside the spherical shell, we have

$$\tilde{P}(\tilde{Z}_\nu(k), \tilde{Z}'_\nu(k)) = J_\nu(k)\,, \qquad P(\nu,z) = I_\nu(z), \tag{9.25}$$

and the degeneracies are given by $d(l) = (2l+d-1)\frac{(l+d-2)!}{l!(d-1)!}$. From the last equation it easily follows that [J.S. Dowker (1994); M. Bordag et al. (1996c)]

$$\zeta_{\mathcal{N}} = \zeta_B\left(2s, \frac{d+1}{2}; d\right) + \zeta_B\left(2s, \frac{d-1}{2}; d\right)\,. \tag{9.26}$$

As a consequence,

$$F(\nu,z) = \frac{1}{\sqrt{2\pi\nu}} \frac{e^{\nu\eta}}{(1+z^2)^{\frac{1}{4}}}\,, \tag{9.27}$$

$$\log \Sigma_1 \sim \sum_{n=1}^{\infty} \frac{D_n(t)}{\nu^n}. \tag{9.28}$$

The asymptotic contributions have been calculated to be [M. Bordag et al. (1996c)]

$$\begin{aligned} A_{-1}(s) &= \frac{1}{4\sqrt{\pi}} \frac{\Gamma\left(s-\frac{1}{2}\right)}{\Gamma(s+1)} \zeta_{\mathcal{N}}\left(s-1/2\right), \\ A_0(s) &= -\frac{1}{4}\zeta_{\mathcal{N}}(s). \end{aligned} \tag{9.29}$$

The zeta function for the present situation is obtained by means of Eqs. (9.17)-(9.21) with the definitions above.

As already anticipated in the previous subsection, for $N = 2$ we have an additional contribution that has to be computed explicitly. With this aim, we recall that, for large z,

$$I_0(z) = \frac{e^z}{\sqrt{2\pi z}} \left\{ 1 + \frac{1}{8z} + \mathcal{O}\left(z^{-2}\right) \right\}, \tag{9.30}$$

and thus, we can write

$$\begin{aligned} Z_0(s) &= \frac{\sin(\pi s)}{\pi} \left\{ \int_0^1 dz\ z^{-2s} \frac{\partial}{\partial z} \log I_0(z) \right. \\ &\quad + \int_1^\infty dz\ z^{-2s} \left[\frac{\partial}{\partial z} \log I_0(z) - 1 + \frac{1}{2z} + \frac{1}{8z^2} \right] \\ &\quad \left. - \frac{1}{4s} + \frac{1}{2(s-1/2)} - \frac{1}{16(s+1/2)} \right\}, \end{aligned} \tag{9.31}$$

where the poles at $s = \pm 1/2$ are shown explicitly. The integrals are now convergent for $s = -1/2$ and can be computed numerically.

For the scalar field outside the spherical shell, we need simply to replace Bessel functions with Hankel functions, ν_l and $d(l)$ being the same. Thus, we have

$$P(\nu, z) = K_\nu(z), \tag{9.32}$$

$$F(\nu, z) = \sqrt{\frac{\pi}{2\nu}} \frac{e^{-\nu\eta}}{(1+z^2)^{\frac{1}{4}}}, \tag{9.33}$$

$$\log \Sigma_3 \sim \sum_{n=1}^{\infty} \frac{D_n(t)}{\nu^n}, \tag{9.34}$$

$$A_{-1}(s) = -\frac{1}{4\sqrt{\pi}} \frac{\Gamma\left(s-\frac{1}{2}\right)}{\Gamma(s+1)} \zeta_{\mathcal{N}}(s-1/2), \tag{9.35}$$

$$A_0(s) = -\frac{1}{4} \zeta_{\mathcal{N}}(s). \tag{9.36}$$

Owing to the particular relation between Σ_1 and Σ_3, the coefficients $D_n(t)$ differ from the corresponding coefficients that one has in the internal case, but just for the trivial factor $(-1)^n$. The same holds also for the quantities $A_n(s)$.

In two dimensions, we have to consider also the contribution due to $\nu = 0$, which can be obtained with the same arguments as in the previous case, Eq. (9.31). The result reads

$$\begin{aligned} Z_0(s) &= \frac{\sin(\pi s)}{\pi} \left\{ \int_0^1 dz\ z^{-2s} \frac{\partial}{\partial z} \log K_0(z) \right. \\ &\quad + \int_1^\infty dz\ z^{-2s} \left[\frac{\partial}{\partial z} \log K_0(z) + 1 + \frac{1}{2z} - \frac{1}{8z^2} \right] \\ &\quad \left. - \frac{1}{4s} - \frac{1}{2(s-1/2)} + \frac{1}{16(s+1/2)} \right\} . \end{aligned} \tag{9.37}$$

For the numerical results corresponding to the zeta functions inside and outside the shell and the total Casimir energy, for all kinds of fields and boundary conditions, we refer the reader to [G. Cognola et al. (2001)].

Let us summarize a little discussion of these methods. The Casimir effect can be analyzed using many different mathematical methods and regularizations, but we think that zeta function is a most natural candidate, since, by definition, it gives the sum over the eigenvalues and directly provides the Casimir energy in simple situations.

The method developed in [M. Bordag et al. (1996a)] and generalized in subsequent works [M. Bordag et al. (1996c); J.S. Dowker and K. Kirsten (1999)], permits the evaluation of the zeta function, also when the eigenvalues are not explicitly known, in a very effective way. Moreover, the results can be given with any desired precision, what is a very remarkable feature of zeta-function regularization. In particular, one is able to treat, at the same time, the big family of the more usual configurations also in quite non-trivial situations, where the boundaries are not flat plates and the boundary conditions can be very general and rather involved.

9.4 A Braneworld Computation

Brane universes were in principle created in order to resolve the hierarchy problem in particle physics [L. Randall and R. Sundrum (1999a); N. Arkani-Hamed et al. (1998); H. Georgi et al. (1974)]. But it was later realized that they can be consistent with observational data, even when the radius of the extra dimension is not so small, and they began to be viewed as feasible descriptions of the universe we live in [L. Randall and R. Sundrum (1999b); P. Binetruy et al. (2000); E.E. Flanagan et al. (2000); G.R. Dvali et al. (2000); J. Garriga and T. Tanaka (2000); C. Csaki et al. (2000)].

For consistency, the five-dimensional braneworld should be stabilized (radion stabilization [W.D. Goldberger and M.B. Wise (1999)]). Then, the idea is that the bulk vacuum energy (Casimir contribution) may be used explicitly for realizing such radion stabilization in a number of models [J. Garriga et al. (2001); S. Nojiri et al. (2000); I. Brevik et al. (2001); B. Grinstein et al. (2001); R. Hofmann et al. (2001); S. Nojiri et al. (2001); I.Z. Rothstein (2001); A. Flachi et al. (2001); H. Kudoh and T. Tanaka (2002); S. Nojiri et al. (2002); W. Naylor and M. Sasaki (2002); A.A. Saharian and M.R. Setare (2002)] (mainly with flat branes). An interesting connection between the bulk Casimir effect and supersymmetry breaking in braneworld [T. Gherghetta and A. Pomarol (2001)] or moving branes [A.L. Maroto (2002)] also exists. On the other hand, the brane Casimir effect may be used for a braneworld realization [S.W. Hawking et al. (2000); S. Nojiri and S.D. Odintsov (2000)] of the anomaly-driven (also called Starobinsky) inflation [A. Starobinsky (1980)].

In the papers mentioned above, the authors discussed mainly the Casimir effect in the situation when the brane is flat space. Furthermore the situation in which the brane is more realistic, say a de Sitter (dS) universe, has been considered in [S. Nojiri et al. (2000); W. Naylor and M. Sasaki (2002)]. It has been shown there that, in an AdS bulk, the Casimir energy for the bulk conformal scalar field in a one-brane configuration is zero. However, in situations where the bulk is different, a non-zero contribution of the Casimir energy is not excluded and even a possibility may exist of gravity trapping on the brane itself.

In the paper [E. Elizalde et al. (2003b)], a study of the bulk Casimir effect for a conformal or massive scalar when the bulk is a five-dimensional AdS or a dS space and the brane is a four-dimensional dS space has been carried out. Presumably, the bulk Casimir effect will play a role in the construction (radion stabilization) of braneworlds. We will here concentrate, in what follows, in the calculation of the Casimir energy density in such a situation, in particular, its comparison with the one obtained in the classical Casimir effect, and on the relevance of the zeta-regularization techniques, at its full power [E. Elizalde et al. (1994); E. Elizalde (1995); A.A. Bytsenko et al. (1996)], in performing such difficult calculation, in a quite general setting and a very elegant way. One interesting result one gets is that, for both bulks (AdS and dS) under discussion with one brane, the bulk effective potential is still zero for a conformal field. And also, the results can be used for the stabilization of the radion and have interesting implications as concerning the brane dynamics too [E. Elizalde et al. (2003b)].

9.4.1 *Casimir energy density for a dS brane in a 5D AdS background*

We recall that the action for a conformally invariant massless scalar with scalar-gravitational coupling is given by

$$\mathcal{S} = \frac{1}{2}\int d^5x\sqrt{g}\left[-g^{\mu\nu}\partial_\mu\phi\partial_\nu\phi + \xi_5 R^{(5)}\phi^2\right] , \tag{9.38}$$

where $\xi_5 = -3/16$, $R^{(5)}$ being the five-dimensional scalar curvature. This action is conformally invariant under the conformal transformations

$$g_{\mu\nu} = e^{\alpha\sigma(x^\mu)}\hat{g_{\mu\nu}} , \qquad \phi = e^{\beta\sigma(x^\mu)}\hat{\phi} , \tag{9.39}$$

where $-3\alpha/4 = \beta$. The Euclidean metric of the five-dimensional AdS bulk is

$$ds^2 = g_{\mu\nu}dx^\mu dx^\nu = \frac{l^2}{\sinh^2 z}\left(dz^2 + d\Omega_4^2\right) , \tag{9.40}$$

$$d\Omega_4^2 = d\xi^2 + \sin^2\xi d\Omega_3^2 , \tag{9.41}$$

where l is the AdS radius which is related to the cosmological constant of the AdS bulk, and $d\Omega_3$ is the metric on the 3-sphere. Two dS branes, which are four-dimensional spheres, are placed in the AdS background (Fig. 9.4.1). If we put one brane at z_1, which is fixed, and the other brane at z_2, the distance between the branes is given by $L = |z_1 - z_2|$. When z_2 tends to ∞, namely $L = \infty$, the two-brane configuration becomes a one-brane configuration.

To calculate the effective potential, we use zeta-function regularization [E. Elizalde et al. (1994); E. Elizalde (1995); A.A. Bytsenko et al. (1996)], as in [J. Garriga et al. (2001); S. Nojiri et al. (2000); W. Naylor and M. Sasaki (2002)]. The first step in this procedure consists of the introduction of a mass parameter in order to work with dimensionless eigenvalues; thus, we should write at every instance L_5/μ^2, etc. However, as is often

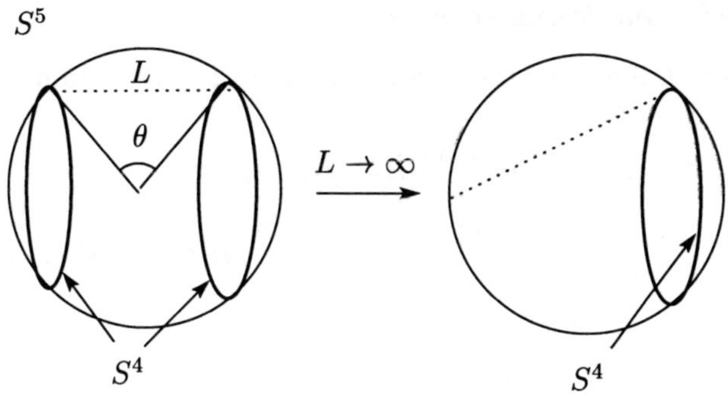

Fig. 9.1 Two dS_4 branes in a dS_5 background. This configuration becomes a one-brane configuration as $L \to \infty$.

done for the sake of the simplicity of the notation, we will just keep in mind the presence of this μ factor, to recover it explicitly only in our final formulae.

Let us assume that the eigenvalues of L_1 and L_4 are of the form $\lambda_n^2, \lambda_\alpha^2 \geq 0$ (with $n, \alpha = 1, 2, \cdots$) respectively. In terms of these eigenvalues, $\log\det L_5$ can be rewritten as follows:

$$\log\det L_5 = \mathrm{Tr}\log L_5 = \mathrm{Tr}\log(L_1 + L_4) = \sum_{n,\alpha}\log(\lambda_n^2 + \lambda_\alpha^2)\,. \tag{9.42}$$

L_1 is a one-dimensional Laplace operator, and the Dirichlet boundary conditions imposed on the two branes result in that the brane separation L can be taken as the width of a one-dimensional potential well. As a consequence, the eigenvalues of L_1 are immediately given by

$$\lambda_n^2 = \left(\frac{\pi n}{L}\right)^2, \tag{9.43}$$

for finite L. The heat kernel for L_5 is written in terms of $K_t(L_1)$ and $K_t(L_4)$, as

$$K_t(L_5) = K_t(L_1)K_t(L_4). \tag{9.44}$$

9.4.2 *The one-brane case*

In the large L limit, for the zeta function $\zeta(s|L_5)$ we obtain:

$$\begin{aligned}\zeta(s|L_5) &= \frac{1}{\Gamma(s)}\int_0^\infty dt t^{s-1} K_t(L_1)K_t(L_4) \\ &\sim \frac{L}{2\sqrt{\pi}}\frac{\Gamma\left(s-\frac{1}{2}\right)}{\Gamma(s)}\frac{1}{\Gamma\left(s-\frac{1}{2}\right)}\int_0^\infty dt t^{\left(s-\frac{1}{2}\right)-1}K_t(L_4) + \mathcal{O}\left(\frac{1}{L}\right) \\ &= \frac{L}{2\sqrt{\pi}}\frac{\Gamma\left(s-\frac{1}{2}\right)}{\Gamma(s)}\zeta\left(s-\frac{1}{2}|L_4\right) + \mathcal{O}\left(\frac{1}{L}\right), \end{aligned} \tag{9.45}$$

and the effective potential is

$$\begin{aligned} V &= -\frac{1}{2L\mathrm{Vol}(M_4)}\left[\zeta'(0|L_5/\mu^2) + \log\mu^2\zeta(0|L_5/\mu^2)\right] \\ &= \frac{1}{2L\mathrm{Vol}(M_4)}\zeta\left(-\frac{1}{2}|L_4/\mu^2\right) + \mathcal{O}\left(\frac{\mu}{L}\right). \end{aligned} \tag{9.46}$$

Note that the μ^2 factor has to be taken into account for obtaining the derivative and, as discussed before, it is in fact everywhere present in each Lagrangian and its eigenvalues (although it is usually not written down in order to simplify the notation). For the spherical brane S_4 whose radius is $\mathcal{R}$, the four-dimensional zeta function $\zeta\left(s|L_4\right)$ is given by

$$\zeta\left(s|L_4\right) = g(s)\mathcal{R}^{2s}, \tag{9.47}$$

where

$$g(s) = \frac{1}{6}\sum_{l=0}^{\infty}(l+1)(l+2)(2l+3)\left(l^2+3l+\frac{9}{4}\right)^{-s}. \tag{9.48}$$

This formula can be rewritten in terms of the values of the Hurwitz zeta-function $\zeta_H(s,a)$ at negative integer argument

$$\zeta_H(-m,a) = -\frac{B_{m+1}(a)}{m+1}, \tag{9.49}$$

where the $B_m(a)$ are Bernoulli polynomials. Thus, $\zeta\left(s|L_4\right)$ is given by

$$\zeta\left(s|L_4\right) = \frac{\mathcal{R}^{2s}}{3}\left[\zeta_H\left(2s-3,\frac{3}{2}\right) - \frac{1}{4}\zeta_H\left(2s-1,\frac{3}{2}\right)\right], \tag{9.50}$$

which leads to

$$\zeta\left(-\frac{1}{2}|L_4\right) = \frac{1}{3\mathcal{R}}\left[\zeta_H\left(-4,\frac{3}{2}\right) - \frac{1}{4}\zeta_H\left(-2,\frac{3}{2}\right)\right] = 0\,. \tag{9.51}$$

As a result, the effective potential Eq. (9.46) becomes zero [E. Elizalde et al. (2003b)] (as first observed in [S. Nojiri et al. (2000)] and confirmed in [W. Naylor and M. Sasaki (2002)] for flat branes in a one-brane configuration).

9.4.3 *The two-brane case*

We will now concentrate on the two-brane situation for the not-so-large distance case (a much more involved one). Using the full power of the zeta-regularization formulae [E. Elizalde et al. (1994); E. Elizalde (1995); A.A. Bytsenko et al. (1996); E. Elizalde (1999)], a precise (albeit not simple) calculation can be carried out which respects at every step the complete structure of the five-dimensional zeta function. That is, the full zeta function is preserved till the end, and the final expression is given in terms of an expansion on the brane distance L over the brane compactification radius $\mathcal{R}$, valid for $L/\mathcal{R} \leq 1$, which complements the one for large brane distance.

As to the specific zeta formulae employed, adhering to the classification that has been given in [E. Elizalde (1999)], the case at hand is indeed to be found there (even if at first sight it would not seem so). It corresponds to a two-dimensional quadratic plus linear form with truncated spectrum. In fact, this is clear from the structure of the spectrum yielding the zeta function

$$\zeta(s|L_5) = \mu^{-2s} \sum_{n,l=0}^{\infty} (\lambda_n^2 + \lambda_l^2)^{-s}, \tag{9.52}$$

where μ is a dimensional regularization scale that renders the argument of the zeta function dimensionless. In the case of the four-dimensional spherical brane of radius $\mathcal{R}$ considered above, this reduces to

$$\begin{aligned}\zeta(s|L_5) &= \frac{\mu^{-2s}}{6} \sum_{n,l=0}^{\infty} (l+1)(l+2)(2l+3) \\ &\quad \times \left[\left(\frac{\pi n}{L}\right)^2 + \mathcal{R}^{-2}\left(l^2 + 3l + \frac{9}{4}\right)\right]^{-s}.\end{aligned} \tag{9.53}$$

This zeta function looks awkward, at first sight. But after some reshuffling it can be brought to exhibit the standard structure mentioned. Specifically,

$$\begin{aligned}\zeta(s|L_5) &= \frac{\mathcal{R}^{2s}}{6\mu^{2s}} \sum_{n,l=0}^{\infty} 2\left(l+\frac{3}{2}\right) \left\{ \left[\left(l+\frac{3}{2}\right)^2 + \frac{\pi^2 n^2 \mathcal{R}^2}{L^2} \right]^{1-s} \right. \\ &\qquad \left. - \left(\frac{\pi^2 n^2}{L^2} + \frac{1}{4}\right) \left[\left(l+\frac{3}{2}\right)^2 + \frac{\pi^2 n^2 \mathcal{R}^2}{L^2} \right]^{-s} \right\} \\ &\equiv \frac{\mathcal{R}^{2s}}{6\mu^{2s}} \left[Z_1(s) + Z_2(s) \right], \end{aligned} \tag{9.54}$$

where both $Z_1(s)$ and $Z_2(s)$ are obtained by taking derivatives (see [E. Elizalde (1994c)] for a discussion of this issue, which is a nontrivial one when asymptotic expansions are involved), with respect to x at $x = 3/2$, of a zeta function of the class just mentioned, e.g.

$$\sum_{l=0}^{\infty} \left((l+x)^2 + q \right)^{-s}, \qquad q \equiv \frac{\pi^2 n^2 \mathcal{R}^2}{L^2}. \tag{9.55}$$

In [E. Elizalde (1999)], explicit formulae for the analytical continuation of this class of zeta functions are given. To be brief (and forgetting for the moment about the sum over n, for simplicity), we just have to recall the celebrated asymptotic expansion [E. Elizalde (1995)] [a particular case of Eqs. (3.1) and (3.2)]

$$\begin{aligned} &\sum_{n=0}^{\infty} \left[(n+c)^2 + q \right] \\ &\quad \sim \left(\frac{1}{2} - c \right) q^{-s} + \frac{q^{-s}}{\Gamma(s)} \sum_{n=1}^{\infty} \frac{(-1)^n \Gamma(n+s)}{n!} q^{-n} \zeta_H(-2n, c) \\ &\quad + \frac{\sqrt{\pi}\,\Gamma(s-1/2)}{2\Gamma(s)} q^{1/2-s} \\ &\quad + \frac{2\pi^s}{\Gamma(s)} q^{1/4-s/2} \sum_{n=1}^{\infty} n^{s-1/2} \cos(2\pi n c) K_{s-1/2}\left(2\pi n \sqrt{q}\right). \end{aligned} \tag{9.56}$$

After some calculations, we get for $Z_1(s)$ and $Z_2(s)$:

$$Z_1(s) = -\frac{1}{2-s} \left(\frac{\pi^2 \mathcal{R}^2}{L^2} \right)^{2-s} \zeta(2s-4)$$

$$- \frac{1}{\Gamma(s-1)} \sum_{n=1}^{\infty} \frac{(-1)^n \Gamma(n+s-2)}{n!}$$
$$\times \left(\frac{\pi^2 \mathcal{R}^2}{L^2} \right)^{2-n-s} \zeta(2s+2n-4) \zeta'_H(-2n, 3/2), \quad (9.57)$$

$$\begin{aligned} Z_2(s) &= \frac{1}{1-s} \left(\frac{\pi^2 \mathcal{R}^2}{L^2} \right)^{1-s} \left[\frac{\pi^2 \mathcal{R}^2}{L^2} \zeta(2s-4) + \frac{1}{4} \zeta(2s-2) \right] \\ &+ \frac{1}{\Gamma(s)} \sum_{n=1}^{\infty} \frac{(-1)^n \Gamma(n+s-1)}{n!} \left(\frac{\pi^2 \mathcal{R}^2}{L^2} \right)^{1-n-s} \\ \times & \left[\frac{\pi^2 \mathcal{R}^2}{L^2} \zeta(2s+2n-4) + \frac{1}{4} \zeta(2s+2n-2) \right] \zeta'_H(-2n, 3/2). \end{aligned} \quad (9.58)$$

The Casimir energy density (pressure) follows easily

$$\mathcal{E}_{\text{Cas}} = \frac{\hbar c}{2LR^4} \zeta \left(-\frac{1}{2} | L_5 \right) = -\frac{\hbar c \pi^3}{36 L^6} \left[\frac{\pi^2}{315} - \frac{1}{240} \left(\frac{L}{\mathcal{R}} \right)^2 + \mathcal{O} \left(\frac{L}{\mathcal{R}} \right)^4 \right]. \quad (9.59)$$

If we put numbers here, we realize that the actual numerical value of this cosmological Casimir energy density corresponding to a scalar field in the conditions considered (in particular, recall that we have set Dirichlet BC on the two branes) is about ten times larger than the one corresponding to the original Casimir setting for the electromagnetic field, namely

$$\mathcal{E}_{\text{Cas,em}} = -\frac{\hbar c \pi^2}{240\, L^4}, \quad (9.60)$$

this one being about 100 dynes/cm^2 at 100 nm.

For calculating the one-loop effective potential

$$V = \frac{1}{2L \text{Vol}(M_4)} \log \det (L_5/\mu^2), \quad (9.61)$$

where

$$L_5 = -\partial_z^2 - \Delta^{(4)} - \xi_5 R^{(4)} = L_1 + L_4 \quad (9.62)$$

and

$$\log \det L_5 = \sum_{n,\alpha} \log(\lambda_n^2 + \lambda_\alpha^2) = -\zeta'(0|L_5), \quad (9.63)$$

we need to obtain the derivative of the five-dimensional zeta function at $s = 0$ [E. Elizalde (1986); E. Elizalde (1985); E. Elizalde (1993)]. The results are the following.

(1) In the one-brane limit, $L \longrightarrow \infty$, we obtain,

$$K_t(L_1) \sim \frac{L}{2\sqrt{\pi t}}, \tag{9.64}$$

$$\zeta'(0|L_5) = \frac{1}{3\mathcal{R}}\left[\zeta_H\left(-4, \frac{3}{2}\right) - \frac{1}{4}\zeta_H\left(-2, \frac{3}{2}\right)\right] = 0. \tag{9.65}$$

(2) For the small distance expansion

$$\begin{aligned}
\zeta'(0|L_5) &= \frac{\zeta'(-4)}{6}\frac{\pi^4\mathcal{R}^4}{L^4} + \frac{\zeta'(-2)}{12}\frac{\pi^2\mathcal{R}^2}{L^2} \\
&+ \frac{1}{24}\left[\zeta'_H(-4,3/2) - \frac{1}{2}\zeta'_H(-2,3/2)\right]\log\left(\frac{\pi^2\mathcal{R}^2}{L^2}\right) \\
&+ \frac{\zeta'(0)}{6}\left[\zeta'_H(-4,3/2) - \frac{1}{2}\zeta'_H(-2,3/2)\right] + \frac{1}{24}\zeta'_H(-4,3/2) \\
&+ \frac{1}{36}\left[\frac{1}{8}\zeta'_H(-4,3/2) - \frac{1}{3}\zeta'_H(-6,3/2)\right]\frac{L^2}{\mathcal{R}^2} + \mathcal{O}\left(\frac{L^4}{\pi^4\mathcal{R}^4}\right) \\
&\simeq 0.129652\frac{\mathcal{R}^4}{L^4} - 0.025039\frac{\mathcal{R}^2}{L^2} - 0.002951\log\left(\frac{\mathcal{R}^2}{L^2}\right) \\
&-0.017956 - 0.000315\frac{L^2}{\mathcal{R}^2} + \cdots
\end{aligned} \tag{9.66}$$

9.4.4 *The massive case*

The Lagrangian for a massive scalar field with scalar-gravitational coupling is given by:

(1) In an AdS background

$$\mathcal{L} = \phi\left(\partial_z^2 + \Delta^{(4)} - m^2 l^2 \sinh^{-2} z + \xi_5 R^{(4)}\right)\phi. \tag{9.67}$$

(2) In a dS background

$$\mathcal{L} = \phi\left(\partial_z^2 + \Delta^{(4)} - m^2 \cosh^{-2} z + \xi_5 R^{(4)}\right)\phi. \tag{9.68}$$

The corresponding effective potentials for the massive scalar field are, respectively,

$$
\begin{aligned}
V &= \frac{1}{2L\mathrm{Vol}(M_4)} \log\det(L_5/\mu^2), \qquad (9.69)\\
L_5 &\equiv -\partial_z^2 + m^2 l^2 \sinh^{-2} z - \Delta^{(4)} - \xi_5 R^{(4)} = L_1 + L_4 \ (AdS),\\
L_5 &\equiv -\partial_z^2 + m^2 \cosh^{-2} z - \Delta^{(4)} - \xi_5 R^{(4)} = L_1 + L_4 \qquad (dS).
\end{aligned}
$$

The small mass limit (with L not large) is given, from the eigenvalues

$$
\lambda_n^2 \simeq \frac{\pi^2 n^2}{\mu^2 L^2} + m^2 l^2 \frac{\tanh(\mu L/2)}{\mu L/2}, \qquad (9.70)
$$

as the result we had for the massless case plus the following mass correction (de Sitter brane in AdS bulk)

$$
\begin{aligned}
\Delta\zeta'(0|L_5) &\simeq \frac{a\rho + a^2\rho^2}{48} - \frac{\pi^2}{144}\left\{\frac{a\rho^2}{2} + [2\zeta'(-4,3/2) - \zeta'(-2,3/2)]\,\rho\right\}\\
&\quad - \frac{\pi^4}{4370}[2\zeta'(-4,3/2) - \zeta'(-2,3/2)]\,\rho^2 + \mathcal{O}(m^6), \qquad (9.71)
\end{aligned}
$$

$$
a \equiv \frac{\pi^2 \mathcal{R}^2}{L^2}, \qquad \rho \equiv \frac{m^2 l^2}{\pi^2}\,\frac{\tanh(L/2l)}{L/2l}.
$$

For the large mass limit (with L not small), we have the eigenvalues

$$
\lambda_n^2 \simeq \frac{\pi^2 n^2 l^2}{L^2} + \frac{2\arctan(\sinh L/2l)}{\sinh(L/2l)} m^2 l^2 + \frac{\pi n m l^2}{L\sinh(L/2l)} + \cdots \qquad (9.72)
$$

and the dominant contribution in the result turns out to be

$$
\begin{aligned}
\zeta(s|L_5) &= \frac{L}{2l\sqrt{\pi}}\,\frac{\Gamma(s-1/2)}{\Gamma(s)}\\
&\quad \times\zeta\left(s - \frac{1}{2}\middle| L_4 + 2m^2 \frac{\arctan(\sinh(L/2l))}{\sinh(L/2l)}\right) + \cdots \qquad (9.73)
\end{aligned}
$$

and

$$
\zeta'(0|L_5) = -\frac{4m^2 l^3}{3\mathcal{R}}\,\frac{\arctan(\sinh L/2l)}{\sinh(L/2l)} + \cdots \qquad (9.74)
$$

which is now non-zero.

Appendix A

Useful Mathematical Relations

For the reader convenience we collect some definitions and properties of special functions which we used throughout the book.

A.1 The Poisson Summation Formula

This is one of the most useful summation formulae. It can be regarded as a non-Abelian version of the Selberg trace formula. There exist several versions. To begin with, in the sense of distributions we have

$$\sum_{n=-\infty}^{\infty} \delta(x-2\pi n) = \frac{1}{2\pi} \sum_{n=-\infty}^{\infty} e^{inx} \,.$$

In general, for any suitable function $f(\mathbf{x})$, $\mathbf{x} \in \mathbb{R}^N$ and any $\mathbf{q} \in \mathbb{R}^N$, the Poisson summation formula reads

$$\sum_{\mathbf{k}\in\mathbb{Z}^N} f(\mathbf{k}+\mathbf{q}) = \sum_{\mathbf{k}\in\mathbb{Z}^N} \int_{\mathbb{R}^N} d^N x\, f(\mathbf{x}) e^{2\pi i\,\mathbf{k}\cdot(\mathbf{x}-\mathbf{q})} \,.$$

Now, choosing $f(x) = e^{-\alpha x^2}$, one directly gets the well known Jacobi identity

$$\sum_{n=-\infty}^{\infty} e^{-\alpha n^2} = \sqrt{\frac{\pi}{\alpha}} \sum_{n=-\infty}^{\infty} e^{-\pi n^2/\alpha} \,. \tag{A.1}$$

A.2 The Mellin Transform

Let $x^{z-1}f(x)$ belong to $L^1(0,\infty)$ (the space of integrable functions) and let $f(x)$ have bounded variation on every finite interval. Then the Mellin transform is defined by

$$\hat{f}(z) = \int_0^\infty dx\, x^{z-1} f(x)\,.$$

In the case when $f(x)$ is continuous, the Mellin inversion formula is

$$f(x) = \frac{1}{2\pi i}\int_{\mathrm{Re}\,z=c} dz\, x^{-z}\hat{f}(z)\,, \tag{A.2}$$

c being a real number in the strip in which $\hat{f}(z)$ is analytic. For any pair of functions f, g with Mellin transforms $\hat{f}$, $\hat{g}$, we have the useful Mellin-Parseval identity

$$\int_0^\infty dz\; f(x)g(x)\,dx = \frac{1}{2\pi i}\int_{\mathrm{Re}\,z=c} \hat{f}(z)\hat{g}(1-z)\,,$$

where c is in the common strip in which $\hat{f}(z)$ and $\hat{g}(1-z)$ are analytic.

A.3 The McDonald Functions

The following integral representations [I.S. Gradshteyn and I.M. Ryzhik (1980)] are frequently used in the paper:

$$K_\nu(z) = \frac{\left(\frac{z}{2}\right)^\nu \Gamma\left(\frac{1}{2}\right)}{\Gamma\left(\nu+\frac{1}{2}\right)}\int_1^\infty dt\, e^{-zt}(t^2-1)^{\nu-\frac{1}{2}}\,, \tag{A.3}$$

$$\left[\mathrm{Re}\left(\nu+\frac{1}{2}\right) > 0 \text{ and } |\arg z| < \frac{\pi}{2} \quad \text{or } \mathrm{Re}\,z = 0 \text{ and } \nu = 0\right],$$

$$K_\nu(z) = \frac{1}{2}\left(\frac{z}{2}\right)^\nu \int_0^\infty dt\, \frac{e^{-t-\frac{z^2}{4t}}}{t^{\nu+1}}\,, \tag{A.4}$$

$$\left[|\arg z| < \frac{\pi}{2} \text{ and } \mathrm{Re}\,z^2 > 0\right].$$

A.4 The Riemann-Hurwitz Functions

The prototype of the zeta functions is the celebrated Riemann-Hurwitz ζ_H-function. For $\text{Re}\, s > 0$, it may be defined by means of equation (see for example [I.S. Gradshteyn and I.M. Ryzhik (1980)])

$$\zeta_H(s;q) = \sum_{n=0}^{\infty}(n+q)^{-s}\,, \tag{A.5}$$

the sum being extended to all non-negative n such that $n + q \neq 0$. Here q is an arbitrary real number. We have the simple relation $\zeta_H(s;1) = \zeta_R(s)$, where $\zeta_R(s)$ is the usual Riemann function. It satisfies the useful functional equation (reflection formula)

$$\pi^{-\frac{s}{2}}\Gamma(\frac{s}{2})\zeta_R(s) = \pi^{-\frac{1-s}{2}}\Gamma(\frac{1-s}{2})\zeta_R(1-s)\,. \tag{A.6}$$

A.5 The Epstein Z-Function

The Epstein Z-function can be considered as a generalization of the one of Riemann-Hurwitz. We consider three N-dimensional vectors $\mathbf{q}$ and $\mathbf{h}$ in $\mathbb{R}^N$ and $\mathbf{k} \in \mathbb{Z}^N$ and an invertible $N \times N$ matrix $\mathcal{R}$. For $\text{Re}\, s > 1$ the Epstein Z-function is defined by [A. Erderlyi (1955)]

$$Z_{\mathcal{R}}(s;\mathbf{q},\mathbf{h}) = \sum_{\mathbf{k}} e^{2\pi i\mathbf{k}\cdot\mathbf{h}}[(\mathbf{k}+\mathbf{q})\mathcal{R}(\mathbf{k}+\mathbf{q})]^{-Ns/2}\,, \tag{A.7}$$

where the sum run over $\mathbf{k} \in \mathbb{Z}^N$ for which $\mathbf{k} + \mathbf{q} \neq 0$. If all components of $\mathbf{h}$ are not integer, then Eq. (A.7) can be analytically continued to an entire function in the complex plane, otherwise it has a simple pole at $s = 1$ with residue $\pi^{Ns/2} \det \mathcal{R}^{-1/2}\Gamma(1 + N/2)$. $Z_{\mathcal{R}}(s;\mathbf{q},\mathbf{h})$ satisfies the functional equation

$$\pi^{-\frac{Ns}{2}}\Gamma\left(\frac{Ns}{2}\right) Z_{\mathcal{R}}(s;\mathbf{q},\mathbf{h}) =$$

$$\det\mathcal{R}^{-\frac{1}{2}}\pi^{-\frac{N(1-s)}{2}}e^{-2\pi i\mathbf{q}\cdot\mathbf{h}}\Gamma\left(\frac{N(1-s)}{2}\right) Z_{\mathcal{R}^{-1}}(1-s;\mathbf{h},-\mathbf{q})\,.$$

Appendix B

The Wodzicki Residue

For the reader's convenience, we will update in this Appendix some basic information concerning classical pseudo-differential operators (ΨDO) and the Wodzicki residue [M. Wodzicki (1987)] (see also [C. Kassel (1989)] and the references to Wodzicki papers quoted therein) that will be used in the rest of the book.

Let us consider a D-dimensional, smooth (compact) manifold without boundary, $\mathcal{M}_D$, and a classical ΨDO, Q, acting on sections of vector bundles on $\mathcal{M}_D$. To any classical ΨDO, it corresponds to a complete symbol

$$\sigma(Q) = Q(x,k) = e^{ikx} Q e^{-ikx} , \tag{B.1}$$

such that, modulo infinitely smoothing operators, one has

$$(Qf)(x) \sim \int_{R^D} \frac{dk}{(2\pi)^D} \int_{R^D} dy \, e^{i(x-y)k} Q(x,k) f(y) \, . \tag{B.2}$$

The complete symbol of the classical operator Q admits an asymptotic expansion for $|k| \to \infty$, given by

$$Q(x,k) \sim \sum_{j=0}^{\infty} Q_{q-j}(x,k) \, , \tag{B.3}$$

where the coefficients fulfill the *homogeneity property*

$$Q_{q-j}(x,tk) = t^{q-j} Q_{q-j}(x,k),$$

for $t > 0$, being $Q_q(x,k) \neq 0$. The number q is called the order of Q. We shall here deal mainly with self-adjoint operators and thus it will be assumed that their complex powers, besides the semigroup property, also satisfy $(A^c)^{-s} = A^{-cs}$, the c and s being arbitrary complex numbers. As an example that will often be encountered, let us consider $\log A$, where A is the Laplace operator plus a constant C. Then,

$$\sigma(\log A) \sim 2\log|k| + \sum_{j=0}^{\infty} A_{-j}(x,k)\,, \tag{B.4}$$

where

$$A_{-j}(x,k) = \left(\frac{C}{|k|^2}\right)^j\,. \tag{B.5}$$

Observe that $\log A$ is *not* a classical ΨDO operator, since in general its symbol differs from that of a zeroth order operator in the presence of the $\log|k|$ term.‡

In order to introduce the definition of the non-commutative residue of Q, let us consider an elliptic operator A, with $a > q$, and form the family of ΨDOs $A_Q(u) = A + uQ$, u being a real parameter. The associated zeta function reads

$$\zeta(s|A_Q(u)) = \mathrm{Tr}(A+uQ)^{-s}\,. \tag{B.6}$$

The meromorphic structure of the above zeta function can be obtained from the short-t asymptotics of $\mathrm{Tr}\, e^{-A_Q(u)}$ [J. Duistermaat and V. Guillemin (1975)], namely

$$\mathrm{Tr}\, e^{-tA_Q(u)} \simeq \sum_{j=0}^{\infty} \alpha_j(u) t^{(2j-D)/a} + \sum_{k=1}^{\infty} \beta_k(u) t^k \log t\,. \tag{B.7}$$

Recalling that

$$\int_0^1 dt t^{z+j-1} \log t = -\frac{1}{(z+j)^2}\,, \tag{B.8}$$

the presence of logarithmic terms leads —with the help of the Mellin transform— to double poles in the meromorphic expansion of $\zeta(s|A_Q(u))$, i.e.

$$\zeta(s|A_Q(u)) = \frac{1}{\Gamma(s)}\left(\int_0^1 + \int_1^{\infty}\right) dt\; t^{s-1}\,\mathrm{Tr}\, e^{-A_Q(u)}$$

‡Observe however, that log $(I + A)$ is, this fact being common knowledge in the ΨDO community.

$$= \frac{1}{\Gamma(s)} \left[\sum_{j=0}^{\infty} \frac{\alpha_j(u)}{s + \frac{2j-D}{a}} - \sum_{k=1}^{\infty} \frac{\beta_k(u)}{(s+k)^2} + J(s,u) \right] , \quad \text{(B.9)}$$

where $J(s,u)$ is the analytic part. Taking the derivative with respect to u, and then the limit $u \to 0$, one gets

$$\lim_{u\to 0} \frac{d}{du} \operatorname{Tr}(A + uQ)^{-s} = -s \operatorname{Tr}(QA^{-s-1})$$

$$= \frac{1}{\Gamma(s)} \left[\sum_{j=1}^{\infty} \frac{\alpha_j'(0)}{s + \frac{j-D}{a}} - \sum_{k=1}^{\infty} \frac{\beta_k'(0)}{(s+k)^2} + J'(s,0) \right] . \quad \text{(B.10)}$$

By definition, the non-commutative residue of Q is

$$\operatorname{res}(Q) = \operatorname{Res} \left[a \lim_{u\to 0} \frac{d}{du} \operatorname{Tr}(A + uQ)^{-s} \right]_{s=-1} = a\beta_1'(0) , \quad \text{(B.11)}$$

where $\operatorname{Re} s$ is here the ordinary Cauchy residue of complex analysis. It is possible to show that the non-commutative residue, $\operatorname{res}(Q)$, is independent of the elliptic operator A and that it is a trace in the algebra of classical ΨDOs (actually, it is the *only* trace up to multiplicative constants).[§] From the above definition and taking the derivative with respect to u at $u = 0$ of Eq. (B.7), one obtains a possible way to compute the non-commutative residue. In fact

$$\operatorname{Tr}\left(Qe^{-tA}\right) \simeq -\sum_{j=0}^{\infty} \alpha_j'(0) t^{(2j-D)/a-1} - \frac{\operatorname{res}(Q)}{a} \log t + \mathcal{O}(t \log t) \quad \text{(B.12)}$$

and, thus, the non-commutative residue of Q can be read off from the short–t asymptotics of the quantity $\operatorname{Tr}\left(Qe^{-tA}\right)$ —just picking up the coefficient associated with $\log t$. When the manifold is non-compact, this is one of the methods that we have at hand for evaluating the Wodzicki residue, as long as all the traces involved exist.

For the case of a compact manifold, Wodzicki has obtained a very useful *local* form of the non-commutative residue, that is, a *density* which can be integrated to yield the non-commutative residue, namely

$$\operatorname{res}(Q) = \int_{\mathcal{M}_D} \frac{dx}{(2\pi)^D} \int_{|k|=1} dk \, Q_{-D}(x,k) . \quad \text{(B.13)}$$

[§] This makes the non-commutative (or Wodzicki) residue of unmatched importance, as the 'unique' extension of the Dixmier trace to the whole class of ΨDOs.

Here, Q_{-D}, the component of order $-D$ (remember that D is the dimension of the manifold) of the complete symbol appears. In particular, it is immediate that if this component is absent, the Wodzicki residue is zero, a situation that is very easy to check by mere inspection of the symbol of the operator.

Let us now consider an elliptic (self-adjoint) operator B of positive order $b > q$. From Eq. (B.10) and the requirement $\mathrm{Tr}[Q(B^c)^{-z}] = \mathrm{Tr}(QB^{-cz})$, one has the following

Lemma B.1 *In a neighborhood of $z = 0$,*

$$\mathrm{Tr}(QB^{-z}) = \frac{res(Q)}{zb} + \frac{\gamma\, res(Q)}{b} - R_Q(B) + \mathcal{O}(z)\,, \tag{B.14}$$

where γ is the Euler-Mascheroni constant and the quantity $R_Q(B)$ satisfies the relation

$$R_Q(B^c) - R_Q(B) = \frac{\gamma\, res(Q)}{b}\frac{1-c}{c}\,, \qquad c > 0\,. \tag{B.15}$$

The latter equation gives another way to compute the non-commutative residue, namely

$$\mathrm{res}(Q) = b\left[\mathrm{Re}\,\mathrm{Tr}(QB^{-z})\right]_{z=0}\,. \tag{B.16}$$

In particular, for $Q = B^{-n}$, one has

$$\mathrm{res}(B^{-n}) = b\left[\mathrm{Res}\,\mathrm{Tr}(B^{-z-n})\right]_{z=0} = b\left[\mathrm{Res}\,\zeta(n+z|B)\right]_{z=0}\,. \tag{B.17}$$

In these cases, when the non-commutative residue is non-vanishing, it is proportional to heat-kernel coefficients associated with B. Furthermore, when $D + b$ is even, one has

$$R_Q(B) = \alpha'_{\frac{D+b}{2}}\,. \tag{B.18}$$

Thus, in principle, the quantity $R_Q(B)$ can be computed evaluating the related heat-kernel expansion.

Making use of Lemma B.1 one also obtains the following

Corollary B.1

$$\begin{aligned}\mathrm{Tr}\left[Q(A^{-c_1 z} - B^{-c_2 z})\right] &= \frac{res(Q)}{z}\left(\frac{1}{ac_1} - \frac{1}{bc_2}\right) + \gamma\, res(Q)\left(\frac{1}{a} - \frac{1}{b}\right) \\ &\quad - R_Q(A) + R_Q(B) + \mathcal{O}(z)\,. \end{aligned} \tag{B.19}$$

Here $c_1 > 0$ and $c_2 > 0$.

In particular, if $ac_1 = bc_2 = 1$, there is no pole at $z = 0$, and then

$$\mathrm{Tr}\left[Q(A^{-\frac{z}{a}} - B^{-\frac{z}{b}})\right] = \gamma\,\mathrm{res}(Q)\left(\frac{1}{a} - \frac{1}{b}\right) - R_Q(A) + R_Q(B) + \mathcal{O}(z)\,.$$

On the other hand, we also have [M. Kontsevich and S. Vishik (1993)]

Lemma B.2 *It turns out that*

$$\mathrm{Tr}\left[Q(A^{-\frac{z}{a}} - B^{-\frac{z}{b}})\right] = -res\left[Q\left(\frac{\log A}{a} - \frac{\log B}{b}\right)\right] + \mathcal{O}(z)\,. \quad \text{(B.20)}$$

This is a direct consequence of Lemma B.1. In fact, if C is a suitable elliptic operator, making use of the relation $C^z = I + z\log C + \mathcal{O}(z^2)$, one obtains

$$\begin{aligned}\mathrm{Tr}\left[Q(A^{-\frac{z}{a}} - B^{-\frac{z}{b}})\right] &= \mathrm{Tr}\left[Q(A^{-\frac{z}{a}} - B^{-\frac{z}{b}})C^zC^{-z}\right] \\ &= -z\,\mathrm{Tr}\left[Q\left(\frac{\log A}{a} - \frac{\log B}{a}\right)C^{-z}\right] + \mathcal{O}(z^2)\,. \quad \text{(B.21)}\end{aligned}$$

Now $Q(\frac{\log A}{a} - \frac{\log B}{a})$ is a classical ΨDO and we can make use of Lemma B.1 to obtain the desired result. As a consequence, we also get

$$\mathrm{res}\left[Q\left(\frac{\log A}{a} - \frac{\log B}{b}\right)\right] = -\gamma\,\mathrm{res}(Q)\left(\frac{1}{a} - \frac{1}{b}\right) + R_Q(A) - R_Q(B)\,. \quad \text{(B.22)}$$

It is easy to show that Lemmas B.1 and B.2 allow us to rewrite Eq. (B.19) as

$$\begin{aligned}\mathrm{Tr}\left[Q(A^{-z} - B^{-z})\right] \quad &= \quad \frac{\mathrm{res}(Q)}{z}\left(\frac{1}{a} - \frac{1}{b}\right) \\ &\quad -\mathrm{res}\left[Q\left(\frac{\log A}{a} - \frac{\log B}{b}\right)\right] + \mathcal{O}(z)\,. \quad \text{(B.23)}\end{aligned}$$

And, as a result, we have

Lemma B.3 *In the case considered, the search for limit is given by*

$$\lim_{z\to 0}\frac{d}{dz}\left\{z\,\mathrm{Tr}\left[Q(A^{-z} - B^{-z})\right]\right\} = -res\left[Q\left(\frac{\log A}{a} - \frac{\log B}{b}\right)\right]\,. \quad \text{(B.24)}$$

Let us consider the case $b + D$ even, then Lemma B.1 leads to

$$\lim_{z\to 0}\frac{d}{dz}\left\{z\,\mathrm{Tr}\left[QA^{-z})\right]\right\} = +\frac{\gamma}{a}\mathrm{res}Q - \alpha'_{\frac{D+b}{2}}\,. \quad \text{(B.25)}$$

Alternatively, Lemma B.24 gives

$$\lim_{z\to 0}\frac{d}{dz}\left\{z\,\mathrm{Tr}\left[QA^{-z}\right]\right\}=\lim_{z\to 0}\frac{d}{dz}\left\{z\,\mathrm{Tr}\left[QB^{-z}\right]\right\}-\mathrm{res}\left[Q\left(\frac{\log A}{a}-\frac{\log B}{b}\right)\right].$$

Definitions and Conventions

Basic Notations

Tr	the trace of an operator in a Hilbert space
tr	the trace on internal indices (i.e. Dirac matrices)
Res	Cauchy Residue of a function
res	non-commutative (Wodzicki) residue of an operator
∂_a	partial derivative with respect to x^a
∇_a	covariant derivative with respect to x^a
A, $\mathcal{A}$	generic elliptic operator and the chosen self-adjoint extension
Δ, Δ_D,	Laplace-Beltrami or D'Alembert operator
L_D, $\mathcal{L}$	Laplacian-like Hermitian operator on $\mathcal{M}^D$
γ^μ	Dirac matrices
$\not{V} = \gamma^\mu V_\mu$	for a generic vector V
$(\mathcal{M}, \mathbf{g})$, $\mathcal{M}^D$	smooth Riemannian or Lorentzian (D-dimensional) manifold
$\mathbf{g}$	Lorentzian or Euclidean metric
$g_{\mu\nu}$, g	components and determinant of $\mathbf{g}$
$\mathcal{H}$	Hilbert space equipped with the scalar product (,)
$\mathcal{D}(A) \subset \mathcal{H}$	generally nonclosed subspace of $\mathcal{H}$

$K_t(A) = \mathrm{Tr}\, e^{-tA}$	heat-kernel trace of A
$K_n(A)$, $A_n(A)$	spectral coefficients related to A
$\zeta(s\vert A)$	zeta function related to A
$\zeta_R(s)$	Riemann zeta function
$\zeta_H(s,a)$	Hurwitz zeta function
$\zeta_B(s,a;d)$	Barnes zeta function
$Z_{\mathcal{R}}(z,\mathbf{g},\mathbf{h})$	Epstein Z-function
$Z(s)$, $\Xi(s)$	Selberg functions
$\Gamma(z)$	Euler gamma function
$K_\nu(z)$	McDonald functions (Bessel of III kind)
γ	Euler-Mascheroni constant

Operators on Hilbert Spaces

In a Hilbert space $\mathcal{H}$ equipped by the scalar product $(\ ,\ )$, the linear operator $A : \mathcal{D}(A) \to \mathcal{H}$ is said

Hermitian: if $(u, Av) = (Au, v)$ for all $u, v \in \mathcal{D}(A)$;
symmetric: if it is Hermitian and $\overline{\mathcal{D}(A)} = \mathcal{H}$;
self-adjoint: if it is symmetric and $A = A^\dagger$, where the latter is the adjoint of A;
essentially self-adjoint: if it is symmetric and $A^\dagger$ is self-adjoint. In this case $\overline{A} = (A^\dagger)^\dagger = A^\dagger$.

Riemannian Geometry as in [L.D. Landau and E.M. Lifshits (1970)]

$R^a_{bcd} = \partial_c \Gamma^a_{db} - \partial_d \Gamma^a_{cb} + \Gamma^a_{ce}\Gamma^e_{db} - \Gamma^a_{de}\Gamma^e_{cb}$	Riemannian tensor components
$R_{bd} = R^a_{bad}$	Ricci tensor components
$R = R_{ab} g^{ab}$	scalar curvature

Differential Geometry

smooth: throughout the book, by *smooth* we mean C^∞. A manifold and its metric are assumed to be smooth whenever it is not explicitly stated further and similarly, a manifold has *no boundary* if it is not explicitly stated further.

vector fields: as usual, if X is a vector field on a manifold $\mathcal{M}$ and $f \in C^\infty(\mathcal{M})$, $X(f)$ denotes the smooth function given by $X^a \frac{\partial f}{\partial x^a}$ in each local coordinate patch.

geodesically convex [M.P. do Carmo (1992)]: if $(\mathcal{M}, \mathbf{g})$ is a smooth Riemannian or Lorentzian manifold, an open set $C \subset \mathcal{M}$ is said to be *geodesically convex* (or *convex normal*) if there is an open set $W \subset T\mathcal{M}$, $W = \{(q,v) \mid q \in C, v \in S_q\}$, $S_q \subset T_q\mathcal{M}$ being a starshaped open neighborhood of the origin, such that $exp \upharpoonright_W$: $(q,v) \mapsto exp_q v$ is a diffeomorphism onto $C \times C$. It is clear that C is connected and there is only one geodesic segment joining any pair $q, q' \in C$ completely contained in C, i.e., $t \mapsto \exp_q(t((\exp_q)^{-1}q'))$ $t \in [0,1]$. Moreover if $q \in C$, $\{e_\alpha|_q\} \subset T_q\mathcal{M}$ is a base, $t = t^\alpha e_\alpha|_q \mapsto \exp_q(t^\alpha e_\alpha|_q)$, $t \in S_q$ define a *global* normal coordinate system onto C centered on q which is called *normal Riemannian coordinate system* with respect to (or centered on) q.

signature: the Lorentzian and Euclidean signatures of a metric read $(-,+,+,\cdots,+)$ and $(+,\cdots,+)$ respectively.

indices: in a Lorentzian D-dimensional manifold with local coordinates $x^0, x^1, \ldots, x^{D-1}$, x^0 is always a temporal coordinate (i.e., $\partial/\partial x^0$ is timelike); Greek indices $(\alpha, \beta, \ldots)$ take the values $0, 1, 2, \ldots, D-1$ while Latin indices $i, j, k, \ldots$ take the values $1, 2, \ldots, D-1$. In either Euclidean or Lorentzian D-dimensional manifolds, Latin indices a, b, c take the values in $1, 2, \ldots, D$.

Wick Rotation, Action and Stress-Energy Tensor

Wick rotation: from Lorentzian to Euclidean, referred to static coordinates, is assumed to be $t \to -i\tau$.

action: Euclidean and Lorentzian actions of a Klein-Gordon scalar real field are defined respectively by

$$S_E \stackrel{\text{def}}{=} \int_{\mathcal{M}} \frac{1}{2}(\nabla_a\phi\nabla^a\phi + m^2\phi^2 + \xi R\phi^2)\, d\mu_{\mathbf{g}}\,,$$

$$S_L \stackrel{\text{def}}{=} -\int_{\mathcal{M}} \frac{1}{2}(\nabla_\mu\phi\nabla^\mu\phi + m^2\phi^2 + \xi R\phi^2)\, d\mu_{\mathbf{g}}\,.$$

stress-energy tensor: Euclidean and Lorentzian stress-energy tensors are defined respectively by

$$T_{ab}(x) \stackrel{\text{def}}{=} \frac{2}{|g(x)|^{1/2}}\frac{\delta}{\delta g_{ab}(x)} S_E[g]\,,$$

$$T_{\mu\nu}(x) \stackrel{\text{def}}{=} -\frac{2}{|g(x)|^{1/2}}\frac{\delta}{\delta g_{\mu\nu}(x)} S_L[g]\,.$$

Units

In all physical applications, we shall use units in which the speed of light, the Planck and the Boltzmann constants have the values $c = \hbar = k = 1$.

Bibliography

Abers E.S. and Lee B.W. (1973), *Phys. Rep.* **9**, 1.

Abramowitz M. and Stegun I.A. (1970), *Handbook of Mathematical Functions* (Dover, New York).

Actor A. (1985), *Phys. Lett.* **B157**, 53.

Actor A. (1986), *Nucl. Phys.* **B256**, 53.

Actor A. (1989), *Fortschr. Phys.* **35**, 793 (1987); *Fortschr. Phys.* **43**, 141.

Adler S. (1969), *Phys. Rev.* **177**, 1426.

Allen B. (1983), *Nucl. Phys.* **B226**, 228.

Allen B. (1986), *Phys. Rev.* **D33**, 3640.

Ambjørn J. and Wolfram S. (1983), *Ann. Phys.* **147**, 1.

Andrianov A. and Bonora L. (1984), *Nucl. Phys.* **B233**, 232.

Arcioni G. and Vazquez-Mozo M.A. (2000), *JHEP*, **0001**, 028.

Arcioni G., Barbon J.L.F., Gomis J. and Vazquez-Mozo M.A. (2000), *JHEP* **0006**, 038.

Ardalan F., Arfaei H. and Sheikh-Jabbari M.M. (1999), *JHEP* **9902**, 016.

Arkani-Hamed N., Dimopoulos S. and Dvali G.R. (1998), *Phys. Lett.* **B429**, 263.

Asada A. (1999), "Spectral invariants and geometry of mapping spaces", in *Geometric aspects of partial differential equations*, B. Booss-Bavnbek and K. Wojticiechowski Eds., *Contemporary Math.* **242**, 1. And private communication.

Atiyah M., Bott R. and Patodi V.K. (1973a), *Invent. Math.* **19**, 279; Errata, *Invent. Math.* **28** (1975), 277.

Atiyah M., Patodi V. and Singer I. (1973b), *Bull. London Math. Soc.* **5**, 229.

Atiyah M.F., Patodi V.K. and Singer I. M. (1975), *Math. Proc. Cambridge Philos. Soc.* **77**, 43.

Atiyah M.F., Patodi V.K. and Singer I.M. (1976), *Math. Proc. Cambridge Philos. Soc.* **79**, 71.

Avramidi I. G. (1991), *Nucl. Phys.* **B355**, 712; Erratum–ibid. (1998), *Nucl. Phys.* **B509**, 577.

Avramidi I. G. (1999), *Rev. Math. Phys.* **11**, 947.

Axelrod S. and Singer I.M. (1994), *J. Diff. Geom.* **39**, 173.

Baacke J. and Igarashi Y. (1983), *Phys. Rev. D* **27**, 460.

Babb J. and Mohideen U. (2002), "Casimir Forces: Recent Developments in Experiment and Theory", Harvard-Smithsonian ITAMP. Recordings are available on-line at the web page *http://itamp.harvard.edu/casimir.html.*

Balachandran A.P., Marmo G., Nair V.P. and Trahern C.G (1982), *Phys. Rev.* **D25**, 2713.

Balakrishnan J. and Toms D.J. (1992), *Phys. Rev.* **D46**, 4413.

Ball R.D. (1989), *Phys. Rep.* **182**, 1.

Barbasch D. and Moscovici H. (1983), *J. Funct. Anal.* **53**, 151.

Bardeen W.A. (1969), *Phys. Rev.* **184**, 1848.

Barnes E.W. (1903), *Trans. Camb. Philos. Soc.* **19**, 374.

Barton G. (1991), *J. Phys. A: Math. Gen.* **24**, 991; *J. Phys. A: Math. Gen.* **24**, 5533.

Barton G. (1999), *J. Phys. A: Math. Gen.* **32**, 525.

Barvinsky A.O., Kamenshchik A.Yu. and Karmazin I.P. (1991), *Ann. Phys.* **219**, 201.

Bateman Manuscript Project (1955), "Higher Transcendental Functions", vol. **3** (McGraw-Hill, New York).

Bekenstein J.D. and Parker L. (1981), *Phys. Rev.* **D23**, 2850.

Bell J. and Jackiw R. (1969), *Nuovo Cimento* **A60**, 47.

Bender C.M. and Orszag S.A. (1978), *Advanced Mathematical Methods for Scientists and Engineers* (McGraw-Hill, New York).

Benson K., Bernstein J. and Dodelson S. (1991), *Phys. Rev.* **D44**, 2480. Bernstein J. and Dodelson S. *Phys. Rev. Lett.* **66**, 683 (1991).

Berline N., Getzler E. and Vergne M. (1996), *Heat Kernel and Dirac Operators*, 2nd Edition (Springer, Berlin).

Bernard C. (1974), *Phys. Rev.* **D9**, 3312.

Besse A. L. (1987), *Eistein Manifolds* (Springer, Berlin).

Bezerra V.B., G.L. Klimchitskaya G.L. and Romero C. (2000), *Phys. Rev.* **A65**, 012111.

Binetruy P., Deffayet C., Ellwanger U. and Langlois D. (2000), *Phys. Lett.* **B477**, 285.

Binosi D. and Zerbini S. (1999), *J. Math. Phys.* **40**, 5106.

Binosi D., Moretti V., Vanzo L. and Zerbini S. (1999a), *Phys. Rev.* **D59**, 104017.

Birrell N.D. and Davies P.C.W. (1982), *Quantum Fields in Curved Space* (Cambridge University Press, Cambridge).

Bismut J.M. and Freed D.S. (1986a), *Commun. Math. Phys.* **106**, 159.

Bismut J.M. and Freed D.S. (1986b), *Commun. Math. Phys.* **107**, 103.

Bismut J.M. and Zhang M. (1992), Asterisque **205**.

Bjorken J.D. (2003), *Phys. Rev.* **D67**, 043508.

Blanchet L. and Faye G. (2000), *J. Math. Phys.* **41**, 7675.

Blau S.K., Visser M. and Wipf A. (1988), *Nucl. Phys.* **B310**, 163.

Bordag M., Elizalde E. and Kirsten K. (1996a), *J. Math. Phys.* **37**, 895.
Bordag M., Elizalde E., Geyer B. and Kirsten K. (1996b), *Commun. Math. Phys.* **179**, 215.
Bordag M., Kirsten K. and Dowker J.S. (1996c), *Commun. Math. Phys.* **182**, 371.
Bordag M., Elizalde E., Kirsten K. and Leseduarte S. (1997), *Phys. Rev.* **D56**, 4896.
Bordag M., Mohideen U., Mostepanenko V.M. (2001), *Phys. Rep.* **353**, 1.
Bowers M.E. and Hagen C.R. (1999), *Phys. Rev.* **D59**, 025007.
Boyer T.H. (1968), *Phys. Rev.* **174**, 1764.
Brandenberger R.H. (1985), *Rev. Mod. Phys.* **57**, 1.
Branson T.P. and Gilkey P.B. (1990), *Commun. Part. Diff. Equations* **15**, 245.
Bressi G., Carugno G., Onofrio R. and Uoso G. (2002), *Phys. Rev. Lett.* **88**, 041804.
Brevik I. and Elizalde E. (1994), *Phys. Rev.* **D49**, 5319.
Brevik I., Marachevsky V.N. and Milton K.A. (1999), *Phys. Rev. Lett.* **82**, 3948.
Brevik I., Milton K.A., Nojiri S. and Odintsov S.D. (2001), *Nucl. Phys.* **B599**, 305.
Brown G.E. and Rho M. (1979), *Phys. Lett. B* **82**, 177.
Brown G.E., Jackson A.D., Rho M. and Vento V. (1984), *Phys. Lett. B* **140**, 285.
Brown M.R. and Ottewill A.C. (1985), *Phys. Rev.* **D31**, 2514.
Brown M.R., Ottewill A.C. and Page D. (1986), *Phys. Rev.* **D33**, 2840.
Brunetti R. and Fredenhagen K. (2002), *Comm. Math. Phys.* **208**, 623; ibid **180** (1996) 633.
Bruning J. (1982), *Manuscripta Math.* **39**, 126.
Bruning J. and Seeley R. (1987), *J. Funct. Anal.* **73**, 3699.
Buchbinder I.L., Odintsov S.D. and Shapiro I.L. (1992), *Effective Action in Quantum Gravity* (A. Hilger, Bristol).
Butzer P.L., Kilbas A., Trujillo J.J. (2002), *J. Math. Anal. Appl.* **269**, 387.
Bytsenko A.A., Vanzo L. and Zerbini S. (1992), *Mod. Phys. Lett.* **A7**, 397.
Bytsenko A.A., Cognola G., Vanzo L. and Zerbini S. (1996), *Phys. Rep.* **266**, 1.
Bytsenko A.A., Cognola G. and Zerbini S. (1997a), *J. Phys.* **A30**, 3543.
Bytsenko A.A., Cognola G. and Zerbini S. (1997b), *Class. Quantum Grav.* **14**, 615.
Bytsenko A.A., Vanzo L. and Zerbini S. (1997c), *Nucl. Phys.* **B505**, 641.
Bytsenko A.A., Gonçalves A.E. and Williams F.L. (1998), *Mod. Phys. Lett.* **A13**, 99.
Bytsenko A.A. and Williams F.L. (1998), *J. Math. Phys.* **38**, 1075.
Bytsenko A.A., Vanzo L. and Zerbini S. (1999), *Phys. Lett.* **B459**, 535.
Bytsenko A.A. and Williams F.L. (1999), *J. Phys. A: Math. Gen.* **32**, 5773.
Bytsenko A.A., Gonçalves A.E. and Williams F.L. (2000), *Mod. Phys. Lett.* **A15**, 1031.
Bytsenko A.A., Elizalde E. and Zerbini S. (2001a), *Phys. Rev.* **D64**, 105024.
Bytsenko A.A., Gonçalves A.E. and Zerbini S. (2001b), *Mod. Phys. Lett.* **A16**, 1479.

Bytsenko A.A. (2002a), *Int. J. Mod. Phys.* **A17**, 95.
Bytsenko A.A. (2002b), *Nucl. Phys. Suppl.* **104**, 127.
Cai R.G. and Ohta N. (2000), *Phys. Rev.* **D61**, 124012.
Caldarelli M.M. (1999), *Nucl. Phys.* **B549**, 499.
Callan C.G., Giddings S.B., Harvey J.A. and Strominger A. (1992), *Phys. Rev.* **45**, 1005.
Callias C. (1983), *Comm. Math. Phys.* **88**, 357.
Camporesi R. (1990), *Phys. Rep.* **196**, 1.
do Carmo M.P. (1992), *Riemannian Geometry* (Birkhäuser, Boston, MA).
Carroll S.M. (2001), *Living Rev. Rel.* **4**, 1.
Casimir H.B.G. (1948), *Proc. K. Ned. Akad. Wet.* **60**, 793.
Casimir H.B.G. and Polder D. (1948), *Phys. Rev.* **73**, 360.
Chan H.B., Aksyuk V.A., Kleiman R.N., Bishop D.J., Capasso F. (2001a), *Phys. Rev. Lett.* **87**, 211801.
Chan H.B., Aksyuk V.A., Kleiman R.N., Bishop D.J., Capasso F. (2001b), *Science* **291**, 1941.
Chavel I. (1984) *Eigenvalues in Riemannian Geometry* (Academic Press, Orlando).
Cheeger J. (1983), *J. Diff. Geom.* **18**, 1856.
Chernoff P. (1973), *J. Funct. Anal.* **12**, 282.
Chern S.S. (1966), *Bull. Amer. Math. Soc.* **72**, 167.
Cheung Y.-K.E. and Krogh M. (1998), *Nucl. Phys.* **B528**, 185.
Chodos A., Jaffe R.L., Johnson K. and Thorn C.B. (1974a), *Phys. Rev. D* **10**, 2599.
Chodos A., Jaffe R.L., Johnson K., Thorn C.B. and Weisskopf V. (1974b), *Phys. Rev. D* **9**, 3471.
Chowla S. and Selberg A. (1949), *Proc. Nat. Acad. Sci. U.S.A.* **35**, 317.
Chu C.S. and Ho P.M. (1999), *Nucl. Phys.* **B550** 151.
Cognola G. and Zerbini S. (1988), *Mod. Phys. Lett.* **3**, 599.
Cognola G. and Giacconi P. (1989), *Phys. Rev.* **D39**, 2987.
Cognola G., Vanzo L. and Zerbini S. (1989a), *Phys. Lett.* **B223**, 416.
Cognola G., Giacconi P. and Soldati R. (1990a), *J. Math. Phys.* **31**, 2699.
Cognola G., Vanzo L. and Zerbini S. (1990b), *Phys. Lett.* **B241**, 381.
Cognola G., Vanzo L. and Zerbini S. (1992), *J. Math. Phys.* **33**, 222.
Cognola G. and Vanzo L. (1992), *Mod. Phys. Lett.* **A7**, 3677.
Cognola G., Vanzo L. and Zerbini S. (1996), *Phys. Rev.* **D54**, 2699.
Cognola G. and Shapiro I.L. (1998), *Class. Quant. Grav.* **15**, 787.
Cognola G. and Zerbini S. (1999), *Lett. Math. Phys.* **48**, 375.
Cognola G., Elizalde E. and Kirsten K. (2001), *J. Phys.* **A34**, 7311.
Connes A., Douglas M.R. and Schwarz A. (1998), *JHEP* **9802**, 003.
Criscuolo G. (1997), *J. Comput. Appl. Math.* **78**, 255.
Csaki C., Graesser M., Randall L. and Terning J. (2000), *Phys. Rev.* **D62**, 045015.
Dai X. and Freed D.S. (1994), *J. Math. Phys.* **35**, 5155.
Davies B. (1972), *J. Math. Phys.* **13**, 1324.

Davies E.B. (1989), *Heat Kernel and Spectral Theory* (Cambridge University Press, Cambridge).
Delbourgo R. (1995), "A square root of the harmonic oscillator", e-Print arXiv: hep-th/9503056.
Deser S., Griguolo L. and Seminara D. (1998), *Phys. Rev.* **D57**, 7444.
DeWitt B.S. and Brehme R.W. (1960), *Ann. Phys.* **9**, 220.
DeWitt B.S. (1965), *The Dynamical Theory of Groups and Fields* (Gordon and Breach, New York).
DeWitt B.S. (1979), "Quantum gravity: the new synthesis", in *General Relativity. An Einstein Centenary Survey.* S.W. Hawking and W. Israel, editors (Cambridge University Press, Cambridge).
Diaconescu D.E., Moore G. and Witten E. (2000), "E_8 Gauge Theory, and a Derivation of K–Theory from M–Theory", e-Print arXiv: hep-th/0005090.
Dolan L. and Jackiw R. (1974), *Phys. Rev.* **D9**, 3320.
Douglas M.R. and Hall C. (1998), *JHEP* **9802**, 008.
Dowker J.S. and Critchley R. (1976), *Phys. Rev.* **D13**, 3224.
Dowker J.S. and Kennedy G. (1978), *J. Phys.* **A11**, 895.
Dowker J.S. and Schofield J.P. (1988), *Phys. Rev.* **D38**, 3327.
Dowker J.S. and Schofield J.P. (1989), *Nucl. Phys.* **B327**, 267.
Dowker J.S. (1994), *J. Math. Phys.* **35**, 4989.
Dowker J.S. (1996a), *Class. Quantum Grav.* **13**, 1.
Dowker J.S. (1996b), *Phys. Lett.* **B366**, 89.
Dowker J.S. (1998), "On the relevance of the multiplicative anomaly", e-Print arXiv: hep-th/9803200.
Dowker J.S. and Kirsten K. (1999), *Comm. in Analysis and Geometry*, in press.
Duistermaat J. and Guillemin V. (1975), *Invent. Math.* **29**, 39.
Dunne G. and Hall T.L. (1999), *Phys. Rev.* **D60**, 065002.
Dvali G.R., Gabadadze G. and Porrati M. (2000), *Phys. Lett.* **B485**, 208.
Dzyaloshinskii I.E., Lifshitz E.M. and Pitaevskii L.P. (1961), *Advances Phys.* **10**, 165.
Eberlein C. (1992), *J. Phys. A: Math. Gen.* **25**, 3015; *J. Phys. A: Math. Gen.* **25**, 3038.
Ederth T. (2000), *Phys. Rev.*, **A62**, 062104.
Einhorn M.B. and Jones D.R.T. (1984), *Phys. Rev.* **D29**, 331.
Elizalde E. (1984), *Nucl. Phys.* **B243**, 398.
Elizalde E. (1985), *J. Phys.* **A18**, 1637.
Elizalde E. and Soto J. (1985a), *Ann. Phys. (NY)* **162**, 192.
Elizalde E. and Soto J. (1985b), *Nucl. Phys.* **B260**, 136.
Elizalde E. (1986), *Math. Computation* **47**, 175.
Elizalde E. (1989a), *J. Phys.* **A22**, 931.
Elizalde E. (1989b), *Nuovo Cimento* **B104**, 685.
Elizalde E. and Romeo A. (1989a), *J. Math. Phys.* **30**, 1133.
Elizalde E. and Romeo A. (1989b), *Phys. Rev.* **D40**, 436.
Elizalde E. and Romeo A. (1991), *Am. J. Phys.* **59**, 711.

Elizalde E. (1993), *J. Math. Phys.* **34**, 3222.

Elizalde E., Odintsov S.D., Romeo A., Bytsenko A.A. and Zerbini S. (1994), *Zeta Regularization Techniques with Applications* (World Scientific, Singapore).

Elizalde E. (1994a), *J. Math. Phys.* **35**, 3308.

Elizalde E. (1994b), *J. Math. Phys.* **35**, 6100.

Elizalde E. (1994c), *J. Phys.* **A27**, 3775.

Elizalde E. (1994d), *J. Phys.* **A27**, L299.

Elizalde E., Kirsten K. and Zerbini S. (1995), *J. Phys.* **A28**, 617.

Elizalde E. (1995), *Ten Physical Applications of Spectral Zeta Functions* (Springer, Berlin).

Elizalde E., Lygren M. and Vassilevich D.V. (1996), *J. Math. Phys.* **37**, 3105.

Elizalde E. (1997), *J. Phys.* **A30**, 2735.

Elizalde E., Lygren M. and Vassilevich D.V. (1997), *Commun. Math. Phys.* **183**, 645.

Elizalde E. (1998), *Commun. Math. Phys.* **198**, 83.

Elizalde E., Bordag M. and Kirsten K. (1998a), *J. Phys.* **A31**, 1743.

Elizalde E, Cognola G. and Zerbini S. (1998b), *Nucl. Phys.* **B532**, 407.

Elizalde E., Filippi A., Vanzo L. and Zerbini S. (1998c), "Is the multiplicative anomaly relevant?", e-Print arXiv: hep-th/9804072.

Elizalde E., Filippi A., Vanzo L. and Zerbini S. (1998d), *Phys. Rev.* **D57**, 7430.

Elizalde E., Vanzo L. and Zerbini S. (1998e), *Commun. Math. Phys.* **194**, 613 .

Elizalde E. (1999), *JHEP* **9907**, 015.

Elizalde E. and Vassilevich D.V. (1999), *Class. Quant. Grav.* **16**, 813.

Elizalde E. (2000), *J. Comput. Appl. Math.* **118**, 125.

Elizalde E. (2001a), *J. Phys.* **A34**, 3025.

Elizalde E. (2001b), *Phys. Lett.* **B516**, 143.

Elizalde E., Santos F.C. and Tort A.C. (2002a), *J. Phys.* **A35**, 7403.

Elizalde E. and Tort A.C. (2002b), *Phys. Rev.* **D66**, 045033.

Elizalde E., Cognola G. and Zerbini S. (2003a), *Commun. Math. Phys.* **237**, 207.

Elizalde E., Nojiri S., Odintsov S.D. and Ogushi S. (2003b), *Phys. Rev.* **D67**, 063515.

Elizalde E., Santos F.C. and Tort A.C. (2003c), *Int. J. Mod. Phys.* **A18**, 1761.

Elizalde E. and Tort A.C. (2003d), *Phys. Rev.* **D67**, 124014.

Elizalde E. and Tort A.C. (2003e), "A note on the Casimir energy of a massive scalar field in positive c urvature space", e–Print arXiv: hep-th/0306049.

Elliott J.D. (1995), *J. Comput. Appl. Math.* **62**, 267.

Emig T., Hanke A., Golestanian R. and Kardar M. (2001), *Phys. Rev. Lett.* **87**, 260402.

Epstein P. (1903), *Math. Ann.* **56**, 615; ibid. **65**, 205 (1907).

Erderlyi A. (1955), *Higher Trascendental Functions* (McGraw-Hill Book Company, New York). Staff of the Bateman Manuscript Project; based on the notes of Harry Bateman.

Estrada R. and Fulling S.A. (1999), *Electronic J. Diff. Eqs.* **7**, 1.

Estrada R. and Kanwal R. (2002), *A distributional approach to asymptotics: the-*

ory and applications (Birkhäuser, Boston, MA).

Evans T.S. (1999), *Phys. Lett.* **B457**, 127.

Faddeev L.D. and Popov V.N. (1967), *Phys. Rev. Lett.* **25**, 30.

Feller W. (1971), *An Introduction to the Theory of Probability*, 2nd ed., Vol II (Wiley, New York).

Feynman R.P. and Hibbs A.R. (1965), *Quantum Mechanics and Path Integrals* (McGraw-Hill, New York).

Filippi A. (1997), "Inclusion of Chemical Potential for Scalar Fields", e-Print arXiv: hep-ph/9703323.

Fischler W., Gorbatov E., Kashani-Poor A., McNees R., Paban S. and Pouliot P. (2000a), *JHEP* **0006**, 032.

Fischler W., Gorbatov E., Kashani-Poor A., Paban S., Pouliot P. and Gomis J. (2000b), *JHEP* **0005**, 024.

Flachi A., Moss I.G. and Toms D.J. (2001), *Phys. Rev.* **D64**, 105029.

Flanagan E.E., Tye S.H.H. and Wasserman I. (2000), *Phys. Rev.* **D62**, 044039.

Ford L.H. (1993), *Phys. Rev.* **47**, 1; **51**, 1962 (1995).

Fosco C.D., Gamboa Saravi R.E. and Schaposnik F.A. (1999), "On the two-dimensional fermion determinant at finite temperature", e-Print arXiv: hep-th/9902002.

Freed D.S. (1997), "Determinant line bundles revisited", e-Print arXiv: dg-ga/9505002, "Geometry and Physics (Aarhus, 1995)", 187195; Lecture Notes in Pure and Appl. Math. **184** (Marcel Dekker, New York, 1997).

Freed D.S. and Witten E. (1999), "Anomalies in String Theory with D–Branes", e-Print arXiv: hep-th/9907189.

Freedman B. and McLerran L. (1977), *Phys. Rev.* **D16**, 1130.

Fried D. (1977a), *Ann. Sci. E'cole Norm. Sup.* **10**, 133.

Fried D. (1977b), *Invent. Math.* **84**, 523.

Friedlander F.G. (1975), *The Wave Equation on a Curved Space–Time* (Cambridge University Press, Cambridge).

Frolov V.P. and Fursaev D.V. (1998), *Class. Quant. Grav.* **15**, 2041.

Fujikawa K. (1975), *Phys. Rev. Lett.* **35**, 1560.

Fujikawa K. (1979), *Phys. Lett.* **42**, 1195.

Fujikawa K. (1980), *Phys. Rev. Lett.* **44**, 1733.

Fulling S.A. (1973), *Phys. Rev.* **D7**, 2850.

Fulling S.A. and Ruijsenaars S.N.M. (1987), *Phys. Rep.* **152**, 135.

Fulling S.A. (1991), *Aspects of Quantum Field Theory in Curved Spacetime* (Cambridge University Press, Cambridge).

Fulling S.A. (1999), *Electronic J. Diff. Eqs.* **6**, 1.

Gamboa Saravi R.E., Muschietti M.A., Schaposnik F.A. and Solomin J.E. (1984), *Ann. Phys.* **157**, 360.

Gamboa Saravi R.E., Muschietti M.A., Schaposnik F.A. and Solomin J.E. (1985), *J. Math. Phys.* **26**, 2045.

Gamboa Saravi R.E., Rossini G.L. and Schaposnik F.A. (1996), *Int. J. Mod. Phys.* **A11**, 2643.

Gangolli R. (1977), *Illinois J. Math.* **21**, 1.

Gangolli R. and Warner G. (1980), *Nagoya Math. J.* **78**, 1.

Garriga J. and Tanaka T. (2000), *Phys. Rev. Lett.* **84**, 2778.

Garriga J., Pujolas O. and Tanaka T. (2001), *Nucl. Phys.* **B605**, 192.

Gelfand I.M. and Chilov G.E. (1972), *Les Distributions*, tome 1 (Dunod, Paris).

Genet C., Lambrecht L. and Reynaud S. (2002), *Int. J. Mod. Phys.* **A17**, 761.

Georgi H., H. Quinn R. and Weinberg S. (1974), *Phys. Rev. Lett.* **33**, 451.

Gherghetta T. and Pomarol A. (2001), *Nucl. Phys.* **B602**, 3.

Gibbons G.W. and Perry M.J. (1978), *Proc. R. Soc. Lond* **A358**, 467.

Gilkey P.B. (1975), *J. Diff. Geometry* **10**, 601.

Gilkey P.B. (1995), *Invariance Theory, the Heat Equation and the Atiyah-Singer Index Theorem* (CRC Press, Boca Raton).

Gilkey P.B. and and Grubb G. (1998), *Comm. Partial Diff. Eqs.* **23**, 777.

Goldberger W.D. and Wise M.B. (1999), *Phys. Rev. Lett.* **83**, 4922.

Gomis J., Landsteiner K. and Lopez E. (2000a), *Phys. Rev.* **D62** 105006.

Gomis J., Mehen T. and M.B. Wise M.B. (2000b), *JHEP* **0008**, 029.

Gradshteyn I.S. and Ryzhik I.M. (1980), *Table of Integrals, Series and Products* (Academic Press, New York).

Graham N., Jaffe R.L., Khemani V., Quandt M., Scandurra M. and Weigel H. (2002), *Nucl. Phys.* **B645**, 49; ibid "Casimir Energies in Light of Quantum Field Theory", e–Print arXiv: hep-th/0207205.

Grib A.A, Mamayev S.G. and Mostepanenko V.M. (1994), *Vacuum quantum Effects in Strong Fields*, (Friedmann Laboratory Publishing, St. Petersburg).

Grinstein B., Nolte D.R. and Skiba W. (2001), *Phys. Rev.* **D63**, 105016.

Gross D.J. and Jackiw R. (1972), *Phys. Rev.* **D6**, 477.

Grubb G. (2002), *Nucl. Phys. Suppl.* **B104**, 71.

Gusynin V.P. and Roman'kov V.V. (1987), *Sov. J. Nucl. Phys.* **46**, 1097.

Guz N.A. and Zozulya V.V. (2001), *Int. J. Nonlinear Sci. Numer. Simul.* **2**, 173.

Haber H.E. and Weldon H.A. (1981), *Phys. Rev. Lett.* **46**, 1497.

Haber H.E. and Weldon H.A. (1982a), *J. Math. Phys.* **23**, 1852.

Haber H.E. and Weldon H.A. (1982b), *Phys. Rev.* **D25**, 502.

Halamos P.R. (1969), *Measure Theory* (Van Nostrand R., Hamburg).

Hardy G.H. (1949), *Divergent Series* (Oxford University Press, Oxford).

Harris B.W., Chen F. and Mohideen U. (1998), *Phys. Rev.* **A69** 052109.

Hawking S.W. and Ellis G.F. (1973), *The Large Scale Structure of Space-Time* (Cambridge University Press, Cambridge).

Hawking S.W. (1975), *Commun. Math. Phys.* **43**, 199.

Hawking S.W. (1977), *Commun. Math. Phys.* **55**, 133.

Hawking S.W. (1978), *Phys. Rev.* **D18**, 1747.

Hawking S.W. (1979), "The path-integral approach to quantum gravity", in *General Relativity. An Einstein Centenary Survey.* S.W. Hawking and W. Israel, editors (Cambridge University Press, Cambridge).

Hawking S.W., Hertog T. and Reall H.S. (2000), *Phys. Rev.* **D62**, 043501.

Helgason S. (1962), *Differential Geometry and Symmetric Spaces*, Pure and Ap-

plied Math. Ser. **12** (Academic Press).
Hochberg D., Molina-Paris C., Juan Perez-Mercader J. and Visser M. (1998), *Phys. Rev. Lett.* **81**, 4802.
Hofmann R., Kanti P. and Pospelov M. (2001), *Phys. Rev.* **D63**, 124020.
Hollands S. and Wald R.M. (2001), *Commun. Math. Phys.* **223**, 289.
Hormander L. (1968), *Acta Math.* **121**, 193.
Huang W.H. (2001), *Phys. Lett.* **B497**, 317.
Hu B.L. (1981), *Phys. Lett.* **B103**, 331.
Hu B.L. (1982), *Phys. Lett.* **B108**, 19.
Hu B.L. (1983), *Phys. Lett.* **B123**, 189.
Hu B.L. and Phillips N.G. (2000), *Int. J. Theor. Phys.* **39**, 1817.
Hu S.H. and Young B.L. (1984), *Phys. Rev.* **D30**, 836.
Iellici D. and Moretti V. (1996), *Phys. Rev.* **D54**, 7459.
Iellici D. (1998), "Aspects and Applications of Quantum Field Theory on Spaces with Conical Singularity", Ph.D. dissertation, Trento University; e-Print arXiv: gr-qc/9805058.
Iellici D. and Moretti V. (1998), *Phys. Lett.* **B425**, 33.
Iyanaga S. and Kawada Y. (1977), *Encyclopedic dictionary of mathematics*, Vol. II, 1372 (The MIT Press, Cambridge).
Kapusta J.I. (1981), *Phys. Rev.* **D24**, 426.
Kapusta J.I. (1989), *Finite-Temperature Field Theory* (Cambridge University Press, Cambridge).
Karoubi M. (1978), "K–Theory. An Introduction", *Grundlehren der Mathematischen Wissenschaften* **226**, A Series of Comprehensive Studies in Mathematics (Springer, Berlin).
Kassel C. (1989), *Asterisque* **177**, 199, Sem. Bourbaki.
Kay B.S. and Wald R. M. (1991), *Phys. Rep.* **207**, 49.
Kiem Y. and Lee S. (2000), *Nucl. Phys.* **B586**, 303.
Kirsten K. (1991), *Class. Quantum Grav.* **8**, 2239.
Kirsten K. and Toms D.J. (1995), *Phys. Rev.* **D51**, 6886.
Kirsten K. and Elizalde E. (1996a), *Phys. Lett.* **B365**, 72.
Kirsten K. and Cognola G. (1996b), *Class. Quant. Grav.* **13**, 633.
Kirsten K. and Toms D.J. (1996c), *Phys. Lett.* **B368**, 119.
Kirsten K. and Toms D.J. (1996d), *Phys. Rev.* **A54**, 4188.
Kirsten K. and Toms D.J. (1997), *Phys. Rev.* **D55**, 7797.
Kirsten K. (1998), *Class. Quant. Grav.* **15**, L5.
Kirsten K. (2002), *Spectral Functions in Mathematics and Physics* (Chapman and Hall, CRC, London).
Knopp K. (1931), *Theorie und Andwendungen der Unendlichen Reihen* (Springer, Berlin).
Kobayashi S. (1972), *Transformation Groups in Differential Geometry* (Springer, Berlin).
Kontsevich M. and Vishik S. (1993), "Geometry of determinants of elliptic operators", in *Functional Analysis on the Eve of the 21st Century* (Eds. Gindikin,

S. et al.) *Progress in Mathematics*, vol. 1, 173.
Kudoh H. and Tanaka T. (2002), *Phys. Rev.* **D65**, 104034.
Kuo C.I. and Ford L.H. (1993), *Phys. Rev.* **47**, 4510.
Lamoreaux S.K. (1997), *Phys. Rev. Lett.*, **78**, 5; erratum, **81**, 5475 [New York Times, Jan. 21, 1997, p. C1].
Landau L.D. and Lifshits E.M. (1970), *Theorie des Champs* (Moscow 1970).
Landsteiner K., Lopez E. and Tytgat M.H.G. (2000), *JHEP* **0009**, 027.
Langmann E. and Mickelsson J. (1996), *Lett. Math. Phys.* **36**, 45.
Lang S. (1985), $SL(2,\mathbb{R})$ (Springer, Berlin).
Lawson H.B. and Michelson M.L. (1989), *Spin Geometry*, Princeton Mathematical Series, **38** (Princeton University Press, Princeton).
Leseduarte S. and Romeo A. (1996), *Ann. Phys.* **250**, 448.
Liberati S., Belgiorno F., Visser M. and Sciama D.W. (2000), *J. Phys.* **A33**, 2251.
Linde A.D. (1983), *Phys. Lett.* **B123**, 185.
Maroto A.L. (2002), e–Print arXiv: hep-th/0207207.
Mastronardi N. and Occorsio D. (1996), *J. Comput. Appl. Math.* **70**, 75.
Mathai V., Melrose R.B. and Singer I.M. (2002), "The index of projective families of elliptic operators", e–Print arXiv: math.DG/0206002.
McKenzie-Smith J.J and Toms D.J. (1998), *Phys. Rev.* **D58**, 105001.
McKeon D.G.C. and Schubert C. (1998), *Phys. Lett.* **B440**, 101.
Miatello R. (1976), "The Minakshisundaram–Pleijel Coefficients for the Vector–Valued Heat Kernel on Compact Locally Symmetric Spaces of Negative Curvature", PhD Thesis (Rutgers University).
Miatello R. (1979), *Manuscripta Math.* **29** 249.
Miatello R. (1980), *Trans. Am. Math. Soc.* **260**, 1.
Mickelsson J. (1993), "Wodzicki residue and anomalies of current algebras", in *Proceedings of Integrable models and Strings (Helsinki 1994)*, 123, e–Print arXiv: hep-th/9404093.
Mickelsson J. (1996), "Chiral anomalies on lattice", e–Print arXiv: hep-lat/9602030.
Milonni P.W. (1994), *The quantum vacuum: An introduction to quantum electrodynamics* (Academic Press, Boston).
Milton K.A. (1980a), *Ann. Phys.* **127**, 49.
Milton K.A. (1980b), *Phys. Rev.* **D22**, 1441; *Phys. Rev.* **D22**, 1444; Erratum-ibid. **D25**, 3441 (1982).
Milton K.A. (1983), *Ann. Phys.* **150**, 432.
Milton K.A. and Ng Y.J. (1997), *Phys. Rev. E* **55**, 4209.
Milton K.A. and Ng Y.J. (1998), *Phys. Rev. E* **57**, 5504.
Milton K.A. (2001), *The Casimir Effect* (World Scientific, Singapore).
Milton K.A. (2002), "Calculating Casimir Energies in Renormalizable Quantum Field Theory", e–Print arXiv: hep-th/0210081.
Minakshisundaram S. (1949), *Can. J. Math.* **1**, 320.
Minakshisundaram S. and Pleijel A. (1949), *Can. J. Math.* **1**, 242.
Minwalla S., Van Raamsdonk M. and Seiberg N. (2000), *JHEP* **0002**, 020.

Mohideen U. and Roy A. (1998), *Phys. Rev. Lett.* **81**, 4549.
Moretti V. and Iellici D. (1997), *Phys. Rev.* **D55**, 3552.
Moretti V. (1997a), *Class. Quant. Grav.* **14**, L123.
Moretti V. (1997b), *Phys. Rev.* **D56**, 7797.
Moretti V. (1999a), *Commun. Math. Phys.* **201**, 327.
Moretti V. (1999b), *J. Math. Phys.* **40**, 3843.
Moretti V. (1999c), *Commun. Math. Phys.* **208**, 283.
Moretti V. (2000), *Commun. Math. Phys.* **212**, 165.
Moretti V. (2003), *Commun. Math. Phys.* **232**, 189.
Moss I.G. and Dowker J.S. (1989), *Phys. Lett.* **B229**, 261.
Mostepanenko V.M. and Trunov N.N. (1997), *The Casimir Effect and Its Applications* (Oxford Univ. Press, Oxford).
Müller W. (1998), *Commun. Math. Phys.* **192**, 309.
Nam S. (2000), *JHEP* **0010**, 044.
Naylor W. and Sasaki M. (2002), *Phys. Lett.* **B542**, 289.
Nesterenko V.V. and Pirozhenko I.G. (1997), *Phys. Rev.* **D57**, 1284.
Niemi A. and Semenoff G. (1983), *Phys. Rev. Lett.* **51**, 2077;
Niemi A. and Semenoff G. (1983a) *Phys. Rep.* **135**, 93.
Niemi A.J. and Semenoff G.W. (1986), *Nucl. Phys.* **B269**, 131.
Ninomiya M. and Tan C.I. (1985), *Nucl. Phys.* **B245**, 199.
Nojiri S. and Odintsov S.D. (2000), *Phys. Lett.* **B484**, 119.
Nojiri S., Odintsov S.D. and Zerbini S. (2000), *Class. Quant. Grav.* **17**, 4855.
Nojiri S., Odintsov S.D. and Ogushi S. (2001), *Phys. Lett.* **B506**, 200.
Nojiri S., Odintsov S.D. and Sugamoto A. (2002), *Mod. Phys. Lett.* **A17**, 1269.
Nörlund N.E. (1922), *Acta Math.* **43**, 21.
Odintsov S.D. (1990), *Fortschr. Phys.* **38**, 371.
Olver F.W. (1954), *Phil. Trans. Roy. Soc.* **A247**, 328.
O'Neil B. (1983), *Semi-Riemannian Geometry* (Academic Press, New York).
Page D. (1982), *Phys. Rev.* **D25**, 1499.
Park J. and Wojciechowski K.P. (2002), *Nucl. Phys. Suppl.* **B104**, 99.
Parker L. and Fulling S.A. (1974), *Phys. Rev.* **D9**, 341.
Parker L., Fulling S.A. and Hu B.L. (1974), *Phys. Rev.* **D10**, 3905.
Pazy A. (1983), *Semigroups of Linear Operators and Applications to Partial Differential Equations* (Springer, New York).
Phillips N.G. and Hu B.L. (1997), *Phys. Rev.* **55**, 6123.
Phillips N.G. and Hu B.L. (2000), *Phys. Rev.* **62**, 084017.
Planck M. and Verh D. (1911), *Phys. Ges.* **2**, 138.
Plunien G., Müller B. and Greiner W. (1986), *Phys. Rep.* **134**, 88.
Quillen D. (1985), *Funk. Anal. i Prilozen.* **19**, 37.
Radu E. (1998a), *Class. Quant. Grav.* **15**, 2743.
Radu E. (1998b), *Phys. Lett.* **A247**, 207.
Radu E. (1999), *Gen. Rel. Grav.* **31**, 287.
Radu E. and Astefanesei D. (2002), *Int. J. Mod. Phys.* **D11**, 715.
Randall L. and Sundrum R. (1999a), *Phys. Rev. Lett.* **83**, 3370.

Randall L. and Sundrum R. (1999b), *Phys. Rev. Lett.* **83**, 4690.
Ray D.B. (1970), *Adv. in Math.* **4**, 109.
Ray D.B. and Singer I.M. (1971), *Adv. in Math.* **7**, 145.
Ray D.B. and Singer I.M. (1973), *Ann. Math.* **98**, 154.
Redlich A.N. (1984), *Phys. Rev. Lett.* **52**, 18;
Redlich A.N. (1984a), *Phys. Rev.* **D29**, 2366.
Reed M. and Simon B. (1980), *Methods of Modern Mathematical Physics* (Academic Press, London).
Reinhart B.L. (1983), *Differential Geometry of Foliations* (Springer, Berlin).
Reuter M. (1985), *Phys. Rev.* **D31**, 1374.
Riesz F. and Nagy B.S. (1972), *Functional Analysis* (Frederick Ungar Publishing Co., Inc., New York).
Roe J. (1998), *Elliptic Operators, Topology and Asymptotic Methods*, 2nd Edition (Longman, Dorchester UK).
Römer H. and Schroer P.B. (1977), *Phys. Lett.* **21**, 182.
Rothstein I.Z. (2001), *Phys. Rev.* **D64**, 084024.
Rudin W. (1997), *Functional Analysis* (TATA McGraw-Hill, New Delhi).
Saharian A.A. and Setare M.R. (2002), *Phys. Lett.* **B552**, 119.
Sahlmann H. and Verch R. (2001), *Rev. Math. Phys.* **13**, 1203.
Sarnak P. (1987), *Commun. Math. Phys.* **110**, 113.
Schaden M. and Spruch L. (2002), *Phys. Rev.* **A65**, 022108.
Schomerus V. (1999), *JHEP* **9906**, 030.
Schwinger J. (1951a), *Phys. Rev.* **82**, 664.
Schwinger J. (1951b), *Proc. Natl. Acad. Sci.* USA, **37** 452.
Schwinger J., DeRaad L.L. and Milton K.A. (1978), *Ann. Phys.* **115**, 1.
Scott D. (1980), *Math. Ann.* **253**, 177.
Seeley R.T. (1967), *Am. Math. Soc. Prog. Pure Math.* **10**, 172.
Seiberg N. and Witten E. (1999), *JHEP* **9909**, 032.
Selberg A. (1956), *J. Indian Math. Soc.* **20**, 47.
Shubin M.A. (1987), *Pseudodifferential Operators and Spectral Theory* (Springer, Berlin).
Silk J. (1994), "A short history of the universe", *Scientific American Library* (Freeman, New York, 1997).
Simon B. (1972), in *Mathematics of Contemporary Physics*, edited by R.F. Streater (Academic Press, London).
Soldati R. and Sorbo L. (1998), *Phys. Lett.* **B426**, 82.
Spruch L. (1996), *Science* **272**, 1452.
Starobinsky A. (1980), *Phys. Lett.* **B91**, 99.
Toms D.J. (1987), *Phys. Rev.* **D35**, 3796.
Toumboulis E. (1977), *Phys. Lett.* **B70**, 361.
Vassilevich D.V. (2003), "Heat kernel expansion: user's manual", e–Print arXiv: hep-th/0306138.
Venkov A.B. (1987), *Proc. Steklov Math. Inst.* **125**, 3.
Verch R. (1995), XI International Congress of Mathematical Physics (Paris, 1994)

352 (Internat. Press, Cambridge, MA).

Vilkovisky G.A. (1984), *Nucl. Phys.* **B234**, 125.

Voros V. (1987), *Commun. Math. Phys.* **110**, 439.

Wakayama M. (1985), *Hiroshima Math. J.* **15**, 235.

Wald R.M. (1978), *Phys. Rev.* **D17**, 1477.

Wald R.M. (1979), *Commun. Math. Phys.* **70**, 226.

Wald R.M. (1984), *General Relativity* (The University of Chicago Press, Chicago).

Wald R.M. (1994), *Quantum Field theory and Black Hole Thermodynamics in Curved Spacetime*, (The University of Chicago Press, Chicago).

Wallach N. (1976), *Bull. Amer. Math. Soc.* **82**, 171.

Weinberg S. (1974), *Phys. Rev.* **D9**, 3357.

Weinberg S. (2000), "The cosmological constant problems", in *Sources and detection of dark matter and dark energy in the universe* (Dark Matter 2000, Marina del Rey, CA); e-Print arXiv: astro-ph/0005265.

Williams F.L. (1981a), *Osaka J. Math.* **18**, 147.

Williams F.L. (1981b), *Contemp. Math.* **191**, 245.

Williams F.L. (1990), "Lectures on the Spectrum of $L^2(\Gamma\backslash G)$", *Pitman Research Notes in Math.* **242** (Longman House Pub.).

Williams F.L. (1992), *Manuscripta Math.* **77**, 17.

Williams F.L. (1993), *J. Physics* **A26**, 3515.

Williams F.L. (1997), *JMP* **38**, 796.

Williams F.L. (1998), *Pacific J. of Math.* **182**, 137.

Williams F.L. (2003), "Topics in Quantum Mechanics", *Progress in Mathematical Physics*, vol. 27, (Birkhäuser, Boston, MA).

Witten E. (1985), *Commun. Math. Phys.* **100**, 197.

Witten E. (1989), *Commun. Math. Phys.* **121**, 351.

Wodzicki M. (1982), *Invent. Math.* **66**, 115.

Wodzicki M. (1984), *Invent. Math.* **75**, 143.

Wodzicki M. (1987), "Noncommutative Residue Chapter I", in *Lecture notes in Mathematics.* Yu.I. Manin, editor, volume 1289, 320 (Springer, Berlin).

Yam P. (1997), *Scientific American*, **277**, 82.

Yano K. (1965), *Differential Geometry on Complex and Almost Complex Spaces* (Pergamon Press, Oxford).

Yano K. and Kon M. (1984), "Structures on Manifolds", *Series in Pure Math.* **3** (World Scientific, Singapore).

Index

abscissa of convergence, 110, 112, 176, 234
accelerating universe, 295
acts discontinuously, 161
actuator, 290
adiabatic limit, 203
admissible cut, 176
AdS background, 303, 309
Agmon-Nirenberg condition, 112
algebra homomorphism, 198
almost–product structures, 147
analytic continuation, 27, 28, 57, 89, 121, 187
analytic regularization, 6, 216, 218, 223
angular frequencies, 257
anti de Sitter, 13, 16, 84, 87
antiperiodic case, 190
asymptotic behavior, 167, 168, 199, 274, 296, 297
asymptotic contributions, 298, 299, 300
asymptotic expansion, 4, 38, 123, 145, 167, 175, 317
Atiyah-Patodi-Singer, 201, 202, 203
automorphic Laplacian, 162
automorphic relative to the group, 164

Baker-Campbell-Hausdorff formula, 246
Barnes polynomial, 120
Barnes' zeta function, 118, 119, 255
base point, 157
Bernoulli numbers, 167
Bessel function, 113, 132, 295, 296, 299
binomial expansion, 132, 189, 193, 221, 239, 242
Bose-Einstein condensation (BEC), 279
Boson fields, 175
boundary condition, 7, 199, 294
boundary free manifolds, 217
brane, 109, 124, 203, 302-305
  brane solutions, 148
  braneworld, 290, 302, 303
broken phase, 280, 282-284
bulk, 302, 303, 310
bundle map, 198

canonical connection, 202
canonical methods, 210
Cartan subgroup, 204
Cartan type, 204
Casimir effect, 209, 287-290, 292, 301-303
  Casimir energy density, 290, 291, 295, 303, 308
  Casimir force, 210, 287-290

Casimir pressure, 295
Casimir slab, 218, 219, 223
Cauchy-Dunford representation, 212
centralizer, 140
character, 129, 140, 151, 158, 203, 205
characteristic class, 201
Charged Relativistic Bose Gas, 265
charged scalar fields, 3
charge operator, 230, 266, 281
charge renormalization, 283, 284
chemical potential, 14, 18, 29, 267
Chern character of the index, 203
Chern-Simons partition function, 204
chiral anomaly, 184, 260
Chowla-Selberg formula, 110, 111, 113, 117
classical action, 2, 3, 8, 10, 11, 211, 261, 265
classical Euclidean action, 86, 171
classical gravitational field, 209
Clifford multiplication, 198
cohomological torsion, 201
cohomology, 200, 203, 204
coincidence limit, 79, 80, 101, 216, 222
commuting operators, 226, 239
compact locally symmetric space, 142
compact manifold, 25, 31, 32, 48, 246, 319
compact non-commutative manifold, 132
compatible connection, 198
complementary subbundles, 147
complete symbol, 175, 275, 276, 317, 320
complex bundles, 203
complexified Clifford bundle, 198
complexified Lie algebra, 150, 204
complex powers, 176, 177, 187, 212, 293, 318
  complex powers of A, 212
  complex powers of the spectrum, 187
computational point of view, 221
confining coordinate, 218
conformal anomaly, 21, 87, 100, 209, 214, 215
conformal coupling, 85, 173
conformally coupled massless scalar field, 214
conformally invariant case, 211
conformally invariant quantum fields, 210
conformal scalar field, 302
conical singularities, 28, 84, 107
conjecture, 277
conjugacy classes, 158, 162, 165, 204
connected component, 140
conserved currents, 261
consistency rules, 235
consistent anomaly, 260, 262, 263
continuity equations, 261, 264
continuous spectrum, 36, 168, 266
contour integral, 292, 293
cosmological Casimir energy, 308
cosmological constant, 13, 16, 209, 289, 290, 295, 303
  cosmological constant problem, 295
cosmological models, 16, 294
cotangent case, 142, 146, 151, 153
Coulomb interaction, 284
coupling constant, 10, 11, 135, 136, 273, 274
covariant anomaly, 262, 263
covariant derivative, 26, 30, 95, 97, 105, 177, 202, 261
covering transformations, 139
critical limit, 269
critical temperature, 280, 283, 284
curvature forms, 203
curved spacetime, 1, 25, 26, 107, 209, 263
  curved spacetime with torsion, 263
cusp, 162, 164
cutoff function, 292
cut-off regularization, 137

D-dimensional Minkowski spacetime,

25
D-dimensional torus, 220
decomposition of the Lie algebra, 140
defect, 225, 227, 244, 290
definition of determinant, 226, 232
Deformation of elliptic operators, 211, 216
derivative of Selberg's zeta function, 153
derivative of the Barnes zeta function, 123
derivative of the zeta function, 193, 195, 250
de Sitter, 13, 16, 84, 87, 148, 302, 310
determinant line, 201-203
Determinant Line Bundles, 201, 203
determinant of non-hermitian matrices, 258
determinant of the Dirac operator, 193, 231, 259
diagonal subcase, 114, 115, 117
differentiated zeta functions, 77, 78, 80, 92
dimensional arguments, 5, 241
dimensional regularization, 6, 8, 11, 109, 306
dipole interactions, 287
Dirac bundles, 198
Dirac-like operator, 187, 199, 241, 253, 258
Dirac matrices, 177, 182, 184, 261, 264
Dirac operator, 3, 177, 182, 186
Dirac spinor, 3
direct sum, 159
Dirichlet boundaries, 218
Dirichlet boundary conditions, 299, 304
Dirichlet Laplacian, 218
discrete (co-compact) subgroup, 139
discrete, co-compact, torsion free subgroups, 148, 158
discrete group, 161, 162, 165, 166, 171
discrete spectrum, 36, 38, 168, 201, 266
divergent series, 231
double pole, 7-9, 168, 318
dS background, 310
duality operator, 199
Dynkin diagram, 140, 204

effective action, 1, 2, 4, 10, 59, 60, 193
effective field equations, 2
effective Lagrangian, 59, 60, 62, 65, 66, 85, 88, 104
effective mass, 130, 132
effective potential, 172, 304, 305, 308, 310
eigenvalues, 3, 17-19, 106, 168
eigenvalues equation, 164
Einstein manifold, 147
Einstein tensor, 209
Eisenstein–Maass series, 164, 165
electrostatic interactions, 289
elliptic operator, 2, 147, 149, 168, 177, 240, 285
embedding, 139, 200
endomorphisms of the vector bundle, 148
Epstein zeta function, 110, 126, 133
equilibrium states, 279
equilibrium temperature, 228, 266
eta function, 177-179, 182
eta invariant, 177, 179-183, 185, 191
eta-like functions, 196
Euclidean action, 86, 171, 227, 266, 273
Euclidean formulation, 175
Euclidean partition function, 2, 211, 227, 285
Euclidean section, 2, 13, 26, 28, 84, 261
Euler Γ-function, 165
Euler-Mascheroni constant, 6, 165, 240, 320
Euler structure, 201
expectation value of the stress tensor, 223

explicit analytic continuation, 111, 126, 128
exponentially convergent series, 112
exponentially quick convergence, 111
extended complex plane, 161
extension of the original Chowla-Selberg formula, 114, 118
external (constant) sources, 278
external domain, 294
external field, 211

2-form curvature of the determinant line bundle, 207
factorization formula, 151
factorization of the Selberg zeta function, 152
fast convergent series, 115, 127, 133, 222
fermion field, 3, 175
Feynman boundary condition, 2
Feynman propagator, 28, 59, 86, 87
field equation, 2, 293
field fluctuation, 85, 88, 90, 92, 209
finite center, 148
finite part, 6, 7, 9, 111, 125, 213
finite temperature effective potential, 265
finite temperature effects, 1, 13, 16, 129, 227, 265, 266, 273
finite temperature partition function, 21, 129
finite thickness, 291
first variation of the effective action, 210, 211, 215, 222, 223
flat boundaries, 223
flat bundle, 201
flat manifold, 22, 182, 189, 220, 223
flat space, 302
fluctuations of quantum fields, 209, 292
foliations, 147
Fourier transform, 163, 169, 188
framings, 204
free case, 188, 274, 276, 277, 280, 284
free energy, 17, 18, 84, 230, 278, 279, 281-284
Friedrichs' elliptic regularity theorem, 72
Fried's trace formula, 158
Fubini's theorem, 70, 80, 106, 249
Fuchsian group, 162
functional determinant, 1, 4, 135, 166, 193
  functional determinant evaluation, 166
functional equation, 133, 165, 315
functional integral, 2, 3, 26, 29, 87
functional measure, 3
fundamental group, 139, 142, 204
fundamental regions, 162
fundamental solution, 144

gamma function, 125, 127, 128, 181, 222
generalized Dirichlet series, 164
generalized Bernoulli polynomials, 256, 257
generalized Laplace-Beltrami operator plus potential, 30
generating functionals, 278
geometric structures on manifolds, 147
global anomaly formula, 201, 203
grading, 198
grand canonical partition function, 18, 228, 266
Green function, 28, 59, 67-69, 87
  Green function formalism, 292
group of isometries, 27, 31, 144, 171

Haar measure, 140, 158, 206, 207
Hadamard parametrix, 47, 102, 103
Hadamard regularization, 291
half quanta, 288
Hankel functions, 299, 301
Harish - Chandra Plancherel density, 140
harmonic oscillators, 230, 232, 253,

257
heat kernel, 4, 5, 7, 32, 38, 73, 144
heat-kernel asymptotics, 8, 22
heat-kernel coefficients, 5, 10, 39, 45, 248
heat-kernel expansion, 4, 7, 32, 38, 149, 184
heat-kernel parametrix, 39
heat kernel techniques, 175, 180, 185
heat-kernel trace, 7, 132, 166, 172, 195
heat operator, 5, 199
  heat operator trace, 5
hermitian vector bundle, 198
high temperature expansion, 131, 136, 137, 230
Hodge decomposition, 200
holonomy formula, 201
holonomy of the determinant line bundles, 203
homogeneity property, 176, 317
homogeneous component, 182, 275, 276
Hurwitz zeta function, 119, 190, 219, 299, 305
hyperbolic D-space, 157
hyperbolic manifold, 139, 157, 168, 193, 204
hyperbolic space, 157
hyperboloid of two sheets, 157
hypergeometric function, 249
hyperspherical harmonics, 299

identity matrix, 140, 161, 211, 254
identity operator, 31, 188, 233, 234, 235
IIA string theory, 203
imaginary time formalism, 129
index of the Dirac operator, 204
index theorem, 198
induced Chern-Simons term, 177, 182
induced gravity, 241
induced unitary representation, 151
inflation, 302
inhomogeneous Epstein zeta function, 113
inhomogeneous zeta function, 111
injectivity radius, 31, 39
integrated zeta function, 60, 84
interacting case, 189, 267, 273, 284
interacting scalar fields, 130
intermediate temperatures, 284, 285
invariant measure, 164
irreducible representation, 158
isometric circle, 161, 162
isometric fundamental domain, 162
isometry, 103, 104, 107, 161, 199
isotropy group, 157
Iwasawa decomposition, 140, 149, 160

Jacobi identity, 22, 114, 313
Jacobi theta-function identity, 110

Kaluza-Klein spectrum, 109
Killing form, 205
Kleinian group, 162
Kontsevich-Vishik formula, 246, 247
K–theory, 203

lack of a factorization, 225
Laplace–Beltrami operator, 14, 30, 142, 171, 172
Laplace-like operators, 248
Laplacian on forms, 158
Legendre transform, 278, 281
Lie algebra of vector fields, 147
lift, 139
linear connected finite covering, 139
linear–fractional transformation, 161
local diagonal heat-kernel, 218
local isomorphism, 139, 157
locally decomposable manifolds, 148
locally-symmetric spaces, 139, 248
local zeta function, 52, 53, 73, 85, 91, 92
Lorentzian section, 27-29, 87, 104
low energy excitations, 109, 124
lowest eigenvalue, 269

loxodromic, 161, 162
  loxodromic elements, 162

MacDonald function, 269
manifolds with boundaries, 209
manifold with a boundary, 81, 241
manifold without boundary, 5, 10, 175, 185, 218, 241, 248, 317
massive case, 187, 192, 211, 222, 309
massive Dirac operator, 186, 187, 189, 193, 195, 258
massive spinors, 186
massless Euclidean Dirac operator, 177
matrix-valued, 3, 211, 227, 266, 274
matter stress tensor, 209
maximal compact subgroup, 139, 148, 151
maximal parabolic subgroup, 162
mean field, 278
measure, 26, 28, 30, 46, 49, 50, 68, 80
Mellin-Barnes integral, 131
Mellin–Parseval identity, 6, 155, 314
Mellin representation, 8, 18, 179, 219
Mellin transform, 5, 6, 314
  Mellin transform inversion, 181
MEM devices, 288
meromorphic continuation, 112, 151, 152, 153, 156
meromorphic structure, 142, 146, 166, 168, 318
metric induced, 106, 157
Miatello coefficients, 141
Minakshisundaram–Pleijel coefficients, 144, 146
minima, 279
MIT bag model, 294
mixed boundary conditions, 296
modular invariance, 121
modular symmetry, 121
momenta, 266
monodromy, 201
M-theory, 109, 124, 203
multidimensional extension of the Chowla-Selberg formula, 117
multiplet, 3, 210, 211, 223
multiplicative anomaly, 228-231, 239, 240, 242, 248
  multiplicative anomaly formula, 241, 244-246
  multiplicative anomaly term, 284, 285
multiplicative property, 225, 228, 237

nanoscale, 289
nanotechnological devices, 288
negative Casimir pressure, 295
neutral scalar fields, 3, 4, 131
Newton's law, 289
nilpotent factor, 140, 149, 160
non-abelian chiral anomaly, 260
non-analytic dependence, 135
non-commutative dimensions, 110, 124, 137
non-commutative Euclidean space, 132
non-commutative field theory, 111
non-commutative gauge theories, 109, 124
non-commutative residue, 176, 189, 246, 318-320
non-commutative space, 109, 129
non-commutative toroidal manifold, 111
non-commutative torus, 109, 124, 129
non-compact symmetric space, 147
non-diagonal subcase, 114, 115
non trace-class operators, 226
non-trivial extension, 113
non-trivial spacetimes, 209
non-twisted p-forms, 158
non-zero modes, 149, 158
normalization condition, 282, 283
numerical calculations, 120
numerical study, 284

observables, 217
off-diagonal differentiated zeta

function, 77
off-diagonal heat-kernel coefficients, 39, 45
off-diagonal local spectral zeta function, 50
on diagonal differentiated zeta-function, 78
one-brane configuration, 302, 304, 306
one-dimensional potential, 304
one-loop approximated theory, 3
one-loop diagrams, 110, 131, 136
one-loop effective action, 7, 8, 10, 125, 179, 193, 261
one-loop partition function, 14, 15, 17, 21, 129, 273
one-loop quantum corrections, 3
one-loop stress-energy tensor, 86
operator spectrum, 137

parabolic elements, 165-167, 171
parabolic mappings, 162
parabolic transformations, 161
parabolic vertex, 162, 163
parametrix, 39, 47, 102, 103, 187, 188
parity anomaly term, 177
partial differential operators, 148
partially toroidal spacetime, 110
partition function, 2, 3, 29, 211
path-integral functional, 1
Pauli-Villars, 185, 231, 235
  Pauli-Villars regularization, 185, 231
periodic boundary condition, 21, 266
perturbation theory, 6, 230, 241-243, 262, 285
phase of the zeta function, 294
planar boundaries, 218, 220
point separation, 210
point-splitting renormalization, 47, 62
Poisson-Jacobi re-summation formula, 219
Poisson-Jacobi resummation formula, 268
Poisson resummation formula, 113, 114, 191
polar coordinates, 299
pole at the origin, 21, 135, 137, 296
pole structure, 111, 118, 125, 126
positive root system, 140, 206
primitive element, 150
principal symbol, 112, 176
product of two Einstein manifolds, 147
pseudo-differential operators, 317
pseudo-Kähler manifold, 147
(pseudo-) Riemannian structure, 147
pushforward map, 202, 203

quadratic form, 85, 111, 112, 124
quadratic+linear+constant form, 113, 118, 119
quadratic plus affine form, 111
quantum correction, 2, 3, 11, 110, 137, 261
Quantum Field Theory, 16, 25-27, 59, 84, 129
quantum fluctuation, 2, 241
quantum gravity, 1, 209, 211
quantum vacuum, 209, 287, 290
  quantum vacuum effects, 209
quantum variances, 222
Quillen metric, 202
quotient, 139

radion stabilization, 302, 303
rank one, 139, 140, 141, 144, 146, 148, 166
  rank one simple group, 141
Rarita-Schwinger operator, 175
rational cohomology, 203
Ray–Singer norm, 198, 200, 201
real form, 140
Real Hyperbolic Manifolds, 139, 157
recurrence for the Barnes zeta function, 122
recursion relations, 297
Redlich result, 185
reflected formula, 114

reflection formula, 110, 111, 315
regularization method, 110, 111, 235, 291
regularization parameter, 217
Regularization Techniques, 1, 290, 303
regularized one-loop effective action, 284
regularized partition function, 216, 267
regularized vacuum energy, 18, 132, 134, 137
relative variance, 210, 215, 216, 218, 223
relativistic charged Bose gas, 284
remainder, 118, 120, 123
renormalization, 3, 11
  renormalization group equation, 10, 11
  renormalization parameter, 11, 136, 137
renormalized form, 219
representation space, 142
representation theory, 144
representatives, 158, 162
residua, 58, 110-112, 125, 127, 129, 257
  residual forces, 289
resolvents, 187
restricted roots, 140
retarded Van der Waals force, 287
Ricci square invariant, 147
Ricci tensor, 147
Riemannian curvature, 202
Riemannian manifold with boundary, 198
Riemannian symmetric space, 139
Riemann zeta function, 19, 111, 117, 131, 227, 254
Robin boundary conditions, 296
roughness, 287, 289, 291
rules for infinite series, 234, 235

scalar curvature, 14, 15, 171, 210, 241, 303
scalar densities, 3
scalar field, 3, 14, 20
Schwinger-DeWitt's expansion, 59
Schwinger-Like Regularizations, 4
Schwinger's perturbative expansion, 216
second-order approximation, 243
second variation of the complex power, 216
second variation of the effective action, 215, 221
second variation of the functional determinant, 217
second variation of the partition function, 215
second variation of the trace, 216
second variations, 210, 221
Seeley-DeWitt coefficient, 11, 184
Seeley theorem, 20, 179
Selberg's trace formula, 149, 166
Selberg trace formula, 144, 149, 162, 165, 166, 170, 313
Selberg transform, 163
Selberg zeta function, 151-153, 171
self-adjoint factorization, 274
semiclassical approximation, 204, 209
semiclassical gravity, 210
semisimple Lie group, 139
short–time asymptotic expansion, 144, 146
simple complex Lie group, 161
simple poles, 6, 53, 57, 92, 110, 111
slow convergence, 111
small distance expansion, 309
small disturbance operator, 3, 171, 273
small-mass expansion, 222
smeared fields, 210
smooth fiber bundle, 202
smooth p-forms, 199
smooth Riemannian D-dimensional manifold, 29
smooth section, 112, 142, 148, 202

Sobolev's lemma, 31, 72
sonoluminescence, 294
source, 20, 211, 278, 279
spacetime, 2
special Bessel functions of the third kind, 269
spectral analysis, 139
spectral coefficients, 4, 10
spectral cut, 112, 176, 186
spectral decomposition, 38, 232, 242
spectral density, 4, 184
spectral theorem, 36, 68, 73, 183, 191, 192
spectral zeta function, 48, 50
spherical harmonic analysis, 140, 142
spherically symmetric situations, 294
spherical shell, 294, 299, 301
spin bundles, 198
spin connection, 177
spinor field, 3, 175, 193, 241
spinorial operators, 255
spinor Laplacian, 178, 184
spin structure, 202, 203
split rank one Lie groups, 148
splittings of determinants, 238
stabiliser, 162
standard case, 22, 217
standard representation, 158
standard zeta functions, 217
static spacetime, 13
statistical sum contribution, 132, 269
stiction, 289, 290
stress-energy tensor, 76, 85
stress tensor, 89, 91, 94, 210
  stress tensor trace, 210, 211, 213, 214
string theory, 109, 124, 203
strongly elliptic metrics, 2
subdomain, 163
symmetric components, 162
symmetric connection, 147
symmetric phase, 282, 283
symmetry breaking solution, 279
Synge world function, 32

Tauberian argument, 169
theorem due to Bismut and Freed, 180
thermal fluctuation operator, 135
Tolman's temperature, 103
topological invariant, 201, 204
topologically non-trivial spaces, 209
torsion free subgroup, 142, 148, 158
torus volume, 220
totally geodesic, 147
trace anomaly, 15, 100
trace formula, 149, 158, 162
transverse directions, 219
trigonometric duplication formula, 218
trivial representation, 158, 160
Twisted Dirac operator, 205, 206, 208
two-brane case, 306
two factorizations, 271

ultrastatic spacetime, 13, 20, 129, 171, 172, 293
ultraviolet divergences, 7, 274
unbroken phase, 230, 279, 280, 285
uncharged bosons, 283
unitary representation, 142, 148, 151, 158
unrenormalized, 281
upper half plane, 178, 187

vacuum contribution, 131, 134, 269
vacuum energy density, 20, 137, 210, 272, 290
  vacuum energy density fluctuations, 210
vacuum expectation value, 209-211, 213, 215, 219, 222
vacuum fluctuation, 209, 289, 292
vacuum structure, 284
Van der Waals interactions, 287
van Vleck-Morette determinant, 46
ε-symbol for the resolvent, 188
variance, 209, 210, 215, 216, 218, 222, 223

variational principle, 147
vector-axial-vector model, 259, 260
vector bundle, 112, 142, 148, 158
volume factor, 274
Voros theorem, 197

Wald's theorem, 99
warped product, 148
Weyl's asymptotic formulae, 49
Wick rotation, 3, 17, 26, 28, 82
Witten's global anomaly formula, 201
Wodzicki function, 178, 179
Wodzicki residue, 182, 230, 248, 264, 317, 319, 320
Wodzicki's formula, 248, 256
Wodzicki's main theorem, 180

zero coupling, 283
zero eigenvalue, 240
zero modes, 37, 149, 154, 155, 158, 160, 240, 241, 255
zero-point energy, 287
zero temperature, 3, 19, 134, 281, 282, 287
zero temperature partition function, 3
zeta-definition of the determinant, 232
zeta measure, 234, 235
zeta trace, 226, 229, 231-233